T0368341

Lecture Notes in Computer Science 15212

The series Lecture Notes in Computer Science (LNCS), including its subseries Lecture Notes in Artificial Intelligence (LNAI) and Lecture Notes in Bioinformatics (LNBI), has established itself as a medium for the publication of new developments in computer science and information technology research, teaching, and education.

LNCS enjoys close cooperation with the computer science R & D community, the series counts many renowned academics among its volume editors and paper authors, and collaborates with prestigious societies. Its mission is to serve this international community by providing an invaluable service, mainly focused on the publication of conference and workshop proceedings and postproceedings. LNCS commenced publication in 1973.

Luca Maria Aiello · Tanmoy Chakraborty ·
Sabrina Gaito
Editors

Social Networks Analysis and Mining

16th International Conference, ASONAM 2024
Rende, Italy, September 2–5, 2024
Proceedings, Part II

Editors
Luca Maria Aiello
IT University of Copenhagen
Copenhagen, Denmark

Tanmoy Chakraborty
Indian Institute of Technology Delhi
New Delhi, Delhi, India

Sabrina Gaito
Università degli Studi di Milano
Milan, Italy

ISSN 0302-9743 ISSN 1611-3349 (electronic)
Lecture Notes in Computer Science
ISBN 978-3-031-78537-5 ISBN 978-3-031-78538-2 (eBook)
https://doi.org/10.1007/978-3-031-78538-2

This Springer imprint is published by the registered company Springer Nature Switzerland AG
The registered company address is: Gewerbestrasse 11, 6330 Cham, Switzerland

Preface

With ASONAM 2024, the International Conference on Advances in Social Network Analysis and Mining marked its 16th anniversary as the flagship, premier and leading venue in the rapidly growing domain of social network analysis and mining. It has emerged into one of the most well-established and successful conferences. It is a great pleasure that the acceptance rate has stabilized below 20% for full papers since ASONAM was organized in Istanbul in 2012. Indeed, this reflects maturity and stability in terms of number of submissions, acceptance rate and participation, and ASONAM has earned its permanent position among top-tier international conferences.

This year, we moved from ACM/IEEE as sponsors to having Springer as the Sponsor and publisher of the proceedings. Authors of all papers presented at ASONAM and the co-located events are invited to submit expanded versions of their manuscripts to the prestigious SNAM journal, NetMAHIB journal or the LNSN series, which are characterized by their high visibility and fast processing of submissions. Special thanks to Springer Nature for their various publication venues which have been well integrated with ASONAM to the benefit of both parties.

We gathered over four days to witness interesting and exciting research achievements by various authors who present full, short, poster or demo papers covering a wide spectrum of research contributions to the foundations and applications of social networks. This would not have been possible without the dedication of a large team of motivated research leaders working closely together for twelve months to put together the attractive and intensive scientific program. Their great achievements contributed much to the visibility of ASONAM. I would like to heartily thank them all.

We do not forget in particular the generous support received from the operational organizing team who spent considerable time and effort handling daily issues and activities, answering emails, updating the Websites, etc. Special thanks to Min-Yuh Day, Panagiotis Karampelas, Tansel Ozyer, Mehmet Kaya, Deniz Bestepe, Diaylo Steiman and Jalal Kawash, who worked hard to produce the proceedings, communicate with participants/authors and handle the registration; also special thanks go to the local arrangements team from the University of Calabria. Indeed, without their highly appreciated effort it would have been really very hard to maintain the quality of the social program and keep the trend of providing rich meals and breaks during the conference and the excursion trip organized.

Thank you to all organizers including general chairs, program chairs and the chairs of various tracks and workshops, to participants, to authors who submitted papers, to program committee members and to the reviewers who invested their valuable time and effort to provide timely and comprehensive reviews. We encourage researchers and practitioners to submit again next year to get the opportunity and privilege to present their work at ASONAM 2025.

September 2024

Luca M. Aiello
Tanmoy Chakraborty
Sabrina Gaito

Organization

Steering Chair

Reda Alhajj	University of Calgary, Canada

Honorary Chair

Frans N. Stokman	University of Groningen, Netherlands

General Chairs

Andrea Tagarelli	University of Calabria, Italy
Roberto Interdonato	CIRAD, UMR Tetis, France
Jon Rokne	University of Calgary, Canada

Program Committee Chairs

Luca M. Aiello	IT University of Copenhagen, Denmark
Tanmoy Chakraborty	IIT Delhi, India
Sabrina Gaito	University of Milan, Italy

Industry Track Chairs

Francesco Gullo	University of L'Aquila, Italy
Gianmarco De Francisci Morales	CENTAI, Italy

Workshops Chairs

I-Hsien Ting	National University of Kaohsiung, Taiwan
Rajesh Sharma	University of Tartu, Estonia
Lucio La Cava	University of Calabria, Italy

Multidisciplinary Track Chairs

Ester Zumpano	University of Calabria, Italy
Shirin Nilizadeh	University of Texas at Arlington, USA
Carmela Comito	ICAR-CNR, Italy

PhD Forum and Posters Track Chairs

Huzefa Rangwala	George Mason University, USA
Alessia Antelmi	University of Turin, Italy
Domenico Mandaglio	University of Calabria, Italy

Demos and Exhibitions Chairs

Elio Masciari	University of Naples, Italy
Tansel Ozyer	Ankara Medipol University, Turkey

Tutorial Chairs

Pasquale De Meo	University of Messina, Italy
Shady Shehata	YOURIKA Labs, Canada
Davide Vega	University of Uppsala, Sweden

Publicity Chairs

Shang Gao	Jilin University, China
Buket Kaya	Firat University, Turkey
Kashfia Sailunaz	University of Calgary, Canada

Publication Chairs

Min-Yuh Day	National Taipei University, Taiwan
Panagiotis Karampelas	Hellenic Air Force Academy, Greece

Registration Chairs

Jalal Kawash	University of Calgary, Canada
Mehmet Kaya	Firat University, Turkey

Web Chair

Deniz Bestepe	Istanbul Medipol University, Turkey

Additional Reviewers

Abdessamad Benlahbib	FSDM, Morocco
Abdessamad Imine	Loria, France
Abiola Akinnubi	COSMOS-UALR, USA
Adnan Hoq	University of Notre Dame, USA
Aisling Third	Open University, UK
Akira Matsui	Yokohama National University, Japan
Alessandro Visintin	University of Padua, Italy
Alexander Rodriguez	Georgia Institute of Technology, USA
Amrit Poudel	University of Notre Dame, USA
Anastasios Giovanidis	Centre National de la Recherche Scientifique, France
Anatoliy Gruzd	Toronto Metropolitan University, Canada
Anggy Eka Pratiwi	Indian Institute of Technology Jodhpur, India
Ankan Mullick	IIT Kharagpur, India
Anurag Singh	National Institute of Technology Delhi, India
Arlei Silva	Rice University, USA
Ashwin Shreyas Mohan Rao	University of Southern California, USA
B. Aditya Prakash	Georgia Tech, USA
Bailu Jin	Cranfield University, UK
Bijaya Adhikari	University of Iowa, USA
Billy Spann	University of Arkansas at Little Rock, USA
Bing He	Georgia Institute of Technology, USA
Bohan Jiang	Arizona State University, USA
Casey Doyle	Sandia National Laboratories, USA
Charalampos Chelmis	University at Albany State University of New York, USA
Christine Largeron	Université de Lyon, France
Constantine Dovrolis	Georgia Institute of Technology, USA
Courtland Vandam	Massachusetts Institute of Technology, USA

David Skillicorn	Queen's University, Canada
Debanjan Datta	Virginia Tech, USA
Dong Wang	University of Illinois Urbana-Champaign, USA
Eduard Dragut	Temple University, USA
Ehsan Ul Haq	Hong Kong University of Science and Technology, China
Etienne Gael Tajeuna	Laval University, Canada
Fattane Zarrinkalam	University of Guelph, Canada
Fernando Terroso-Saenz	Catholic University of Murcia, Spain
Frank Liu	Southern Illinois University, USA
Fujio Toriumi	University of Tokyo, Japan
George Panagopoulos	École Polytechnique, France
Gita Sukthankar	University of Central Florida, USA
Hadassa Daltrophe	Shamoon College of Engineering, Israel
Hamid R. Rabiee	Sharif University of Technology, Iran
Hanjia Lyu	University of Rochester, USA
Hasan Davulcu	Arizona State University, USA
Hitkul Jangra	Indraprastha Institute of Information Technology, Delhi, India
Huimin Zeng	University of Illinois at Urbana-Champaign, USA
Humayun Kabir	Microsoft, USA
Isabel Murdock	Carnegie Mellon University, USA
Jiaming Cui	Georgia Institute of Technology, USA
Jiamou Liu	University of Auckland, New Zealand
Jiten Sidhpura	Sardar Patel Institute of Technology, India
Jose Luis Fernandez-Marquez	University of Geneva, Switzerland
Julio Cesar Soares dos Reis	Federal University of Viçosa, Brazil
Keith Burghardt	University of Southern California, USA
Kenji Yokotani	Tokushima University, Japan
Keyan Guo	University at Buffalo, USA
Kijung Shin	Korea Advanced Institute of Science and Technology, South Korea
Kshiteesh Hegde	Western Digital, USA
Lanyu Shang	University of Illinois Urbana-Champaign, USA
Lara Quijano-Sanchez	Universidad Autónoma de Madrid, Spain
Lu-An Tang	NEC Labs America, USA
Mainuddin Shaik	University of Arkansas at Little Rock, USA
Mehrdad Jalali	Karlsruhe Institute of Technology, Germany
Michael Smit	Dalhousie University, Canada
Mirela Riveni	University of Groningen, The Netherlands
Muhammad Abulaish	South Asian University, India
Nayoung Kim	Arizona State University, USA

Contents – Part II

Research

Research

Federated Learning-Based Tokenizer for Domain-Specific Language Models in Finance

Farouk Damoun[1,2](✉), Hamida Seba[2,3], and Radu State[1]

[1] University of Luxembourg, Esch-sur-Alzette, Luxembourg
{farouk.damoun,radu.state}@uni.lu
[2] Université Claude Bernard Lyon 1, Lyon, France
{farouk.damoun, hamida.seba}@univ-lyon1.fr
[3] UCBL, CNRS, INSA Lyon, LIRIS, UMR5205, Villeurbanne, France

Abstract. The Federated Byte-level Byte-Pair Encoding (BPE) Tokenizer (FedByteBPE) leverages a Federated Learning (FL) approach for a privacy-preserving approach to train language models tokenizer across distributed datasets. This approach enables entities to train and refine their tokenizer models locally, with vocabulary aggregation performed on a centralized server. This method ensures the creation of a robust, domain-specific tokenizer while preserving privacy. Supported by theoretical analysis and empirical results from experiments on a real-world distributed financial dataset, our findings demonstrate that the federated tokenizer significantly outperforms off-the-shelf and individual local tokenizers in vocabulary coverage. This highlights the potential of federated learning to address training language model tokenizers in a privacy-preserving setting.

Keywords: Federated Learning · Language Models · Tokenizer

1 Introduction

Tokenizers are tools or algorithms used in Natural Language Processing (NLP) to convert text into an organized structure, typically by breaking it down into smaller units known as tokens, which can be words, subwords, characters, or bytes. These tools are fundamental to most state-of-the-art Language Models (LMs) utilized in various downstream NLP tasks, and are mainly trained on a centralized dataset. However, this data-centric LM training approach faces challenges when individual data, especially sensitive and personally identifiable information (PII), cannot be centralized or exposed for processing in industries such as healthcare and finance in a cross-silo setting. Federated learning [28] presents a novel solution by facilitating collaborative model training across multiple clients without the need for direct data sharing, thus offering privacy-preserving guarantees [14]. This method, which contrasts with traditional LM training, allows clients to train local models with initial parameters from a (trusted) central

L. M. Aiello et al. (Eds.): ASONAM 2024, LNCS 15212, pp. 3–20, 2025.
https://doi.org/10.1007/978-3-031-78538-2_1

server and only share back the updated local parameters for global aggregation. These become the new global parameters for all client models. After several rounds, the aggregator server shares the final model parameters. This paradigm adheres to regulations like the General Data Protection Regulation (GDPR) in the European Union by supporting collaborative learning without exposing sensitive individual information, such as bank transactions, geographical locations, and textual communications.

Privacy-preserving language models, focusing on the transformer architecture with its Embedding, Encoder, and Decoder components, are increasingly trained in federated settings for enhanced privacy [25,39]. Despite this progress, these models do not include tokenizers trained in federated settings. This limits achieving a truly end-to-end federated LM. Across the different frameworks proposed [9,25], tokenizers are initialized from an existing public tokenizer, even for domain-specific downstream tasks [45], due to their parameterless structure and training process.

Previous works, such as the federated heavy hitters algorithm [47], adapt distributed frequent sequence mining for word-level discovery but are primarily suited for word/subword WordPieces-based tokenizers. In [3], tokenizers are learned through sampling and auto-regressive text generative federated models, yet these approaches face limitations due to biases and high computational costs. However, these approaches are not applicable to the most widely used state-of-the-art subword/char/byte-level BPE-based tokenizers [36,42], defined as a subword tokenization technique for merging the most frequent pairs of characters or character sequences. Our focus is on BPE, which, unlike WordPieces, ensures no out-of-vocabulary (OOV) terms, and unlike SentencePiece [20], it does not require handling cross-boundary words. However, developing Federated Tokenizer faces several significant challenges: ensuring privacy preservation to prevent leakage of sensitive information during the training process, and achieving model convergence despite asynchronous updates from participants, which can lead to unstable convergence.

Contributions. In this paper, we address these challenges by focusing our experiments on real-world financial data for efficiency, privacy, and robustness. Our contributions are as follows.

- We introduce a Federated Byte-level BPE Tokenizer algorithm to train a tokenizer across distributed datasets in real-world scenarios.
- We compare state-of-the-art generic and domain-specific pretrained tokenizers, highlighting their performances and trade-offs. Our results show that the federated tokenizer outperforms and competes with centralized pretrained tokenizers.
- We analyze the impact of federated tokenizer algorithm parameters, including the number of participants, partial data sharing, and privacy budget, on text compression performance. Our findings indicate that the privacy budget significantly affects performance, while partial data sharing acts as a regularization technique with minimal impact.

Outline. The paper is structured as follows: Sect. 2 formalizes the problem and introduces key concepts. Section 3 reviews related work. Section 4 details our methodology. Section 5 describes the dataset and experiments. Section 6 evaluates our approach. Section 7 concludes with future research directions.

2 Preliminaries and Problem Statement

Tokenizer. Tokenizers transform a string w into sequence of non-empty strings based on the vocabulary $\mathcal{V}$ obtained through the training of a tokenizer algorithm on text corpus D. The function $\tau(w)$ represents the tokenization process, systematically producing all possible token sequences from w by repeatedly applying rules from $\mathcal{V}$ until no further rules apply. Tokenization is defined formally in [8] as follows:

Definition 1. *Given an Σ as a finite set of symbols, the process of tokenization for a string $w \in \Sigma^*$ is the transformation of the string w into a sequence of tokens $u_1, u_2, \ldots, u_l$, where each $u_i \in \Sigma^+$ for $1 \leq i \leq l$ given a collection of tokenization predefined rules $\mathcal{V}$ that guide the segmentation of a string into tokens. The operation of recombining these tokens into the original string w is facilitated by a concatenation function π, such that $w = \pi(u_1, u_2, \ldots, u_l)$.*

Federated Learning. Federated learning trains a shared global model F using an aggregation protocol such as averaging involving N distributed participants (clients), each agreeing to train a local model f starting with the same initial configuration θ. Mathematically, let us assume that all N clients, where the data reside, are available. Let D_i represent the data associated with client i, and n_i the number of samples available, with a total sample size is $\sum_{i=1}^{N} n_i$. Following [28], this setup aims to solve an empirical risk minimization problem of the form:

$$\min_{\theta \in \mathbb{R}^d} F(\theta) = \sum_{i=1}^{N} \frac{n_i}{n} F_i(\theta) \quad \text{where} \quad F_i(\theta) = \frac{1}{n_i} \sum_{x \in D_i} f_i(\theta), \tag{1}$$

where d is the model parameters size to be learned.

Local Differential Privacy. Local Differential Privacy (LDP) is a state-of-the-art privacy preservation technique that addresses the potential risk of an aggregation protocol at the server level to steal, expose, or leak clients' privacy over training rounds. In this setting, each client performs randomized perturbation on local data before sharing model parameters with the server, which then performs the aggregation protocol to obtain effective global model parameters. The same applies to statistics. LDP is defined formally in [17] as follows:

Definition 2. *For a randomized algorithm M, its definition domain and range are $D(M)$ and $R(M)$, respectively. For any two records I and I' in $D(M)$, their same output of M is S, where $S \subseteq R(M)$. If the following inequality holds, then the randomized algorithm M satisfies (ε, δ)-LDP,*

$$\Pr[M(I) = S] \leq e^{\varepsilon} \cdot \Pr[M(I') = S] + \delta. \tag{2}$$

where the parameter ϵ is called the privacy budget, refer to the privacy protection level of the client data can be adjusted through this parameter. And the parameter δ is called sensitivity of M, and represents the tolerance for the probability of ϵ deviating from its expected privacy guarantee, reflecting the maximum impact a single individual can have on the locally trained parameters or statistics. In practice, a larger ϵ means that the lower the privacy protection level M with a higher utility.

Problem Statement. In a federated setting comprising N participants denoted as $\{C_i\}_{i=1}^N$, each with a private local dataset D_i, our goal is to construct a federated tokenizer τ_{Fed}. This tokenizer should efficiently leverage all local datasets without compromising data privacy. The challenge lies in maximizing the similarity between τ_{Fed} and an hypothetical global tokenizer τ_{Global}, which would be constructed using a unified dataset $D = \bigcup_{i=1}^N D_i$. Formally, our goal is to:

$$\text{maximize} \quad sim(\tau_{\text{Global}}, \tau_{\text{Fed}}) \tag{3}$$

Here, $sim(.,.)$ represents the similarity measure between τ_{Global} and τ_{Fed}. Assessing this similarity poses a unique challenge, as tokenizers with divergent training datasets and differing vocabulary sizes can yield equivalent outputs. To effectively assess this similarity, we propose utilizing metrics derived from the tokenization output of both tokenizers across the federation corpus. Specifically, we aim to:

$$\min_{\tau_{\text{Fed}}} \mathcal{F}(\tau_{\text{Fed}}) = \sum_{i=1}^{N} \frac{n_i}{n} \mathcal{F}_i(\tau_{\text{Fed}}) \quad \text{where} \quad \mathcal{F}_i(\tau_{\text{Fed}}) = \mathcal{H}(\tau_{\text{Fed}}(D_i)), \tag{4}$$

where $\mathcal{H}(\cdot)$ denotes a suitable metric computed over the tokenized output from τ_{Fed} applied to each local dataset D_i. This approach ensures the tokenizer efficacy in approximating τ_{Global} while preserving the privacy of the federated learning participants.

3 Related Work

In this paper, we are interested in federated learning, with a specific emphasis on the tokenization algorithms used in language models and the associated privacy concerns.

Tokenizers in Language Models. Tokenization serves as a initial step in language models in both training and inference, transforms raw text into a model-understandable format. Techniques like WordPiece [42], Byte Pair Encoding (BPE) [36], SentencePiece [20], and Byte-level BPE [42] are fundamental in language model pretraining. Recent studies have explored the impact and effectiveness of various tokenization techniques. Studies by [13,40] reveal how tokenization strategies, from BPE merge counts to granularity levels in low-resource languages, influence model performance. Insights from [29] into BERT's word-level tokenization highlight a preference for frequency over semantics, while [24] discusses the selection of optimal tokenizers for multilingual models through extensive experimentation. Furthermore, the works by [31,35], and [7] explore the monolingual performance of multilingual models, the introduction of language biases, and the representation of OOV terms, respectively. However, these studies predominantly focus on the impact of tokenizers in multilingual settings, yet there remains a notable gap in examining their efficacy within domain-specific monolingual contexts.

Tokenizers in Federated Learning. Studies focusing on language models within federated learning frequently employ word-level tokenization. While some research, such as [28], constructs vocabularies from publicly available datasets, it is common to utilize standard pretrained tokenizers, subsequently fine-tuning the entire or partially language model for both generic [1,16,28] and domain-specific downstream tasks [6,38]. Previous research on privately finding vocabulary items, such federated heavy hitters algorithm [47], has been utilized for word-level discovery mainly from a single sequence per participant with a large privacy budget, rather than for training a tokenizer. With the rising attention on federated tokenizers, recent work by [3] introduces a method for learning a tokenizer through sampling, leveraging auto-regressive text generative models. This method applies to both off-the-shelf and federated pre-trained models using a publicly trained tokenizer τ_{pub}, based on a public corpus. Then learns a new tokenizer by prompting the text generative model to generate new tokens. This approach, however, is limited by biases in the prediction head, that reflect the word frequency used to train the public tokenizer in the pretraining corpus [19]. Additionally, the [3] method requires training a language model in a federated learning setting, which is both expensive and unnecessary for training a tokenizer. Different from [3,47], our goal is to develop a bytes-level tokenizer specifically designed for a federated learning setting, by training on the vocabulary corpus instead of raw text itself.

4 Methodology

In this section, we describe our methodology for creating a privacy-preserving tokenizer in a federated environment, aligned with data protection norms. Our approach starts with the local construction of vocabularies across participating entities, crucial for enhancing both privacy and effectiveness [11], and beneficial for domain-specific NLP tasks [40]. We then implement a modified Byte Pair Encoding (BPE) algorithm within a federated learning framework [46], integrated with differential privacy [14]. The aggregation of these local vocabularies forms our federated BPE tokenizer, with further details in the following subsections.

4.1 Privacy-Aware Data Pre-processing

Our Federated Learning-based tokenizer for financial documents adheres to strict privacy standards such as GDPR and CCPA. As illustrated in Fig. 1, the process consists of four main steps ① –④, begins with **Text Pre-Annotations ①**,

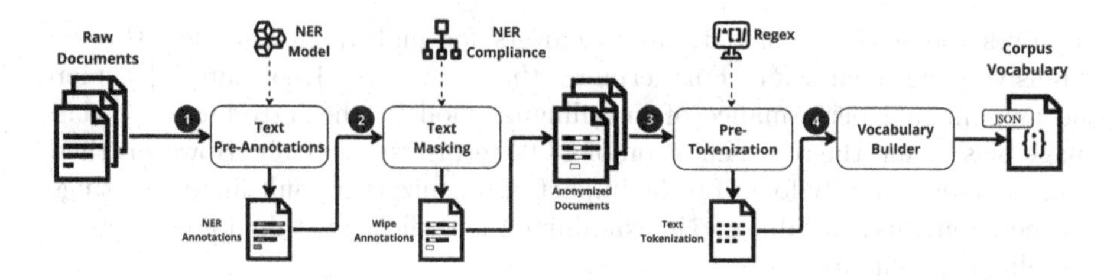

Fig. 1. Automated Pipeline for Privacy-aware Vocabulary Corpus Builder.

employing the specialized Named Entity Recognition (NER) model (NERPII) to identify and tag personal information [27]. The subsequent **Text Masking** step ② anonymizes sensitive data using wiping as an efficient technique based on [21], and in the **Pre-Tokenization** phase ③, we employ an efficient regex pattern expressed as follows:

(?\p{L}+|?\p{N}+|?^\s\p{L}\p{N}|\r\n|\s+(?!\S)|\s+)

In text processing this regex pattern is know as digital- and punctuation-aware regex patterns, demonstrated as an effective pre-tokinization for language models [10,11,30]. The final step, **Vocabulary Builder** ④, constructs a domain-specific corpus vocabulary from pretokenized outputs using a counting function to map each unique token w_i from D to its frequency f_i, forming the tuple set $\mathcal{D} = \{(w_i, f_i)\}$, where each token w_i is transformed into a byte sequence to improve the training of the tokenizer at the byte level, as recommended in the literature [33,42]. This streamlined approach ensures the maintenance of data confidentiality and integrity throughout the tokenization process. This structured approach not only facilitates the systematic construction of a standardized vocabulary but also ensures that our methodology, particularly in the context of byte-level BPE tokenizers.

4.2 Federated BPE Tokenizer

We detail our Federated Byte-level BPE algorithm, an efficient, privacy-preserving tokenizer within a federated framework, in Algorithms 1 and 2. Following the BPE framework [36,42], participants C_i start with datasets $\mathcal{D}_i$ (step 4 outcome). The algorithm initializes the local tokenizer vocabulary of byte values (0–255) and an empty set of merges (Algorithm 1, lines 1–5) then iteratively expands the vocabulary to size k through two main aggregation phases per iteration.

- **Single-Token Aggregation Phase.** (Algorithm 1, lines 7–9): A subset of clients is randomly sampled, and they collectively send their single-token tuples to the server, which aggregates the tuples frequency of single-tokens and sends them back to a subset of clients.

Algorithm 1. Federated Byte-level BPE (1/2)

Require: clients $\mathcal{C} = \{C_i\}_{i=1}^N$, associated datasets $\{\mathcal{D}_i\}_{i=1}^n$
Parameters: **k** - Target vocab size, θ - threshold, ϵ and δ - privacy budget, **p** - Percentage of data from $\mathcal{D}_i$, **K** - Fraction of clients used in each round.
▷Initial local tokenizer vocabulary
1: $V_{size} \leftarrow 256$
2: **for all** C_i in $\mathcal{C}$ **do**
3: $C_i.vocab \leftarrow$ all possible byte values (0 to 255)
4: $C_i.merges \leftarrow \{\}$
5: **end for**
▷Initial tokenizer training
6: **while** $V_{size} \leq k$ **do**
7: $\mathcal{C}' \leftarrow$ RandomSampler($\mathcal{C}, K$) Single-Token Aggregation Phase
8: $L \leftarrow$ {ClientSendTuples($C_i, \emptyset$) | $C_i \in \mathcal{C}'$}
9: $U \leftarrow$ ServerAggregateTuples(L, θ)
10: $\mathcal{C}'' \leftarrow$ RandomSampler($\mathcal{C}, K$) Token-Pair Aggregation Phase
11: $L \leftarrow$ {ClientSendTuples(C_i, U) | $C_i \in \mathcal{C}''$}
12: $B \leftarrow$ ServerAggregateTuples(L, θ)
▷ Update Tokenizer and Local Vocab.
13: **if** $B = \emptyset$ **then**
14: **break** {No Vocabulary Left}
15: **end if**
16: $B_{\max} \leftarrow B[0]$
17: $(t_L, t_R) \leftarrow B_{\max}[0], B_{\max}[1]$
18: **for all** $C_i \in \mathcal{C}$ **do**
19: ClientTupdateDataset(C_i, t_L, t_R)
20: **end for**
21: $V_{size} \leftarrow V_{size} + 1$
22: **end while**

- **Token-Pair Aggregation Phase.** (Algorithm 1, lines 10–12): Following the single-token phase, a different subset of clients is sampled to receive the aggregated tuples from the server then generate token-pair tuples where one of token-pair must be in received single-tokens, and send back to the server to find the most frequent token-pair (Algorithm 2, lines 1–18).

If no vocabulary can be aggregated further (Algorithm 2, line 11) through all participants, indicating that B is empty (Algorithm 1, line 13), the process halts for that iteration and the target tokenizer vocabulary size is not reached. Otherwise, the most frequent tuple is identified (Algorithm 1, lines 16–17), and each client updates its local tokenizer vocabulary and dataset by merging the newly discovered byte pairs accordingly (Algorithm 1, lines 18–20). For a clearer understanding, here are the core procedures for client-server interactions in our system: **RandomSampler**(C, K) selects a subset K from set C, crucial for privacy and efficiency. **ClientSendTuples**(C_i, S) has clients generate tuples from data samples; if S is empty, only single tokens are processed and empty vocabularies return

Algorithm 2. Federated Byte-level BPE (2/2)

1: **Procedure** ClientSendTuples(C_i, S):
2: $\mathcal{D}'_i \leftarrow$ RandomSampler($\mathcal{D}_i, p$)
3: **if** $S = \emptyset$ **then**
4: $\mathcal{U}'_i \leftarrow$ GetUnigrams($\mathcal{D}'_i$)
5: $Singles \leftarrow$ Sort$\{(t_L, f) \mid t_L \in \mathcal{U}'_i\}$
6: $T \leftarrow \{(t_L, f + \mathcal{L}(0, \frac{\delta}{\epsilon})) \mid (t_L, f) \in Singles\}$
7: **else**
8: $\mathcal{B}'_i \leftarrow$ GetPairs($\mathcal{D}'_i$)
9: $Pairs \leftarrow$ Sort$\{((t_L, t_R), f) \mid (t_L, t_R) \in \mathcal{B}'_i, t_L \in S\}$
10: **if** $Pairs = \emptyset$ **then**
11: **return** ("NVL", 0) {No Vocabulary Left}
12: **else**
13: $T \leftarrow \{(pair, f + \mathcal{L}(0, \frac{\delta}{\epsilon})) \mid (pair, f) \in Pairs\}$
14: **end if**
15: **end if**
16: $T_{\max} \leftarrow T[0]$
17: **return** $T_{\max}$
18: **EndProcedure**

19: **Procedure** ServerAggregateTuples(L, θ):
20: $T_{\text{pool}} \leftarrow \bigcup_{(key_i, f_i) \in L} \{(key_i, f_i)\}$
21: $T_{agg} \leftarrow \{(key, \sum_{j \mid key_j = key} f_j)\}_{key \in T_{\text{pool}}}$
22: $T_{filtered} \leftarrow$ Sort$\{key \mid (key, f) \in T_{agg}, f > \theta\}$
23: **return** $T_{filtered}$
24: **EndProcedure**

25: **Procedure** ClientUpdateDataset(C_i, t_L, t_R):
26: $mergedToken \leftarrow t_L + t_R$
27: **for** each w in $\mathcal{D}_i$ **do**
28: **if** w contains t_L followed by t_R **then**
29: $w \leftarrow$ Replace "t_L t_R" with $mergedToken$ in w
30: **end if**
31: **end for**
32: $C_i.vocab \leftarrow$ Extend($C_i.vocab, mergedToken$)
33: $C_i.merges \leftarrow$ Extend($C_i.merges, (t_L, t_R)$)
34: **EndProcedure**

a special tuple. **ServerAggregateTuples**(C_i, θ) involves the server aggregating received tuples and applying a frequency threshold θ to maintain relevant data. **ClientUpdateDataset**(C_i, t_L, t_R) updates local vocabularies to reflect global changes, ensuring synchronization in the progress of federated learning.

After all iterative rounds, each client's tokenizer is generated for the target number of tokens k within $2k$ federated rounds, resulting in a tokenizer size of $256 + k$ unless the algorithm halts (Algorithm 1, line 14). We didn't study the communication overhead due to the limited bytes exchanged per round. However, a limitation of our work is the time complexity of $O(2kN \log N)$, which increases linearly with the number of rounds k, compared to the traditional BPE complexity of $O(N \log N)$ [48].

4.3 Privacy Protection

Here we highlight our efforts in protecting privacy by building a federated tokenizer through two privacy mechanisms.

First, we begin our algorithm with k as the target number of tokenizer vocabulary size, setting K as the number of clients to sample without replacement within a uniform distribution and θ as the minimal frequency for tokens. In every round, we derive two subsets from all the clients $\mathcal{C}$: with (a) uniformly sampled K clients without replacement $\mathcal{C}^{'} \subset \mathcal{C}$, for single token selection, and (b) uniformly sampled K clients without replacement $\mathcal{C}^{''} \subset \mathcal{C}$, for pair token selection predicated on previously selected single tokens to extend the already discovered merges (pair tokens), with respect to the thresholding operation θ. This approach ensure not only k-anonymity properties but also reinforces differential privacy, as demonstrated in [47, Theorem 1], where $k = \theta$.

Given our dual-faceted process, we define $\mathcal{M}_1$ as a global privacy mechanism, inherently deferentially private, as substantiated by [12,47]. In instances where a client $\mathcal{C}_i$ is randomly chosen in both $\mathcal{C}^{'}$ and $\mathcal{C}^{''}$ for the discovery of token pairs, only those pairs that exceed the frequency threshold θ are evaluated. Thus, $\mathcal{M}_1$ improves a form of privacy by locally excluding low-frequent tokens, potentially identifiable.

Finally, the local private token frequency is carried out using a different mechanism $\mathcal{M}_2$ relying on Local Differential Privacy, denoted by (ϵ, δ)-LDP. First, we define $p \in [0, 1]$ as the subset ratio for sampling local client dataset, where we subsample $\mathcal{D'}_i \subset \mathcal{D}_i$, following a uniform distribution without replacement, then we add a controlled random noise sampled from Laplace distribution to the local frequency distribution of single and pair tokens, $F_{D'_i}$. As a result, the $\mathcal{M}_2$ mechanism enhance privacy and retains utility by using both subsampling [5] and for noise addition [17].

Following our second mechanism, let w be a single or pair token in $\mathcal{D}_i$ and due to the randomness inherent in the uniform subsampling process used in $\mathcal{M}_2$, the expected frequency can be expressed as;

$$E[F_{D'_i}(w)] = p \times F_{D_i}(w);$$

where $E[F_{D'_i}(w)]$ is the expected frequency of token w in the D_i. Due to the subsampling perturbation introduced, $\mathcal{D'}_i$ might not perfectly reflect the distribution of $\mathcal{D}_i$, this process can be likened to the effects of BPE-dropout [32]. By integrating the Laplace noise and subsampling, we ensures a deviation in the original distribution that will effectively mask these variations with a privacy budget ϵ.

$$|F_{D'_i}(w) - F_{D'_i,\text{noisy}}(w)| \leq \epsilon$$

Although both mechanisms increase the level of privacy separately, their sequential combination, as described in the composition theorem [14], forms a unified mixture mechanism $\mathcal{M}$. This integration involves employing $\mathcal{M}_1$ to protect against identifying rare token patterns and $\mathcal{M}_2$ to add controlled random noise to the local frequency distributions of tokens. The resultant mechanism $\mathcal{M}$ achieves $\epsilon' = \ln(1 + p(e^\epsilon - 1))$ and $\delta' = p\delta$, according to [4, Theorem 9], while not relying solely on k anonymity for privacy [47].

5 Experimental Settings

In this section, we describe the datasets and metrics used and the experiment set up to assess and evaluate the efficacy of the proposed FedByteLevelBPE algorithm.

5.1 Real Dataset

We utilize the CFPB[1] open dataset, initially for risk monitoring in financial consumer services, now repurposed for federated learning in NLP. Unlike previous studies [23], we distributed the data set between 30 US financial institutions based on the `CompanyName`, encompassing 680,086 complaints from January 2016 to May 2023, totaling about 90 million tokens. This partitioning forms a natural cross-silo setup, mirroring real-world data variations and is processed uniformly as described in Sect. 4.1.

5.2 Evaluation Metrics

Tokenizer performance evaluates the relationship between input text and output lengths to assess how effectively the tokenizer encodes and compresses data. We focus on two metrics: tokenizer's fertility (ψ), which measures the average subtokens per token, reflecting granularity ($\psi \geq 1$ indicates optimal dataset tailoring), given by:

$$\psi(\tau) = \frac{1}{|D|} \sum_{s \in D} \frac{|\tau(s)|}{|s|} \tag{5}$$

and the proportion of continued words (Π), which assesses the tokenizer's tendency to split words, calculated by:

$$\Pi(\tau) = 1 - \frac{|D|}{\sum_{s \in D} |\tau(s)|} \tag{6}$$

These metrics, ψ and Π, scale from individual tokens to datasets by considering D as either concatenated sentences for vocabulary-level or as document collections for broader evaluations.

[1] The Consumer Financial Protection Bureau (CFPB) database: link.

Table 1. Hyperparameter Tuning Space for the FedByteLevelBPE Algorithm

Hyperparameter	Definition	Search Interval
θ	Minimal frequency per tokens	$\{8, 15, 22, 30\}$
ϵ	Privacy budget for (ϵ, δ)-LDP	$\{10^{-3}, 5 \cdot 10^{-3}, 5 \cdot 10^{-2}, 10^{-1}, 5 \cdot 10^{-1}, 1, 2\}$
p	Percentage of data from D_i	$\{0.7, 0.8, 0.9, 0.95, 1.0\}$
K	Fraction of clients in each round	$\{0.25, 0.5, 0.75, 1.0\}$

5.3 Approach for Comparison

We assess our algorithm against recent tokenizer models by classifying tokenizers into three types: **Public Tokenizer** τ_{pub}, using a general corpus; **Local Tokenizer** τ_{loc}, trained on data from a single institution; and **Federated Tokenizer** τ_{fed}, developed in a federated learning context. Each category utilized an identical setup with a vocabulary size of 50257. Specifically, 30 local tokenizers were trained for individual financial institutions, and a federated tokenizer aggregated insights from all datasets, with performance metrics averaged across federated datasets. The details of the hyperparameters of the FedByteLevelBPE algorithm are detailed in Table 1, where the tokenizer vocabulary is fixed at 50257, δ is set to 1 according to [2], and a total of 560 hyperparameter configurations are evaluated across the 30 client nodes. These evaluations were conducted using the `flwr` framework on a server with a 32 core CPU and 128 GB RAM running Ubuntu 22.04.

6 Experimental Results

6.1 Hyperparameter Tuning Results

In this section, we conducted a comprehensive hyperparameter tuning for FedByteBPE, exploring the sensitivity of the algorithm to each parameter through

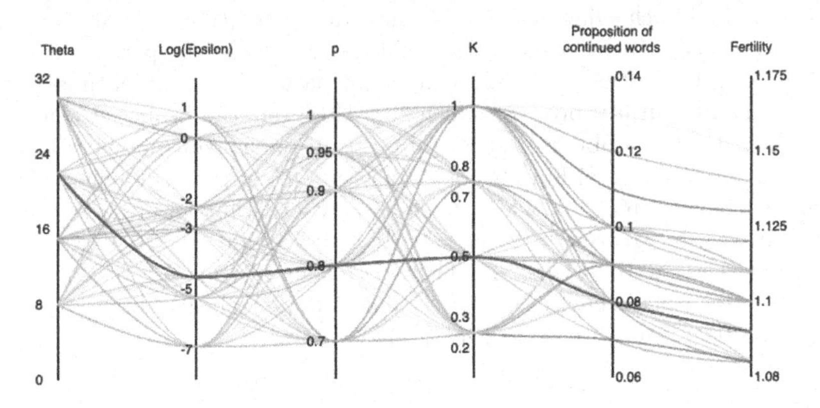

Fig. 2. Parallel Coordinates Analysis of 560 Hyperparameter Combinations Affecting Federated Tokenizer Performance: Fertility (ψ) and Proportion of Continued Words (Π). Optimal parameters are highlighted with a red line. (Color figure online)

a grid search detailed in Table 1. Figure 2 shows the impact of the parameters on the tokenizer's performance at the document level, particularly metrics Π and ψ.

In our analysis of the privacy parameter ϵ, we explored the lower and upper bounds of the privacy impact on the tokenizer's performance relative to our ground truth, the hypothetical centralized tokenizer, with ψ of 1.07 and Π of 0.06 as upper bound benchmark for our comparison.

- **High Privacy Budget** ($\epsilon \geq 10^{-2}$): When ϵ is set higher, less noise is introduced, resulting in better performance metrics that closely align with those of the hypothetical centralized tokenizer. Specifically, the fertility ψ exhibits a minimal decrease ranging from –2.8% to –0.93%, compared to the centralized tokenizer. This indicates high utility with minor privacy trade-offs. Similarly, the proportion of continued words Π also shows slight decrease, ranging from –1.67% to –1.33%, suggesting better preservation of linguistic structure under lower noise conditions. However, this setting carries a potential risk of exposing sensitive information due to minimal data perturbation.
- **Low Privacy Budget** ($\epsilon \leq 5 \cdot 10^{-4}$): A lower ϵ adds more noise, significantly enhancing data privacy but reducing utility. This leads to greater deviation from the centralized tokenizer's performance, with the proportion of continued words Π and fertility ψ showing a decrease, ranging from -3.0% to -2.0% and -7.48% to -3.74% respectively. Under these conditions, the tokenizer tends to generate more subtokens per input token at the document level, reflecting decreased efficiency and a stronger emphasis on privacy.

In a high privacy budget setting, our analysis revealed that sub-sampling of data, parameterized by p, significantly enhances tokenizer construction by introducing diversity or regularization, while the number of clients (K) has a marginal impact, suggesting redundancy in vocabulary does not significantly alter performance beyond a certain participant threshold. Furthermore, increasing θ narrows the vocabulary, enhancing tokenizer efficiency. These optimizations led to a refined performance close to our ground truth emphasizing the robustness of θ and p in achieving optimal tokenizer functionality in federated settings without the need for excessively large numbers of participants per round.

To this end, identifying a universal set of parameters that optimally balance privacy and utility proved challenging. However, the proposed algorithm demonstrated considerable robustness in terms of participant numbers and the volume of shared data, within the constraints of a given privacy budget. Consequently, we have empirically selected the best hyperparameters (marked by the red line in Fig. 2), where $\theta = 24$, $\epsilon = 0.01$, $p = 0.8$, and $K = 0.5$ are used to build the Federated Tokenizer τ_{fed}. The τ_{fed} tokenizer achieve ψ of 1.09 and Π of 0.08, representing decreases of -1.86% and –1.67%, respectively, compared to the hypothetical centralized tokenizer (Table 2).

6.2 Performance Comparison Across Tokenizers

This section provides a comparative analysis of the performance of various tokenizer models. Table 1 show state of the art generic and domain specific pretrained tokenizer, where the reported performance is the average cross 30 distributed datasets, same for local, centralized and federated tokenizers. We focus on two primary metrics in Sect. 5.2 at both the vocabulary and document levels.

At the vocabulary level, public tokenizers pretrained on generic corpora exhibit higher fertility ψ rates and proportion of continued words Π. On the other hand, the performance of domain-specific tokenizers show more tailored performance. However, models like FinGPT and FinMegatronGPT that supplement their training data with public corpora due to limited open-source financial datasets shows a decrease in the Π by approximately 1.9% compared to BERT, reflecting a modest enhancement in maintaining more original word forms. Still GPT3.5-Turbo stands out among generic models with a competitive performance that approaches domain-specific models due to its large-scale training base and the extremely high tokenizer size. Seeking more efficiency and smaller models, FinBERT, FinancialBERT and FLANG-BERT shows a best fertility ψ with a significantly lower proportion of continued words Π. Given that the vocabulary level metrics treat all unique corpus words with equal importance, we augmented our analysis with document level performance metrics, here fertility is weighted by the frequency of

Table 2. Performance Metrics of Various Tokenizer Models at Vocabulary and Document Levels Across Different Domains (Lower is Better). Each Subgroup Underlines the Best Model, with Gray Highlight for the Overall Best Across All Groups.

Model Name	Size	Vocabulary Level		Document Level	
		ψ	Π	ψ	Π
Off-The-Shelf Domain-General Tokenizers					
τ_{BERT}[18]	30,522	2.18 ± 0.21	0.54 ± 0.04	1.13 ± 0.03	0.10 ± 0.02
τ_{BART}[22]	50,265	2.14 ± 0.12	0.53 ± 0.02	1.21 ± 0.02	0.17 ± 0.01
τ_{GPT2}[33]	50,257	2.14 ± 0.12	0.53 ± 0.02	1.21 ± 0.02	0.17 ± 0.01
$\tau_{GPT3.5\text{-}Turbo}$[a]	100,261	$\underline{1.97\pm0.11}$	$\underline{0.49\pm0.03}$	1.16 ± 0.02	0.13 ± 0.01
τ_{T5}[34]	32,100	2.01 ± 0.18	0.50 ± 0.04	$\underline{1.11\pm0.02}$	$\underline{0.10\pm0.01}$
τ_{Llama2}[41]	32,000	2.13 ± 0.15	0.53 ± 0.03	1.15 ± 0.02	0.12 ± 0.01
Average		2.09 ± 0.15	0.52 ± 0.03	1.16 ± 0.02	0.13 ± 0.01
Off-The-Shelf Domain-Specific Tokenizers					
τ_{FinGPT}[45]	32,000	2.13 ± 0.15	0.53 ± 0.03	1.15 ± 0.02	0.12 ± 0.01
$\tau_{FinBERT}$[26]	30,873	1.91 ± 0.13	$\underline{0.30\pm0.05}$	1.08 ± 0.01	0.07 ± 0.01
$\tau_{SecBERT}$[b]	30,000	1.95 ± 0.12	0.36 ± 0.05	$\underline{1.07\pm0.01}$	0.08 ± 0.01
$\tau_{FLANG\text{-}BERT}$[37]	30,522	1.90 ± 0.14	0.37 ± 0.05	1.07 ± 0.01	0.07 ± 0.01
$\tau_{FinancialBERT}$[15]	30,873	$\underline{1.88\pm0.09}$	0.44 ± 0.03	1.11 ± 0.02	0.09 ± 0.01
$\tau_{FinanceDeBERTa}$[43]	128,000	1.91 ± 0.13	0.32 ± 0.06	1.08 ± 0.01	$\underline{0.06\pm0.01}$
$\tau_{FinMegatronGPT}$[44]	50,257	2.14 ± 0.12	0.53 ± 0.02	1.21 ± 0.02	0.17 ± 0.01
$\tau_{FinMegatronBERT}$[44]	30,522	2.11 ± 0.14	0.47 ± 0.05	1.16 ± 0.01	0.15 ± 0.01
$\tau_{FinanceDistilGPT2}$[c]	50,257	2.14 ± 0.12	0.53 ± 0.02	1.21 ± 0.02	0.17 ± 0.01
Average		2.00 ± 0.13	0.43 ± 0.04	1.13 ± 0.01	0.11 ± 0.01
Local Domain-Specific Tokenizers					
$\tau_{Equifax}$	50,257	$\underline{2.08\pm0.11}$	$\underline{0.52\pm0.03}$	$\underline{1.11\pm0.02}$	$\underline{0.10\pm0.02}$
$\tau_{Experian}$	50,257	$\underline{2.08\pm0.11}$	$\underline{0.52\pm0.03}$	$\underline{1.11\pm0.02}$	$\underline{0.10\pm0.02}$
$\tau_{TransUnion}$	50,257	2.10 ± 0.11	0.52 ± 0.03	1.12 ± 0.02	0.11 ± 0.02
$\tau_{BankOfAmerica}$	50,257	2.13 ± 0.12	0.53 ± 0.03	1.14 ± 0.01	0.12 ± 0.01
$\tau_{JPMorganChase}$	50,257	2.13 ± 0.12	0.53 ± 0.03	1.14 ± 0.01	0.12 ± 0.01
$\tau_{Citibank}$	50,257	2.15 ± 0.12	0.53 ± 0.03	1.14 ± 0.02	0.12 ± 0.01
$\tau_{CapitalOne}$	49,724	2.18 ± 0.12	0.54 ± 0.03	1.14 ± 0.02	0.12 ± 0.01
$\tau_{WellsFargo}$	50,257	2.13 ± 0.12	0.53 ± 0.03	1.14 ± 0.01	0.12 ± 0.01
$\tau_{Navient}$	38,790	2.32 ± 0.13	0.57 ± 0.02	1.22 ± 0.02	0.17 ± 0.01
$\tau_{Synchrony}$	40,141	2.29 ± 0.13	0.56 ± 0.02	1.19 ± 0.02	0.15 ± 0.02
τ_{Amex}	35,419	2.36 ± 0.12	0.57 ± 0.02	1.23 ± 0.02	0.18 ± 0.02
$\tau_{U.S.Bank}$	37,226	2.34 ± 0.12	0.57 ± 0.02	1.22 ± 0.02	0.17 ± 0.01
$\tau_{PortfolioRecovery}$	24,654	2.54 ± 0.12	0.60 ± 0.02	1.30 ± 0.04	0.23 ± 0.02
τ_{PayPal}	31,137	2.45 ± 0.13	0.59 ± 0.02	1.27 ± 0.03	0.21 ± 0.02
$\tau_{BreadFinancial}$	28,016	2.48 ± 0.12	0.60 ± 0.02	1.27 ± 0.03	0.21 ± 0.02
$\tau_{Discover}$	32,855	2.39 ± 0.12	0.58 ± 0.02	1.25 ± 0.02	0.19 ± 0.01
$\tau_{Nationstar}$	36,530	2.35 ± 0.12	0.57 ± 0.02	1.23 ± 0.02	0.19 ± 0.01
τ_{AES}	28,364	2.47 ± 0.12	0.59 ± 0.02	1.27 ± 0.02	0.21 ± 0.01
τ_{Ocwen}	36,203	2.36 ± 0.12	0.58 ± 0.02	1.24 ± 0.02	0.19 ± 0.01
$\tau_{EncoreCapital}$	24,206	2.55 ± 0.12	0.61 ± 0.02	1.31 ± 0.04	0.23 ± 0.03
τ_{TDBank}	29,585	2.45 ± 0.12	0.59 ± 0.02	1.27 ± 0.02	0.21 ± 0.01
τ_{PNC}	28,940	2.47 ± 0.12	0.59 ± 0.02	1.28 ± 0.02	0.21 ± 0.01
$\tau_{Barclays}$	26,127	2.51 ± 0.12	0.60 ± 0.02	1.29 ± 0.03	0.22 ± 0.02
$\tau_{Santander}$	25,305	2.54 ± 0.12	0.61 ± 0.02	1.31 ± 0.03	0.23 ± 0.02
τ_{Ally}	24,917	2.55 ± 0.12	0.61 ± 0.02	1.33 ± 0.02	0.24 ± 0.01
$\tau_{ResurgentCapital}$	20,066	2.66 ± 0.12	0.62 ± 0.02	1.36 ± 0.04	0.26 ± 0.02
τ_{USAA}	27,716	2.49 ± 0.12	0.60 ± 0.02	1.30 ± 0.02	0.23 ± 0.01
τ_{ERC}	18,165	2.70 ± 0.12	0.63 ± 0.02	1.38 ± 0.04	0.27 ± 0.02
$\tau_{NavyFederal}$	26,268	2.51 ± 0.12	0.60 ± 0.02	1.31 ± 0.02	0.23 ± 0.01
$\tau_{Coinbase}$	19,605	2.69 ± 0.12	0.63 ± 0.02	1.42 ± 0.03	0.29 ± 0.02
Average		2.38 ± 0.12	0.58 ± 0.02	1.24 ± 0.02	0.19 ± 0.01
Centralized/Federated Domain-Specific Tokenizers					
$\tau_{Centralized}$	50,257	$\underline{2.03\pm0.12}$	$\underline{0.51\pm0.03}$	$\underline{1.07\pm0.01}$	$\underline{0.06\pm0.01}$
$\tau_{Federated}$	50,257	2.04 ± 0.11	$\underline{0.51\pm0.03}$	1.09 ± 0.01	0.08 ± 0.01

[a] GPT3.5-Turbo:link

[b] SecBER:link

[c] FinanceDistilGPT2:link

each word in the corpus. Similarly, the proportion of continued words is adjusted by the extent to which a document's length changes.

This section provides a comparative analysis of the performance of various tokenizer models. Table 1 show state of the art generic and domain specific pretrained tokenizer, where the reported performance is the average cross 30 distributed datasets, same for local, centralized and federated tokenizers. We focus on two primary metrics in Sect. 5.2 at both the vocabulary and document levels.

At the vocabulary level, public tokenizers pretrained on generic corpora exhibit higher fertility rates ψ and the proportion of continued words Π. On the other hand, the performance of domain-specific tokenizers shows a more tailored performance. However, models like FinGPT and FinMegatronGPT that supplement their training data with public corpora due to limited open-source financial datasets shows a decrease in the Π by approximately 1.9% compared to BERT, reflecting a modest enhancement in maintaining more original word forms. Still GPT3.5-Turbo stands out among generic models with a competitive performance that approaches domain-specific models due to its large-scale training base and the extremely high tokenizer size. Seeking more efficiency and smaller models, FinBERT, FinancialBERT and FLANG-BERT shows a best fertility ψ with a significantly lower proportion of continued words Π. Given that the vocabulary level metrics treat all unique corpus words with equal importance, we augmented our analysis with document level performance metrics, here fertility is weighted by the frequency of each word in the corpus. Similarly, the proportion of continued words is adjusted by the extent to which a document's length changes. At the document level, the performance differences between generic and domain-specific tokenizers are even more pronounced. Excluding domain-specific GPT-based models and BPE-based tokenizers, generic models do not outperform their domain-specific counterparts. The T5, as the best generic model, shows a ψ of 1.11 and a Π of 0.10, while FLANG-BERT, the best domain-specific model, outperforms it with a ψ of 1.07 and a Π of 0.07. This represents a 3.6% improvement in fertility and a 30% reduction in word splits by FLANG-BERT compared to T5, highlighting the substantial benefits of domain-specific tokenizers in reducing unnecessary fragmentation and maintaining semantic integrity in specialized fields.

Local trained tokenizers developed for specific financial institutions, such as Bank of America, Citibank, and Wells Fargo, perform slightly better than generic models. However, they are less effective compared to domain-specific models like FLANG-BERT. In a federated learning environment, the Federated tokenizer $\tau_{Federated}$ exhibits competitive capabilities, with a ψ of 1.09 and a Π of 0.08, closely approaching the performance of the hypothetical centralized tokenizer $\tau_{Centralized}$ with ψ of 1.07 and Π of 0.06. This shows that Federated tokenizer not only outperforms all BPE-based and most WordPiece-based tokenizers but also provides a robust privacy-preserving solution that nearly matches the top-performing domain-specific model FLANG-BERT as centralized model in maintaining linguistic integrity in tokenization.

7 Conclusion

This paper presents the Federated Byte-Level BPE Tokenizer (FedByteBPE), a language model tokenizer that combines the robustness of federated learning with the efficiency of byte-level tokenization to ensure privacy without sacrificing performance. Our extensive comparative analysis and parameter studies confirm that FedByteBPE not only matches but often exceeds the performance of centralized pretrained tokenizers, demonstrating its effectiveness in real-world financial distributed datasets. FedByteBPE consistently outperformed local models developed for major banks corpora in terms of document-level fertility and proportion of continued words to federated domain-specific language processing in the financial sector.

Acknowledgment. We acknowledge the invaluable knowledge exchange facilitated by the Data Analysis Lab team and our industry partners' experts. Partial funding for this work was provided by the Luxembourg National Research Fund (FNR) under grant number 15829274, supplemented by ANR-20-CE39-0008.

References

1. Amid, E., et al.: Public data-assisted mirror descent for private model training. In: International Conference on Machine Learning, pp. 517–535. PMLR (2022)
2. Arapinis, M., Figueira, D., Gaboardi, M.: Sensitivity of counting queries. In: International Colloquium on Automata, Languages, and Programming (ICALP) (2016)
3. Bagdasaryan, E., Song, C., van Dalen, R., Seigel, M., Áine Cahill: training a tokenizer for free with private federated learning. In: ACL (2022). https://arxiv.org/abs/2203.09943
4. Balle, B., Barthe, G., Gaboardi, M.: Privacy amplification by subsampling: tight analyses via couplings and divergences. Adv. Neural Inform. Process. Syst. **31** (2018)
5. Balle, B., Barthe, G., Gaboardi, M.: Privacy profiles and amplification by subsampling. J. Priv. Confidential. **10**(1) (2020)
6. Basu, P., Roy, T.S., Naidu, R., Muftuoglu, Z.: Privacy enabled financial text classification using differential privacy and federated learning. arXiv preprint arXiv:2110.01643 (2021)
7. Benamar, A., Grouin, C., Bothua, M., Vilnat, A.: Evaluating tokenizers impact on oovs representation with transformers models. In: Proceedings of the Thirteenth Language Resources and Evaluation Conference, pp. 4193–4204 (2022)
8. Berglund, M., van der Merwe, B.: Formalizing bpe tokenization. arXiv preprint arXiv:2309.08715 (2023)
9. Cai, D., Wu, Y., Wang, S., Lin, F.X., Xu, M.: Efficient federated learning for modern nlp. In: Proceedings of the 29th Annual International Conference on Mobile Computing and Networking. ACM MobiCom 2023, Association for Computing Machinery, New York (2023). https://doi.org/10.1145/3570361.3592505
10. Chowdhery, A., et al.: Palm: scaling language modeling with pathways. J. Mach. Learn. Res. **24**(240), 1–113 (2023)
11. Dagan, G., Synnaeve, G., Rozière, B.: Getting the most out of your tokenizer for pre-training and domain adaptation. arXiv preprint arXiv:2402.01035 (2024)

12. Denning, D.E.: Secure statistical databases with random sample queries. ACM Trans. Database Syst. (TODS) **5**(3), 291–315 (1980)
13. Ding, S., Renduchintala, A., Duh, K.: A call for prudent choice of subword merge operations in neural machine translation. arXiv preprint arXiv:1905.10453 (2019)
14. Dwork, C., Roth, A., et al.: The algorithmic foundations of differential privacy. Foundat. Trends Theoret. Comput. Sci. **9**(3–4), 211–407 (2014)
15. Hazourli, A.: Financialbert-a pretrained language model for financial text mining. Res. Gate **2** (2022)
16. Kairouz, P., McMahan, B., Song, S., Thakkar, O., Thakurta, A., Xu, Z.: Practical and private (deep) learning without sampling or shuffling. In: International Conference on Machine Learning, pp. 5213–5225. PMLR (2021)
17. Kasiviswanathan, S.P., Lee, H.K., Nissim, K., Raskhodnikova, S., Smith, A.: What can we learn privately? SIAM J. Comput. **40**(3), 793–826 (2011)
18. Kenton, J.D.M.W.C., Toutanova, L.K.: Bert: Pre-training of deep bidirectional transformers for language understanding. In: Proceedings of naacL-HLT, vol. 1, p. 2 (2019)
19. Kobayashi, G., Kuribayashi, T., Yokoi, S., Inui, K.: Transformer language models handle word frequency in prediction head. In: Rogers, A., Boyd-Graber, J., Okazaki, N. (eds.) Findings of the Association for Computational Linguistics: ACL 2023, pp. 4523–4535. Association for Computational Linguistics, Toronto, Canada (Jul 2023).https://doi.org/10.18653/v1/2023.findings-acl.276, https://aclanthology.org/2023.findings-acl.276
20. Kudo, T., Richardson, J.: Sentencepiece: A simple and language independent subword tokenizer and detokenizer for neural text processing. arXiv preprint arXiv:1808.06226 (2018)
21. Larbi, I.B.C., Burchardt, A., Roller, R.: Clinical text anonymization, its influence on downstream nlp tasks and the risk of re-identification. In: Proceedings of the 17th Conference of the European Chapter of the Association for Computational Linguistics: Student Research Workshop, pp. 105–111 (2023)
22. Lewis, M., et al.: Bart: denoising sequence-to-sequence pre-training for natural language generation, translation, and comprehension. arXiv preprint arXiv:1910.13461 (2019)
23. Li, Q., Diao, Y., Chen, Q., He, B.: Federated learning on non-iid data silos: an experimental study. In: 2022 IEEE 38th International Conference on Data Engineering (ICDE), pp. 965–978. IEEE (2022)
24. Limisiewicz, T., Balhar, J., Mareček, D.: Tokenization impacts multilingual language modeling: Assessing vocabulary allocation and overlap across languages. arXiv preprint arXiv:2305.17179 (2023)
25. Lin, B.Y., et al.: Fednlp: benchmarking federated learning methods for natural language processing tasks. arXiv preprint arXiv:2104.08815 (2021)
26. Liu, Z., Huang, D., Huang, K., Li, Z., Zhao, J.: Finbert: a pre-trained financial language representation model for financial text mining. In: Proceedings of the Twenty-ninth International Conference on International Joint Conferences on Artificial Intelligence, pp. 4513–4519 (2021)
27. Mazzarino, S., Minieri, A., Gilli, L.: Nerpii: a python library to perform named entity recognition and generate personal identifiable information. In: Proceedings of the Seventh Workshop on Natural Language for Artificial Intelligence (NL4AI 2023) co-located with 22th International Conference of the Italian Association for Artificial Intelligence (AI* IA 2023) (2023)
28. McMahan, H.B., Ramage, D., Talwar, K., Zhang, L.: Learning differentially private recurrent language models. arXiv preprint arXiv:1710.06963 (2017)

29. Nayak, A., Timmapathini, H., Ponnalagu, K., Venkoparao, V.G.: Domain adaptation challenges of bert in tokenization and sub-word representations of out-of-vocabulary words. In: Proceedings of the first workshop on insights from negative results in NLP, pp. 1–5 (2020)
30. Nogueira, R., Jiang, Z., Lin, J.: Investigating the limitations of transformers with simple arithmetic tasks. arXiv preprint arXiv:2102.13019 (2021)
31. Petrov, A., La Malfa, E., Torr, P., Bibi, A.: Language model tokenizers introduce unfairness between languages. Adv. Neural Inform. Process. Syst. **36** (2024)
32. Provilkov, I., Emelianenko, D., Voita, E.: Bpe-dropout: simple and effective subword regularization. arXiv preprint arXiv:1910.13267 (2019)
33. Radford, A., Wu, J., Child, R., Luan, D., Amodei, D., Sutskever, I., et al.: Language models are unsupervised multitask learners. OpenAI blog **1**(8), 9 (2019)
34. Raffel, C., Shazeer, N., et al.: Exploring the limits of transfer learning with a unified text-to-text transformer. J. Mach. Learn. Res. **21**(140), 1–67 (2020). http://jmlr.org/papers/v21/20-074.html
35. Rust, P., Pfeiffer, J., Vulić, I., Ruder, S., Gurevych, I.: How good is your tokenizer? on the monolingual performance of multilingual language models. In: Proceedings of the 59th Annual Meeting of the Association for Computational Linguistics and the 11th International Joint Conference on Natural Language Processing (Volume 1: Long Papers), pp. 3118–3135. Association for Computational Linguistics (Aug 2021).https://doi.org/10.18653/v1/2021.acl-long.243, https://aclanthology.org/2021.acl-long.243
36. Sennrich, R., Haddow, B., Birch, A.: Neural machine translation of rare words with subword units. arXiv preprint arXiv:1508.07909 (2015)
37. Shah, R.Set al.: When flue meets flang: benchmarks and large pretrained language model for financial domain. In: Proceedings of the 2022 Conference on Empirical Methods in Natural Language Processing (EMNLP). Association for Computational Linguistics (2022)
38. Shoham, O.B., Rappoport, N.: Federated learning of medical concepts embedding using behrt. arXiv preprint arXiv:2305.13052 (2023)
39. Tian, Y., Wan, Y., Lyu, L., Yao, D., Jin, H., Sun, L.: Fedbert: when federated learning meets pre-training. ACM Trans. Intell. Syst. Technol. (TIST) **13**(4), 1–26 (2022)
40. Toraman, C., Yilmaz, E.H., Şahinuç, F., Ozcelik, O.: Impact of tokenization on language models: an analysis for Turkish. ACM Trans. Asian Low-Resource Lang. Inform. Process. **22**(4), 1–21 (2023)
41. Touvron, H., et al.: Llama: Open and efficient foundation language models. arXiv preprint arXiv:2302.13971 (2023)
42. Wang, C., Cho, K., Gu, J.: Neural machine translation with byte-level subwords. In: Proceedings of the AAAI conference on artificial intelligence, vol. 34, pp. 9154–9160 (2020)
43. Wang, Y.J., Li, Y., Qin, H., Guan, Y., Chen, S.: A novel deberta-based model for financial question answering task. arXiv preprint arXiv:2207.05875 (2022)
44. Wu, X.: Finmegatron: large financial domain language models. Proc. Second Type Res. Meeting **2021**(FIN-026), 22 (2021)
45. Yang, H., Liu, X.Y., Wang, C.D.: Fingpt: open-source financial large language models. In: FinLLM Symposium at IJCAI 2023 (2023)
46. Yang, Q., Liu, Y., Chen, T., Tong, Y.: Federated machine learning: concept and applications. ACM Trans. Intell. Syst. Technol. (TIST) **10**(2), 1–19 (2019)

47. Zhu, W., Kairouz, P., McMahan, B., Sun, H., Li, W.: Federated heavy hitters discovery with differential privacy. In: International Conference on Artificial Intelligence and Statistics, pp. 3837–3847. PMLR (2020)
48. Zouhar, V., et al.: A formal perspective on byte-pair encoding. arXiv preprint arXiv:2306.16837 (2023)

Robust Stance Detection: Understanding Public Perceptions in Social Media

Nayoung Kim[1](✉), David Mosallanezhad[2], Lu Cheng[3], Michelle V. Mancenido[4], and Huan Liu[1]

[1] Arizona State University, Tempe, AZ 85281, USA
{nkim48,huanliu}@asu.edu
[2] NVIDIA, Santa Clara, CA 95051, USA
dmosallanezh@nvidia.com
[3] University of Illinois Chicago, Chicago, IL 60607, USA
lucheng@uic.edu
[4] Arizona State University West Valley Campus, Glendale, AZ 85306, USA
mmanceni@asu.edu

Abstract. The abundance of social media data has presented opportunities for accurately determining public and group-specific stances around policy proposals or controversial topics. In contrast with sentiment analysis which focuses on identifying prevailing emotions, stance detection identifies precise positions (i.e., supportive, opposing, neutral) relative to a well-defined topic, such as perceptions toward specific global health interventions during the COVID-19 pandemic. Traditional stance detection models, while effective within their specific domain (e.g., attitudes towards masking protocols during COVID-19), often lag in performance when applied to new domains *and* topics due to changes in data distribution. This limitation is compounded by the scarcity of domain-specific, labeled datasets, which are expensive and labor-intensive to create. A solution we present in this paper combines counterfactual data augmentation with contrastive learning to enhance the robustness of stance detection across domains and topics of interest. We evaluate the performance of current state-of-the-art stance detection models, including a prompt-optimized large language model, relative to our proposed framework succinctly called STANCE-C3 (domain-adaptive Cross-target STANCE detection via Contrastive learning and Counterfactual generation). Empirical evaluations demonstrate STANCE-C3's consistent improvements over the baseline models with respect to accuracy across domains and varying focal topics. Despite the increasing prevalence of general-purpose models such as generative AI, specialized models such as STANCE-C3 provide utility in safety-critical domains wherein precision is highly valued, especially when a nuanced understanding of the concerns of different population segments could result in crafting more impactful public policies.

Keywords: Stance Detection · Contrastive Learning · Counterfactual Generation · Domain Adaptation · Cross-target Stance Detection

L. M. Aiello et al. (Eds.): ASONAM 2024, LNCS 15212, pp. 21–37, 2025.
https://doi.org/10.1007/978-3-031-78538-2_2

1 Introduction

The influence of public perceptions on policy development and societal outcomes is well-established in the context of public health [44], legislation [33], environmental sustainability [9], and other community-focused sectors. While policymakers from previous generations relied on expensive surveys or town halls to gauge public opinion toward a specific issue, the meteoric rise of social media has facilitated the swift and cost-effective collection of user-generated insights, which are voluntarily and freely provided by netizens. At the height of COVID-19, for example, social media data and natural language processing (NLP) models were leveraged to establish associations between public support for non-pharmaceutical interventions and the decreased spread of the virus [1].

NLP models that infer the collective attitude or stance toward specific targets using corpora of user-provided data from online forums (e.g., X) are called *stance detection* models. Unlike sentiment analysis, which detects general emotional responses toward a topic, stance detection models identify positions toward a target by analyzing the context, linguistic nuances, and even the implied meanings within the text. For instance, consider an online post stating, "*I hate having to wear a mask every day, but I totally agree it's essential to stop the spread of COVID-19*." Sentiment analysis might classify this as negative due to the expressions of dislike and discomfort. However, despite the negative emotional tone, the stance towards the target (mask-wearing as a COVID-19 health measure) is clearly supportive. Consequently, stance detection offers significant opportunities to leverage social media data for accurately identifying public stances towards various initiatives. This can substantially enhance policy-making, refine strategies that impact the public, and contribute to a more informed public dialogue [2].

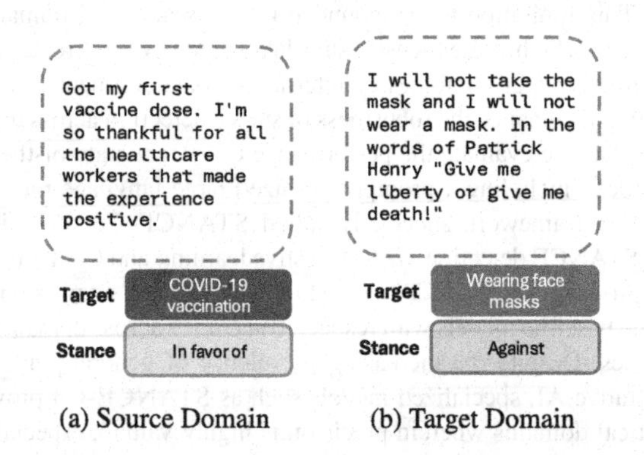

Fig. 1. Examples of domain-adaptive cross-target stance detection. (a) Tweets from source domain are collected using target-related keywords/hashtags (e.g., *get vaccinated*) from January 1st, 2020 to August 23rd, 2021, whereas (b) tweets from target domain are collected with target-related hashtags (e.g., *#MasksSaveLives*) from February 27th, 2020 to August 20th, 2020.

One of the significant challenges in leveraging social media data for stance detection is the limited adaptability of existing models across different domains. This limitation is

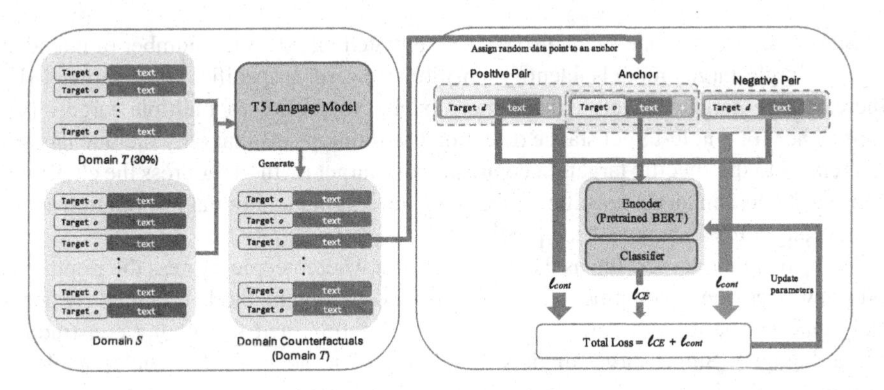

Fig. 2. The STANCE-C3 architecture consists of two key components: a counterfactual data generation network and a contrastive learning network. The counterfactual network (left) is built on a T5-based model and generates training examples that maintain sentence structure but has increased diversity in semantic context. The contrastive learning network (right) then uses the augmented dataset to learn cross-target representations by minimizing distances between examples with the same stance (positive pairs) and maximizing distances between examples with different stance (negative pairs). This approach helps acquire domain-invariant features, improving stance detection across targets.

particularly acute in rapidly changing situations, where shifting public opinion across various domains (e.g., different time points) can significantly influence responses to crises, such as public health emergencies. The social cost of misinformation and the urgent need for responsive public opinion analysis across diverse domains highlight a pressing challenge: developing stance detection models that are both robust and adaptable to different contexts.

Current state-of-the-art (SOTA) stance detection models are largely focused on improving the model's performance within a single domain. Yet, these models often underperform when applied to domains distinct from their training datasets [20]. We attribute this performance degradation to several factors: (1) Attention bias toward events related to the given target, a phenomenon we will illustrate with experiments in later sections; (2) The absence of target words in stance prediction during the training phase; and (3) The model's reliance on auxiliary words (i.e., words that are not necessarily the target or stance) specific to a given domain. Further, the limited availability of training data restricts these models' effectiveness across multiple domains and targets. Although Large Language Models (LLMs) have gained popularity, their resource-intensive and general-purpose nature pose issues in their usefulness for stance detection. LLMs demand substantial computational resources for training and inference, making them financially and environmentally costly. For specific tasks, smaller models can achieve comparable or even better performance without the resource burden [17,39]. Additionally, human factors significantly influence the inference process when using LLMs; for example, it is well-established that different prompting techniques could yield widely varying results [43].

Stance detection models can also be differentiated based on the number of intended targets. Single-target models identify positions toward a specific, fixed target [31], whereas cross-target models extend the model's versatility to multiple targets [4]. Approaches for single-target stance detection use training sets that only include labeled data related to the specific target. In contrast, cross-target methods address the challenge of the sparseness in target-specific labeled data, leveraging additional information about the new target for accurate stance predictions.

With social media serving as the main platform where people express their opinions and grievances, the development of robust stance detection methods is crucial for accurately gauging public opinion, deterring misinformation, and facilitating constructive online dialogue. Single-target models often struggle to adapt to the quickly evolving landscape of public discourse, where topics such as climate change, social justice, and public health are deeply interwoven in their influence and digital presence. Cross-target stance detection models offer a more comprehensive understanding of public opinions across related topics, eliminating the need to rebuild models for each new issue. This not only makes AI tools more responsive and adaptable but also improves their ability to provide real-time insights across a wide range of issues. Such capabilities are particularly valuable in fields like policy making and public health, where timely and informed decision-making is crucial. For example, a cross-target model could analyze opinion on health interventions like vaccination and mask-wearing, despite varying domains and data sources, enabling more effective responses to public health crises. Figure 1 shows an example of a domain-adaptive cross-target stance detection problem, illustrating differences in domains by search query and data acquisition time periods, while targets differ in topical content (e.g., *vaccination* vs. *mask-wearing*). Other attributes of the sampling process could account for distinctions among domains and targets, such as data sources (e.g., social media vs. news), content genre (e.g., political vs. gossip news), or publication type (e.g., opinion pieces vs. reportage).

In this paper, we introduce STANCE-C3 as a domain- and target-robust model that uses Counterfactual Data Augmentation (CDA) [30], an approach that has been shown to decrease bias and improve the robustness of neural network-based models [40]. We combine CDA with text style transfer methods [19] to facilitate effective text transfer from one domain to another. These language model-based data augmentation techniques are expected to enrich existing datasets, improving the performance of stance detection models across multiple domains.

Previous work on stance detection addressed domain robustness using adversarial training and incorporating the target word as an input to the model. However, this approach often led to overfitting on the specific target, restricting its robustness in performance for other targets. In STANCE-C3, we modify a self-supervised technique called contrastive learning [7] to extract both shared and distinct high-level features to capture specific characteristics of statement-target pairs. We incorporate components such as cosine similarity and negative pair loss to direct the model to minimize the distance between pairs with the same stance labels and maximize the distance between pairs with different stance labels. By requiring fewer manually annotated data points and combining CDA during the pre-processing stage, STANCE-C3 ensures that the training set is optimally constructed for a cross-domain and cross-target setting (i.e., a sufficient balance between data points that exhibit similarities and differences, allowing the model to infer salient features that are useful for stance detection) while avoiding overfitting.

The major contributions of this work are threefold: firstly, it addresses the problem of domain-adaptive cross-target stance detection, offering a novel approach to train models effectively even when data on the target domain and subject are limited. Secondly, it tests the potential of a modified contrastive loss learning method as a more effective alternative for improving model robustness in classification tasks. Lastly, it demonstrates STANCE-C3's efficacy and improvement over existing SOTA stance detection models through rigorous experiments and ablation studies on real-world datasets.

2 Related Work

STANCE-C3 is proposed in this paper to improve existing models for stance detection and robustness across domains. In this section, we review current NLP SOTA models for stance detection and domain adaptation.

2.1 Stance Detection

There are two general categories of distinctions among stance detection models. Firstly, models are distinguished on how stance features are modeled. *Content-level* approaches use linguistic features such as topics [15] or targets [42] alongside sentiment information [37]. *User-level* approaches focus on user-related attributes interactions [35], information [5], and timelines [45]. Hybrid models combine content- and user-level features for post representations [22]. In this paper, we focus on content-level modeling with target embedding using BERT due to its practicality in data acquisition.

Secondly, stance detection models differ concerning the specificity of the target of interest. While many previous works focus on specific single-target scenarios [8], recent studies have considered cross-target [42], multi-target [24], and few-shot or zero-shot stance detection [28]. While both cross-target and zero-shot stance detection models are trained on one or more source targets and used on previously unseen targets [26], zero-shot methods can predict stances for unseen targets that are not necessarily related to the target in the training set [3]. Both attempt to extract transferable knowledge from sources using methods such as hierarchical contrastive learning [25]. STANCE-C3 uses a similar but modified approach to the contrastive loss during training.

2.2 Domain Adaptation

As a category of transfer learning, domain Adaptation (DA) leverages agnostic or common information between domains with the target task remaining constant [10]. There are several DA setups, including unsupervised domain adaptation (UDA) where both labeled source data and unlabeled target data are available [29], semi-supervised domain adaptation (SSDA) where labeled source data and a small number of labeled target data are accessible [23], and supervised domain adaptation (SDA) where both labeled source and target data are available during training [27]. Our focus is on SSDA, where a smaller set of labeled target data is used in the training process.

Domain adaptive models are categorized by their focus on the model, training data, or both. Model-centric approaches aim to construct domain-agnostic model structures, such as in Deng et al. [11]. Data-centric methods enhance robustness by leveraging on data attributes using techniques such as the incorporation of loss functions based on inter-domain distances [13] or the employment of pre-training [14]. Finally, hybrid approaches combine model- and data-centric approaches [16]. Many of these DA methods grapple with spurious correlations from differing training and test dataset distributions, thus limiting their effectiveness. Our hybrid approach addresses this by using domain counterfactual examples, which minimizes the impact of domain-invariant features in cross-domain classification.

3 Problem Statement

Suppose we have two sets of instances $\mathcal{D}_{\mathcal{S}}$ and $\mathcal{D}_{\mathcal{T}}$ from source domain $\mathcal{S}$ and target domain $\mathcal{T}$, respectively. The set $\mathcal{D}_{\mathcal{S}} = (r_{\mathcal{S}}, t^o, y_{\mathcal{S}})^N$ contains N annotated instances from a source domain $\mathcal{S}$ where each instance $r_{\mathcal{S}}$ is a text (e.g., tweet) consisting of a sequence of k words, and $y_{\mathcal{S}}$ is the stance toward a target t^o associated with each annotated instance. Another set $\mathcal{D}_{\mathcal{T}} = (r_{\mathcal{T}}, t^d)^{N'}$ is a set of N' instances from target domain $\mathcal{T}$ with the destination target t^d. The goal of the domain-adaptive cross-target stance detection is to learn the stance classifier F that uses the features of the source domain $\mathcal{S}$ and target words t^o in $\mathcal{D}_{\mathcal{S}}$ and predict the stance label $y_{\mathcal{T}}$ of each text $r_{\mathcal{T}}$ in $\mathcal{D}_{\mathcal{T}}$. We formally define the problem as:

Definition. Given statements from two separate datasets $\mathcal{D}_{\mathcal{S}}$ and $\mathcal{D}_{\mathcal{T}}$, and corresponding targets t^o and t^d from the source $\mathcal{S}$ and target $\mathcal{T}$ domains, respectively, learn a domain and target-agnostic text representation using $\mathcal{D}_{\mathcal{S}}$ and a small portion of labeled $\mathcal{D}_{\mathcal{T}}$ that can be classified correctly by the stance classifier F.

4 Proposed Model

In this section, we describe our proposed model STANCE-C3. As illustrated in Fig. 2, the architecture consists of two main components: (1) a domain counterfactual generator, which is trained with input representations of the labeled samples from the source domain and a small portion from the target domain, to generate a set of domain counterfactuals. These generated counterfactuals enrich the knowledge of the target domain during the stance classification training phase. (2) These samples are then combined with the original dataset (i.e., source domain dataset with a small portion of the target domain dataset) to train a BERT-based stance classifier. This training utilizes contrastive learning to develop a domain-adaptive, cross-target stance detection classifier. We partition a small subset (i.e., 50) of randomly selected samples from the combined datasets into positive and negative pairs based on their stance, regardless of the target. This contrastive loss approach enhances the method's ability to generalize stance detection to unseen targets. The following subsections provide a detailed explanation of these two stages.

4.1 Domain-Adaptive Single-Target

One major challenge in domain-adaptive stance detection is that collecting samples for a target domain can be expensive and usually requires human annotations. Even with annotated data, modeling a domain-adaptive stance detection is difficult due to the complexity of the dataset which contains far more distinct features over domains [36]. These features often form spurious correlations within a specific domain and make the models brittle to domain shift. For instance, the presence of question marks in a specific stance or highly negative sentiment associated with a particular stance has been demonstrated to lead to a performance decrease [20].

To address the aforementioned challenges and control these variants on the domain-adaptive stance detection problem, we adapt counterfactual concepts to the source domain. A domain-counterfactual example is defined as a result text of intervening on the domain variable of the original example and converting it to the target domain while keeping other features (e.g., overall structure of the sentence) fixed. Given an example data from the COVID-19 vaccination domain (source domain), the generator first recognizes the terms related to the domain. Then, it intervenes on these terms, replacing them with text that links the example to the wearing face mask domain (target domain) while keeping the stance. We utilize source domain instances in combination with a small portion of target domain instances to incorporate target domain information in generating the counterfactuals specific to the target domain. These counterfactuals are then combined as input for the subsequent stage. Following the work of [6], we adopt a T5-based language model to generate counterfactuals for a given target domain $\mathcal{T}$. In this approach, we train the T5-based LM to convert samples from source domain $\mathcal{S}$ to a given target domain $\mathcal{T}$.

Generating domain counterfactuals consists of two main steps - domain corruption and reconstruction. For domain corruption, we get the masking score of all n-grams w ($n \in \{1, 2, 3\}$) in a dataset $\mathcal{D}_{\mathcal{S}}$ following [6]. The affinity of w to domain $\mathcal{S}$ is defined as $\rho(w, \mathcal{S}) = P(\mathcal{S}|w)(1 - \frac{H(\mathcal{S}|w)}{\log N})$, where $H(\mathcal{S}|w)$ is the entropy of $\mathcal{S}$ given each n-gram w and N is the number of unlabeled domains we know. The final masking score given source domain $\mathcal{S}$ and target domain $\mathcal{T}$ is $mask(w, \mathcal{S}, \mathcal{T}) = \rho(w, \mathcal{S}) - \rho(w, \mathcal{T})$. A higher score implies the n-gram is highly related to the source domain and distant from the target domain.

The next step involves reconstructing the masked examples $M(x)$ by concatenating them with a trainable embedding matrix, or domain orientation vectors. These vectors are embedded with domain-specific information including the nuances, vocabulary, and style of a particular domain. They are initialized with the embedding vector of each domain name and representative words from that domain. The concatenated matrix is then trained using a T5 architecture. Note that $\mathcal{T} = \mathcal{S}$ and $mask(w, \mathcal{S}, \mathcal{T}) = 0$ during the training process. Given the target domain and its orientation vectors in the test phase, the trained model generates domain counterfactual x'.

4.2 Domain-Adaptive Cross-Target

One of the main challenges in detecting stances across different targets is the variation in data distribution for each target, even within the same domain. For instance, the word *WHO* may appear more frequently in the domain where the target is *COVID-19 vaccination* than in another domain where the target is *Donald Trump* [41]. The frequent co-occurrences of specific target words with certain instances make the model biases the model's learning process. Therefore it is necessary to identify an effective way to model transferable knowledge. To learn target-invariant features in instances from both the origin t^o and destination t^d targets, we use a simple yet effective supervised contrastive learning [21] approach. This loss function aims to enhance the separability of different classes by maximizing the distance between an anchor and negative samples.

In our problem setting, considering samples with different stance targets t^o and t^d, the goal is to reduce the distance between samples sharing the same stance label and increase it for those with differing labels. Note that contrastive learning is applied to all pairs, regardless of whether the targets in the pair are equivalent or not. This approach creates a representation for the stance classifier that indicates the relation between the statement and the stance target t. We introduce a modified version of the supervised contrastive learning tailored to better address the nuances of our problem setting:

$$\ell_{\text{cont}} = \sum_{i \in I} \frac{-1}{|P(i)|} \sum_{p \in P(i)} \log \frac{\exp(z_i \cdot z_p/\tau) \cdot sim(z_i, z_p)}{\sum_{a \in A(i)} \exp(z_i \cdot z_a/\tau)} + \sum_{i \in I} \frac{1}{|N(i)|} \sum_{n \in N(i)} \log \frac{\exp(z_i \cdot z_n/\tau) \cdot sim(z_i, z_n)}{\sum_{a \in A(i)} \exp(z_i \cdot z_a/\tau)}$$

The loss takes a batch of samples $i \in I$ as an "anchor" from the training set where each sample is mapped to a vector z_i. Each anchor allows for multiple positive vectors z_p that have the same label as an anchor ($y_j = y_i$) in addition to multiple negative vectors z_n ($z_n \in A(i)$, $y_n \neq y_i$). $P(i)$ and $N(i)$ represent the indices of all positive and negative vectors, respectively. τ shows the temperature parameter that affects the distance of instance discrimination. The notation $sim(\cdot)$ refers to the cosine similarity. Here, we introduce the second term to balance the weight of positive and negative pairs. According to our observations on experiments involving different variants of the modified contrastive loss, we discovered that weighing the original loss with text similarity improves the model's generalization ability across different targets. Note that anchor z_i is from target t^o but z_p and z_n are from both t^d and t^o for target-agnostic feature learning.

Optimization Algorithm. The training process of STANCE-C3 for domain-adaptive cross-target scenario has two stages[1]. In the first stage, the parameters of the T5-based LM are trained for conversion between source $\mathcal{S}$ and the target $\mathcal{T}$ domain following the work of [6]. This stage enriches the input dataset to include more samples similar to the target domain. The next stage uses the contrastive learning approach to create a cross-target sentence representation. We randomly selected 50 anchors from $\mathcal{D}_\mathcal{S}$ and its domain counterfactuals, along with their positive and negative pairs from across all dataset except for generated counterfactuals. For each instance x_i, the overall loss is formulated as $\ell_{\text{total}} = \lambda\ell_{\text{cont}} + (1 - \lambda)\ell_{\text{CE}}$ where ℓ_{CE} stands for the

[1] Source code and dataset is available at https://github.com/Davood-M/RobustStanceDet.

Table 1. Datasets' statistics: For COVID-19-Stance dataset, a significant number of tweets could not be downloaded due to Twitter restrictions and deletion of the tweets. In this context, the number of samples mentioned on the left refers to the actual sample size whereas the original data count is on the right.

Dataset	Target	Platform	Labels
CoVaxNet	COVID-19 Vaccination	Twitter	Pro (1,495,991), Anti (335,229)
COVID-19-Stance	Wearing a face mask, Anthony S. Fauci, M.D., Keeping schools closed, Stay at home orders	Twitter	In-favor (2,421/43,799), Against (1,577/29,693)

cross-entropy loss for classification and (ℓ_{cont}) represents the contrastive loss. λ determines the balance between the two loss components. Note that while the classification loss is computed over the entire training dataset, the contrastive loss is derived solely from the selected pairs mentioned previously. The cross-entropy loss is calculated as $CE = -\sum_{i=1}^{k} y_i \log(p_i)$ for k classes based on the true stance label y and predicted stance probability p for the ith class. Similar to [21] we set the temperature parameter $\tau = 0.08$.

5 Experiments

The experiments conducted in this study investigate how changing the domains and targets affects the performance of several stance detection models, including STANCE-C3. We also examine the effects of data augmentation and contrastive learning on performance accuracy in cross-domain and cross-target scenarios. More specifically, we aim to answer the following research questions: **Q1.** Does STANCE-C3 achieve better performance in comparison to other SOTA when trained on source domain $\mathcal{S}$ with some data from target domain $\mathcal{T}$, and tested on target domain $\mathcal{T}$ for target words in the source domain $\mathcal{S}$? **Q2.** Does STANCE-C3 achieve better performance in comparison to other SOTA when trained on a source domain $\mathcal{S}$ and tested on a target domain $\mathcal{T}$ for target words not in the source domain? **Q3.** What is the effect of changes in the model's parameters and components on its performance in the stance detection task? We consider the following scenarios in the experimental design: (1) single domain + single target, (2) cross-domain + single target, and (3) cross-domain + cross-target, where the targets and domains of the datasets have clear distributional differences. We address **Q1** by training and testing the baseline and proposed models under scenarios (1) and (2) where targets are the same for training and testing sets. For **Q2**, the models are evaluated under scenario (3) to compare and contrast the models' domain-adaptive and cross-target performance. Finally, we address **Q3** by performing ablation studies and parameter analysis. Notably, domain counterfactual generation is applied in all scenarios, and contrastive learning is employed only for scenario (3).

5.1 Baselines

We consider the following baselines. Table 2 shows the difference between baselines and the proposed approach. Notably, we have implemented baselines for both cross-domain and cross-target tasks regardless of their specific task targets. This decision

stems from the unique nature of the task, which presents limitations in terms of available baselines. We chose BERT as our base model instead of larger language models due to its demonstrated effectiveness in text classification, accessibility to task-specific pretrained models, and computational efficiency.
MoLE [16]**:** Utilizes mixture-of-experts models where each domain expert represents each domain and produces probabilities for all the target labels. Then uses domain adversarial training to learn domain-invariant representations.
MTL [36]**:** A benchmarking framework for evaluating the robustness of stance detection systems. The proposed method leverages a diverse set of challenging datasets with varying levels of noise, bias, and adversarial attacks to evaluate the performance and robustness of stance detection systems. For each dataset, this model adds a domain-specific projection layer to the final layer of a pre-trained language model and freezes other layers during training.
RLFND [32]**:** A domain-adaptive reinforcement learning framework for fake news detection. The proposed method leverages a deep reinforcement learning algorithm to learn the optimal policy for fake news detection while adapting to the domain shift between the source and target domains. **MTSD** [24]**:** A multi-task framework that performs stance detection on multiple targets. The framework leverages a shared encoder to capture the common features of the text and a task-specific decoder to predict the stance toward each target.
Llama-2-7B [38]**:** A pre-trained and fine-tuned large language model released by Meta AI, designed for a variety of NLP tasks. We use the smallest variant for our problem setting. Due to Llama-2's incapability of performing stance detection on our dataset with an acceptable performance, NeMo framework was used to perform Supervised Fine-Tuning (SFT) on this model.

Table 2. Baselines targeted goals indicate the difference between the proposed approach STANCE-C3 and the baselines.

Model	Cross-Domain	Cross-Target
MoLE [16]	✓	✓
MTL [36]	✓	
REAL-FND [32]	✓	
MTSD [24]		✓
Llama-2-7B [38]	✓	✓
STANCE-C3	✓	✓

5.2 Datasets

We use two representative datasets related to COVID-19, each from different domains and targets (details provided in Table 1). The large-scale Twitter dataset CoVaxNet [18] was employed for training across both domains and targets. CoVaxNet was compiled

Table 3. Cross-domain, single-target performance results in accuracy and AUC (in parentheses). In this experiment we only use the datasets that have the same target word (p-value < 0.05 for all McNemar's tests).

Source Domain	Target Domain	BERT	MoLE	MTL	MTSD	RLFND	Llama-2-7B	STANCE-C3
CoVaxNet_pre	CoVaxNet_post	0.713 (0.697)	0.646 (0.621)	0.734 (0.736)	0.734 (0.731)	0.752 (0.747)	0.721 (0.725)	**0.765 (0.753)**
	CoVaxNet_pre	0.762 (0.727)	**0.901 (0.876)**	0.823 (0.789)	0.845 (0.831)	0.843 (0.835)	0.812 (0.807)	0.863 (0.858)
Performance Degradation		**0.049 (0.030)**	0.255 (0.255)	0.089 (0.053)	0.111 (0.100)	0.091 (0.088)	0.091 (0.082)	0.098 (0.105)
CoVaxNet_post	CoVaxNet_post	0.826 (0.796)	0.821 (0.835)	0.811 (0.823)	**0.841 (0.839)**	0.812 (0.784)	0.761 (0.719)	0.831 (0.837)
	CoVaxNet_pre	0.698 (0.655)	0.713 (0.702)	0.623 (0.611)	0.712 (0.692)	0.674 (0.632)	0.688 (0.641)	**0.723 (0.723)**
Performance Degradation		0.128 (0.141)	0.108 (0.133)	0.188 (0.212)	0.129 (0.147)	0.138 (0.152)	**0.073 (0.078)**	0.108 (0.114)

throughout the pandemic period (Jan 1, 2020 - Dec 31, 2021) and annotated with stances towards *COVID-19 vaccination*. To evaluate the performance of STANCE-C3 in a single-target context, we categorized the *domains* as distinct time intervals distinguished by significant events, following common practice in social media analysis [34]. We divided CoVaxNet into two periods, with the full approval of the Pfizer-BioNTech vaccine by the US Food and Drug Administration (FDA) on Aug 23, 2021, serving as the dividing event. Thus, the source and target domains consist of 10,000 tweets each, selected randomly from January 1, 2020 to Aug 22, 2021 ("CoVaxNet_pre"; source domain) and from Aug 23, 2021 to Dec 31, 2021 ("CoVaxNet_post"; target domain).

For the cross-target scenario, CoVaxNet_pre and CoVaxNet_post were separately used as training sets for STANCE-C3. Model performance was evaluated on the targets (*"Anthony S. Fauci, M.D."*, *"Wearing a Face Mask"*, *"Keeping Schools Closed"*, *"Stay at Home Orders"*) from the COVID-19-Stance [12] dataset[2]. It includes different targets and keywords of interest (e.g., #lockdown), thereby justifying an adequate distributional shift between the source and target. Domain counterfactuals for this dataset were generated in the same manner as the cross-domain setting to augment the training set. Importantly, we labeled the generated examples the same as the original example as we observed a positive correlation between the targets in the source and target domain. Future works should explore scenarios where the stances towards targets in source and target domains are conflicting.

We excluded tweets that were too short or contained excessive hashtags, emojis, and URLs from both datasets to preserve their quality. Additionally, we balanced the training and test set across the different labels. It should be noted that congruent labels from the two datasets were treated as equivalent; that is, *pro* is equivalent to *in-favor* and *anti* is the same as *against*. In the context of stance detection, the terms in these pairs essentially convey the same stance. Both CoVaxNet and COVID-19-Stance datasets are publicly available for research purposes [18].

5.3 Evaluation and Results

Table 3 and Table 4 summarize the comparative performance of the baseline and proposed models concerning the accuracy and AUC (in parentheses).

[2] Dataset collected between Feb 27, 2020 and Aug 20, 2020.

Table 4. Cross-domain, cross-target performance results in accuracy and AUC (in parentheses). In this experiment, we evaluate the model's performance on a dataset that has a different target word in comparison to the source domain's target (p-value < 0.05 for all McNemar's tests).

Source Domain	Target Domain	BERT	MoLE	MTL	MTSD	RLFND	Llama-2-7B	STANCE-C3
CoVaxNet_pre	Face Masks	0.623 (0.568)	0.732 (0.711)	0.741 (0.712)	0.789 (0.746)	0.742 (0.625)	0.862 (0.815)	**0.868 (0.823)**
	Fauci	0.741 (0.706)	0.725 **(0.719)**	0.691 (0.687)	0.732 (0.698)	0.699 (0.625)	0.673 (0.635)	**0.743** (0.714)
	School Closures	0.832 (0.627)	0.778 (0.761)	0.711 (0.651)	0.872 (0.731)	0.834 (0.804)	0.894 **(0.796)**	**0.920** (0.734)
	Stay at Home	0.641 (0.625)	0.734 (0.738)	0.621 (0.613)	0.738 (0.732)	0.665 (0.658)	**0.761 (0.759)**	0.736 (0.742)
Average Performance		0.709 (0.631)	0.742 (0.732)	0.691 (0.665)	0.782 (0.726)	0.735 (0.678)	0.797 (0.751)	**0.816 (0.753)**
CoVaxNet_post	Face Masks	0.570 (0.555)	0.801 **(0.786)**	0.698 (0.676)	0.794 (0.754)	0.734 (0.712)	0.808 (0.758)	**0.819** (0.769)
	Fauci	0.548 (0.624)	0.721 (0.740)	0.709 (0.683)	**0.763 (0.778)**	0.713 (0.694)	0.678 (0.644)	0.743 (0.714)
	School Closures	0.578 (0.543)	0.718 (0.743)	0.621 (0.620)	0.687 (0.701)	0.694 (0.675)	0.725 (0.736)	**0.739 (0.772)**
	Stay at Home	0.623 (0.637)	0.672 (0.691)	0.669 (0.681)	0.699 (0.704)	0.675 (0.662)	0.705 (0.711)	**0.725 (0.727)**
Average Performance		0.579 (0.589)	0.728 (0.740)	0.674 (0.665)	0.735 (0.734)	0.704 (0.685)	0.729 (0.712)	**0.756 (0.745)**

Q1: In this experiment, the proposed approach and the domain-adaptive baselines used a fixed portion (30%) of the target domain data during counterfactual data generation and maintained the same ratio in the training set. The models were trained on a single domain and tested on both source and target domains for stance detection for a single target of interest from the source domain. Our results provide some evidence that even when the target of interest is contained in the source domain, performance in the target domain generally degrades for all models. The *Performance Degradation* in the table highlights how a model's performance changes when shifting from the source domain to the target domain. We observe that all models exhibit performance degradation, confirming the difference between domains.
Q2: The results of our experiments on the cross-domain cross-target setting, as presented in Table 4, demonstrate the performance of our approach and the baselines. In this setting, we augment the training set of baselines with a small portion (30%) of data from the target domain. The results of our experiments suggest that the proposed approach is more effective than the domain-adaptive baselines. Specifically, our approach surpasses all baselines regarding average accuracy and AUC score, underscoring its suitability for the cross-domain cross-target context. Furthermore, these results underscore the robustness of our approach to variations in the target domain, a strength attributable to our counterfactual data generation technique which is tailored to produce data that better represents the target domain.
Parameter Analysis. To address Q3, we analyze the impact of hyperparameters on the performance of STANCE-C3. To demonstrate the effect of contrastive loss and the balance between contrastive and cross-entropy loss, we vary the λ value in the range of $\{0.0, 0.25, 0.5, 0.75, 1.0\}$ during training. As illustrated in Fig. 3a, the AUC score varies significantly for different λ values. The results suggest that using $\lambda = 0.5$ yields the best performance.

We further investigate the impact of the target domain data portion γ utilized for counterfactual data generation by varying its value within the range of $\{5\%, 15\%, 30\%, 45\%\}$. The results, as illustrated in Fig. 3b, demonstrate that incorporating a larger portion of target domain data leads to enhanced performance. However, in practical scenarios, access to substantial data in the target domain is often

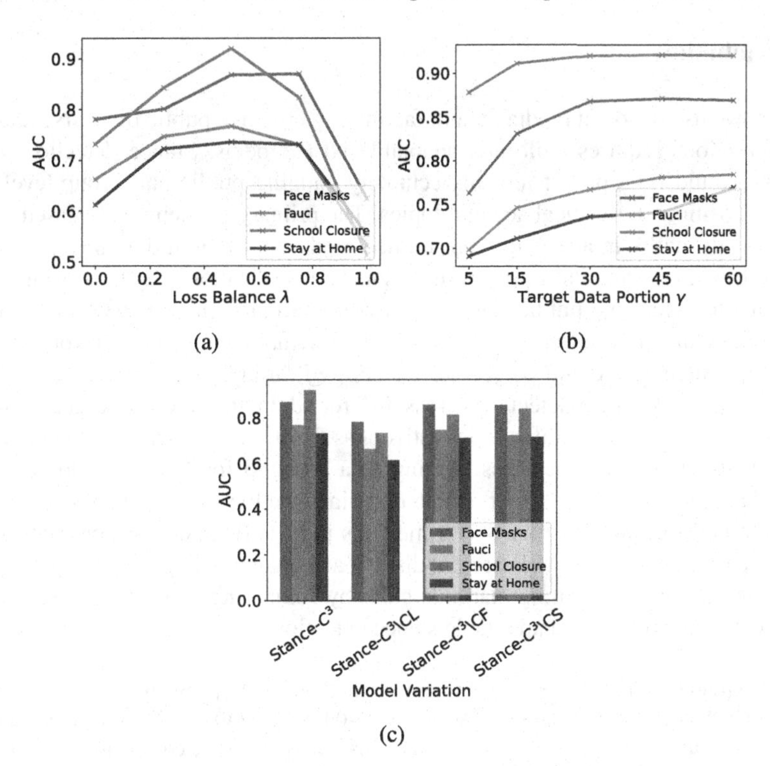

Fig. 3. Impact analysis of model's parameters and components. Figures (a) and (b) show the impact of the target domain data portion and the balance between the loss values, respectively. Figure (c) shows the impact of different components - removing modified contrastive loss (STANCE-C3\CL), removing counterfactual data generation component (STANCE-C3\CF), and using simple contrastive loss (STANCE-C3\CS) - on the model's performance.

constrained. Considering there is no statistically significant performance difference observed between $\gamma = 30\%$ and $\gamma = 45\%$, we conclude that employing 30% of the target domain data enables an acceptable level of performance.

Ablation Study. We investigate the impact of each component in STANCE-C3by testing different model variants: (1) **STANCE\CL** where we remove the contrastive loss component by setting $\lambda = 1.0$, (2) **STANCE-C3\CF** where we remove the counterfactual data generation component and only use 30% of target domain data ($\gamma = 30\%$) in the training set, and (3) **STANCE\CS** where we replace the modified contrastive loss with unmodified triplet loss. Figure 3c shows AUC degradation when domain counterfactual examples are removed, and when contrastive loss is either removed or modified during the training process. While removing or replacing either component (counterfactuals and contrastive loss) yielded worse AUC's, the contrastive loss component seems to have the greatest impact on STANCE-C3's performance. Further comparisons with the unmodified triplet loss show gains with respect to the AUC, suggesting that using the constrastive loss function is a step in the right direction for enhancing the robustness of stance detection models.

6 Conclusion

The effective use of social media data in facilitating effective public discourse and policy interventions requires agile and adaptable NLP models. Stance detection, as an NLP task, could be a potent tool in accurately gauging public and group-level support or oppositions for critical societal topics. The primary challenge addressed in this work is the difficulty in adapting a pre-trained model to different domains and targets of interest, which limits its utility in time-sensitive applications, such as using social media data to determine public opinions on non-pharmaceutical interventions during a pandemic. Our proposed approach uses counterfactual data generation for extended contexts in which data could be sparse. We showed through ablation studies that this approach improves stance detection across different domains and new targets. Further, we showed that using a modified contrastive loss function complements the data augmentation strategy with much less training data required for the target domain. The proposed architecture in this paper, which combines the two strategies, has shown evidence of outperforming existing SOTA methods in COVID datasets. For applications wherein specialized NLP models present distinct advantages over general purpose models, we showed that aggregating marginal gains by combining strategies is a step in the right direction for robust and adaptable stance detection.[3]

Acknowledgments. This work received support from the U. S. Department of Homeland Security under Grant Award Number 17STQAC00001-05-00 and the Office of Naval Research under Grant Award Number N00014-21-1-4002. Opinions, interpretations, conclusions, and recommendations within this article are solely those of the authors.

Disclosure of Interests. The authors have no competing interests to declare that are relevant to the content of this article.

References

1. Agusto, F.B., et al.: Impact of public sentiments on the transmission of covid-19 across a geographical gradient. PeerJ (2023). https://doi.org/10.7717/peerj.14736
2. Agusto, F.B., et al.: Impact of public sentiments on the transmission of covid-19 across a geographical gradient. PeerJ (2023)
3. Allaway, E., McKeown, K.: Zero-shot stance detection: a dataset and model using generalized topic representations. In: EMNLP. Association for Computational Linguistics (2020). https://doi.org/10.18653/v1/2020.emnlp-main.717
4. Augenstein, I., Rocktäschel, T., Vlachos, A., Bontcheva, K.: Stance detection with bidirectional conditional encoding. In: EMNLP. Association for Computational Linguistics (2016). https://doi.org/10.18653/v1/D16-1084
5. Benton, A., Dredze, M.: Using author embeddings to improve tweet stance classification. In: Proceedings of the 2018 EMNLP Workshop. Association for Computational Linguistics, Brussels (2018). https://doi.org/10.18653/v1/W18-6124

[3] Disclaimer: "The views and conclusions contained in this document are those of the authors and should not be interpreted as representing the official policies, either expressed or implied, of the Department of Homeland Security.".

6. Calderon, N., Ben-David, E., Feder, A., Reichart, R.: DoCoGen: domain counterfactual generation for low resource domain adaptation. In: ACL. Association for Computational Linguistics (2022). https://doi.org/10.18653/v1/2022.acl-long.533
7. Chen, T., Kornblith, S., Norouzi, M., Hinton, G.: A simple framework for contrastive learning of visual representations. In: ICML, pp. 1597–1607. PMLR (2020)
8. Darwish, K., Stefanov, P., Aupetit, M., Nakov, P.: Unsupervised user stance detection on twitter. In: Proceedings of the International AAAI Conference on Web and Social Media **14** (2020)
9. Dash, G., Sharma, C., Sharma, S.: Sustainable marketing and the role of social media: an experimental study using natural language processing (NLP). Sustainability **15**(6), 5443 (2023)
10. Daume, H., III., Marcu, D.: Domain adaptation for statistical classifiers. J. Artif. Intell. Res. **26**, 101–126 (2006)
11. Deng, R., Panl, L., Clavel, C.: Domain adaptation for stance detection towards unseen target on social media. In: ACII. IEEE (2022)
12. Glandt, K., Khanal, S., Li, Y., Caragea, D., Caragea, C.: Stance detection in covid-19 tweets. In: ACL-IJCNLP, vol. 1 (2021)
13. Guo, H., Pasunuru, R., Bansal, M.: Multi-source domain adaptation for text classification via distancenet-bandits. In: Proceedings of the AAAI Conference on Artificial Intelligence, vol. 34, pp. 7830–7838 (2020)
14. Gururangan, S., et al.: Don't stop pretraining: adapt language models to domains and tasks. In: ACL. Association for Computational Linguistics (2020). https://doi.org/10.18653/v1/2020.acl-main.740
15. Gómez-Suta, M., Echeverry-Correa, J., Soto-Mejía, J.A.: Stance detection in tweets: a topic modeling approach supporting explainability. Expert Syst. Appl. **214**, 119046 (2023). https://doi.org/10.1016/j.eswa.2022.119046
16. Hardalov, M., Arora, A., Nakov, P., Augenstein, I.: Cross-domain label-adaptive stance detection. In: EMNLP. Association for Computational Linguistics (2021). https://doi.org/10.18653/v1/2021.emnlp-main.710
17. Hoffmann, J., et al.: An empirical analysis of compute-optimal large language model training. Adv. Neural Inf. Process. Syst. (2022)
18. Jiang, B., Sheth, P., Li, B., Liu, H.: Covaxnet: an online-offline data repository for covid-19 vaccine hesitancy research. arXiv preprint arXiv:2207.01505 (2022)
19. Jin, D., Jin, Z., Hu, Z., Vechtomova, O., Mihalcea, R.: Deep learning for text style transfer: a survey. Comput. Linguist. **48**(1), 155–205 (2022). https://doi.org/10.1162/coli_a_00426
20. Kaushal, A., Saha, A., Ganguly, N.: TWT–WT: a dataset to assert the role of target entities for detecting stance of tweets. In: NAACL (2021)
21. Khosla, P., et al.: Supervised contrastive learning. Adv. Neural. Inf. Process. Syst. **33**, 18661–18673 (2020)
22. Lai, M., et al.: #brexit: leave or remain? the role of user's community and diachronic evolution on stance detection. J. Intell. Fuzzy Syst. **39**(2), 2341–2352 (2020). https://doi.org/10.3233/JIFS-179895
23. Li, D., Hospedales, T.: Online meta-learning for multi-source and semi-supervised domain adaptation. In: Computer Vision–ECCV 2020: 16th European Conference, Glasgow, 23–28 August 2020, Proceedings, Part XVI, pp. 382–403. Springer (2020)
24. Li, Y., Caragea, C.: A multi-task learning framework for multi-target stance detection. In: ACL-IJCNLP, pp. 2320–2326 (2021)
25. Liang, B., Chen, Z., Gui, L., He, Y., Yang, M., Xu, R.: Zero-shot stance detection via contrastive learning. In: Proceedings of the ACM Web Conference 2022, pp. 2738–2747 (2022)
26. Liang, B., et al.: Target-adaptive graph for cross-target stance detection. In: Proceedings of the Web Conference 2021, pp. 3453–3464 (2021)

27. Lin, B.Y., Lu, W.: Neural adaptation layers for cross-domain named entity recognition. In: EMNLP. Association for Computational Linguistics, Brussels (2018). https://doi.org/10.18653/v1/D18-1226
28. Liu, R., Lin, Z., Ji, H., Li, J., Fu, P., Wang, W.: Target really matters: target-aware contrastive learning and consistency regularization for few-shot stance detection. In: COLING, pp. 6944–6954 (2022)
29. Liu, X., et al.: Deep unsupervised domain adaptation: a review of recent advances and perspectives. APSIPA Trans. Signal Inf. Process. **11**(1) (2022)
30. Lu, K., Mardziel, P., Wu, F., Amancharla, P., Datta, A.: Gender bias in neural natural language processing. In: Logic, Language, and Security: Essays Dedicated to Andre Scedrov on the Occasion of His 65th Birthday, pp. 189–202 (2020)
31. Mohammad, S., Kiritchenko, S., Sobhani, P., Zhu, X., Cherry, C.: Semeval-2016 task 6: detecting stance in tweets. In: Proceedings of the 10th International Workshop on Semantic Evaluation (SemEval-2016), pp. 31–41 (2016)
32. Mosallanezhad, A., Karami, M., Shu, K., Mancenido, M.V., Liu, H.: Domain adaptive fake news detection via reinforcement learning. In: Proceedings of the ACM Web Conference 2022, pp. 3632–3640 (2022)
33. Nababan, A.H., Mahendra, R., Budi, I.: Twitter stance detection towards job creation bill. Procedia Comput. Sci. **197**, 76–81 (2022)
34. Pak, A., Paroubek, P.: Twitter as a corpus for sentiment analysis and opinion mining. In: Proceedings of the Seventh International Conference on Language Resources and Evaluation (LREC 2010). European Language Resources Association (ELRA), Valletta (2010)
35. Rashed, A., Kutlu, M., Darwish, K., Elsayed, T., Bayrak, C.: Embeddings-based clustering for target specific stances: the case of a polarized turkey. Proc. Int. AAAI Conf. Web Soc. Media **15**(1), 537–548 (2021). https://doi.org/10.1609/icwsm.v15i1.18082
36. Schiller, B., Daxenberger, J., Gurevych, I.: Stance detection benchmark: how robust is your stance detection? KI-Künstliche Intelligenz **35**(3) (2021)
37. Sobhani, P., Mohammad, S., Kiritchenko, S.: Detecting stance in tweets and analyzing its interaction with sentiment. In: Proceedings of the Fifth Joint Conference on Lexical and Computational Semantics. Association for Computational Linguistics, Berlin (2016). https://doi.org/10.18653/v1/S16-2021
38. Touvron, H., et al.: Llama: open and efficient foundation language models. arXiv preprint arXiv:2302.13971 (2023)
39. Turc, I., Chang, M.W., Lee, K., Toutanova, K.: Well-read students learn better: on the importance of pre-training compact models. In: Computation and Language (2019)
40. Wang, Z., Culotta, A.: Robustness to spurious correlations in text classification via automatically generated counterfactuals. In: Proceedings of the AAAI Conference on Artificial Intelligence, vol. 35 (2021)
41. Wang, Z., Wang, Q., Lv, C., Cao, X., Fu, G.: Unseen target stance detection with adversarial domain generalization. In: 2020 International Joint Conference on Neural Networks (IJCNN), pp. 1–8. IEEE (2020)
42. Wei, P., Mao, W.: Modeling transferable topics for cross-target stance detection. In: Proceedings of the 42nd International ACM SIGIR Conference on Research and Development in Information Retrieval (2019)
43. Zamfirescu-Pereira, J., Wong, R.Y., Hartmann, B., Yang, Q.: Why Johnny can't prompt: how non-AI experts try (and fail) to design LLM prompts. In: Proceedings of the 2023 CHI Conference on Human Factors in Computing Systems, pp. 1–21 (2023)

44. Zhang, S., Qiu, L., Chen, F., Zhang, W., Yu, Y., Elhadad, N.: We make choices we think are going to save us: Debate and stance identification for online breast cancer cam discussions. In: Proceedings of the 26th International Conference on World Wide Web Companion (2017)
45. Zhu, L., He, Y., Zhou, D.: Neural opinion dynamics model for the prediction of user-level stance dynamics. Inf. Process. Manag. **57**(2), 102031 (2020). https://doi.org/10.1016/j.ipm.2019.03.010

Impacts of Personalization on Social Network Exposure

Nathan Bartley(✉), Keith Burghardt, and Kristina Lerman

Information Sciences Institute, Marina Del Rey, CA 90292, USA
nbartley@usc.edu

Abstract. Algorithms personalize social media feeds by ranking posts from the inventory of a user's network. However, the combination of network structure and user activity can distort the perceived popularity of user traits in the network well before any personalization step. To measure this "exposure bias" and how users might perceive their network when subjected to personalization, we conducted an analysis using archival X (formerly Twitter) data with a fixed inventory. We compare different ways recommender systems rank-order feeds: by recency, by popularity, based one the expected probability of engagement, and random sorting. Our results suggest that users who are subject to simpler algorithmic feeds experience significantly higher exposure bias compared to those with chronologically-sorted, popularity-sorted and deep-learning recommender models. Furthermore, we identify two key factors for bias mitigation: the effective degree-attribute correlation and session length. These factors can be adjusted to control the level of exposure bias experienced by users. To conclude we describe how this framework can extend to other platforms. Our findings highlight how the interactions between social networks and algorithmic curation shape—and distort—user's online experience.

Keywords: Recommender Systems · Exposure bias · Social Networks

1 Introduction

Online social networks (OSNs) serve as a critical conduit for the spread of news and information in everyday life as well as in times of emergency, including natural disasters [26]. As OSNs have grown in popularity, the volume of user-generated content has exploded. To deal with information overload, OSNs use recommender systems to create personalized feeds for users; the systems sort content and bring useful content to the top of the user's feed. While personalized feeds have great utility, evidence has emerged in recent years that recommender systems can narrow the types of information users are exposed to as they interact with the system, creating echo chambers [25] and potentially increasing partisanship and affective polarization [7,28]. While there is conflicting evidence as to the extent that these recommender systems cause these problems versus how

L. M. Aiello et al. (Eds.): ASONAM 2024, LNCS 15212, pp. 38–53, 2025.
https://doi.org/10.1007/978-3-031-78538-2_3

much user choices are responsible [2,29], it is nonetheless acknowledged that a problem exists.

The problem of affective polarization in particular is theorized to have multiple interacting causes, one of which is the use of other peoples' identity signals by users as a cognitive shortcut to establish the credibility or popularity of the information being presented [15]. As algorithmic curation in aggregate may result in an ideological asymmetric environment where algorithmic amplification may be over-representing one ideology over another [19], it is essential to understand how visibility and biased exposure can affect the perception of social reality online (which ultimately may create a self-fulfilling prophecy).

As a simple example, some accounts a user follows may be much more active than others, which can make their opinions over-represented in a user's feed. In addition, users themselves vary in how much time they spend on Twitter: some users may have short sessions where they read just a few of all posts, while others may have longer sessions where they read a much larger fraction of new posts. These factors, combined with how the recommender algorithm orders the posts within the user's feed, will affect the information the user sees.

We use the term *exposure bias* described previously in Bartley, Burghardt and Lerman 2023 to refer to the disproportionate representation of certain types of content, users, or items in general in a user's feed as influenced by algorithmic design and user behavior patterns [5]. While studies have explored various facets of algorithmic personalization, there remains a gap in how different ranking mechanisms—such as recency, popularity, and deep learning based mechanisms— can affect what users get exposed to, and through this their perception of their network. This paper aims to bridge this gap by detailing a framework for measuring exposure bias, incorporating elements of network structure and user-platform interaction dynamics.

In this work we address the following research questions:

RQ1. How do different feed recommendation algorithms affect exposure bias?
RQ2. Do different length sessions experience different levels of exposure bias?

We explore these questions on an X follower network that contains all tweets posted by the users in the sample. We emulate different feed ranking algorithms and then proceed to measure what each user is exposed to. Moreover, we use three measures of bias that measure a different component of that user's network exposure. We also vary the number of tweets the user sees, i.e., the length of the feed. We find that users under simpler model-based feeds exhibit more exposure bias than deep-learning based, chronological, and popularity-ranked feeds. We also describe the relationship that measures of exposure bias have to the degree-attribute correlation of a network, i.e., how degree of a node correlates with the probability of having an attribute. We finally show that exposure bias can be both affected by the time users spend online as well as the algorithm, which implies the benefits of some algorithms over others at reducing echo chambers should they account for what tweets users observe[1].

[1] In an effort for reproducibility we provide a public repository with the simulation and analysis scripts: https://github.com/bartleyn/laughing-train.

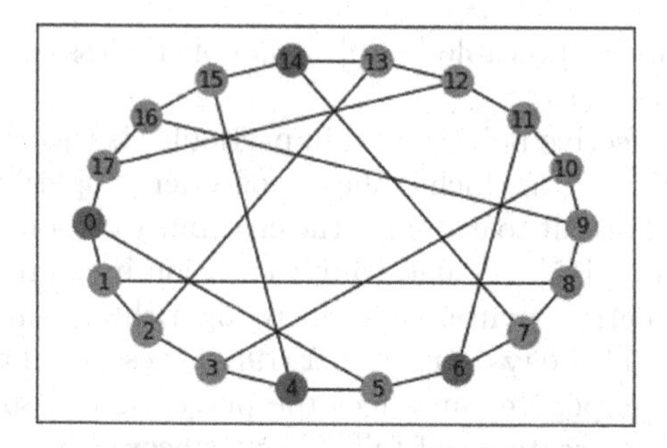

Fig. 1. Majority Illusion. Plotted is a graph with 18 nodes and 27 edges, with 4 nodes (0, 4, 14, 6) having an "active" binary trait. Nodes 15, 7, and 5 would experience the majority illusion in that the majority of their connections have the active trait, whereas the minority of the global population has the trait.

2 Related Work

This study can be situated at the interface between social psychology, network science, and the study/auditing of recommender systems.

2.1 Cognitive Biases and Information Diffusion

The cognitive sciences has long held the idea that humans are prone to multiple cognitive biases. Of particular interest is the salience bias: that humans are prone to paying more attention to stimuli that are remarkable or prominent, i.e., those stimuli that are irregular or unexpected [21]. In social psychology this salience bias can manifest in identifying minority groups as standing out, often resulting in an overestimation of their size. This is similar to the *majority illusion*, a statistical bias in which network structure distorts a user's local view of the network [21]. For illustrative purposes, we present an example network in Fig. 1 where some nodes experience the illusion due to how the minority trait is distributed across the network.

There is a similar assessment of cognitive biases from the social sensing literature: while people have the neurological capacity to observe and infer properties of their social network and mental states of others, their judgments may be differentially accurate depending on the populations being asked about. Galesic, Olsson and Rieskamp, 2012 detail a social sampling model where individuals sample from their immediate social network to estimate characteristics like average household wealth [13]. Individuals surveyed tended to be accurate within their immediate network, but were less accurate for some measurements when judging larger populations. This speaks to the importance of understanding how OSNs mediate our exposure to our networks largely through recommender systems: distorted exposures to one's network can plausibly affect inferences about broader populations.

In the study of social networks these perception and cognitive biases are well understood to be integral to how information diffuses through a network. Kooti, Hodas, and Lerman, 2014, investigated the origins of network paradoxes and found that they have behavioral origins, suggesting that individuals have distorted perceptions of their networks [22]. Rodriguez, Gummadi, and Schoelkopf, 2014, studied the effects of cognitive overload on information diffusion and found that exposure to information is less likely to infect a user if they are processing information from an ordered "queue" at higher rates. Our paper focuses on comparing different feed "queue" strategies and is not explicitly concerned with information diffusion/overload [30]. Our framework for comparing personalized feeds can be used for assessing impact on information diffusion.

The most relevant work to our study is that of Alipourfard et al., 2020, where they both introduce a primary measure we use in our work and study the local perception bias of various hashtags in Twitter data from 2014 [1]. They identify various hashtags that were overrepresented to users relative to the hashtags' global prevalence. A key difference in our work is that we study how constructing feeds in different ways can affect the perception of user traits. We also are more concerned with understanding how sensitive different feeds are to the distribution of the trait itself.

2.2 Algorithmic Audits of OSN Recommender Systems

Beginning with Sandvig et al., 2014, there has been a steady and growing interest in auditing algorithms in online systems for discriminatory behavior [31]. Algorithmic audits have focused on a wide variety of sectors, ranging from e-commerce platforms [20] to search queries [32,34] and OSNs [3,4].

Both Beattie, Taber, and Cramer, 2022 and Ramaciotti Morales and Cointet, 2021 analyzed the friend recommendation engine on Twitter, where the former presents a method for breaking echo chambers via user embeddings [6] and the latter is focused on incorporating ideological positions into similar user embeddings [27]. Our work is not concerned with friend recommendations, however they are a very valuable tool for intervention.

From within Twitter, Huszár et al., 2022 looked at the algorithmic amplification of political parties across different countries on Twitter with proprietary data on users [19]. They identify right-wing ideological bias under the algorithmic condition, suggesting that users in aggregate may be unduly influenced in how they perceive politics (at least on the platform).

Guess et al., 2023 study Facebook and Instagram feed data to assess the impact of personalization on user attitudes and political behaviors around the 2020 US election. They found no significant impact on behaviors but a significant difference in on-platform exposure to untrustworthy and uncivil content on the platforms [16]. Gonzalez-Bailon et al., 2023 study the impact of algorithmic and social amplification in spreading ideological URLs on Facebook, identifying ideological segregation taking place and both the algorithmic/exposure stage as well as the social amplification [14]. While these studies do not identify changes in beliefs or behaviors pertaining to the US 2020 presidential election, it is important to note that the effects may not be generalizable across different platforms and are focused on political-related outcomes from 2020 that may not apply in other domains given the relationship to Covid-19.

Donkers and Ziegler, 2021 studied both "epistemic" and "ideological" echo chambers in OSNs and how diversifying recommendations can potentially depolarize discussions [9]. They constructed a recommender system based on knowledge graph embeddings, allowed for users with varying propensities for accepting new information. However the utility of their framework, utilizing knowledge graph embeddings, might be limited as they do not consider different methods of ranking the tweets that are proposed to users in their evaluations.

2.3 User Perception of Ranked Feeds

Understanding user perceptions of their ranked feeds has been a focus of human-computer interaction work for the past decade. Notably this has been focused primarily on Facebook, where prominent work done by Eslami et al., 2016 identified several "folk theories" for how Facebook's personalized News Feed curated the information users in the study saw [10,11]. FeedVis, the tool they developed, empowers users by presenting which friends they will never see, rarely see, often see alongside other information about their feeds [10]. Our study differs from this vein of research as we focus on comparing different recommender systems directly.

3 Data and Methods

3.1 Twitter Data

Starting in March 2014, Alipourfard et al., 2020 [1] queried Twitter to identify accounts followed by each of 5,599 seed users. These followed accounts are known as *follower graph friends* or *friends* for short. The authors continued to query Twitter for the followed accounts daily through September 2014 to identify any new friends of seed users. This subset of the Twitter follower graph has over 4M users and more than 17M edges.

In addition to follow relations, the authors also collected messages posted by seed users and their friends over this time period, roughly 81.2M tweets. For this study we consider tweets from May 2014 to September 2014. At the time of data collection, Twitter created a feed for each user by aggregating all messages posted by the user's (follower graph) friends and ranking them in reverse chronological order. Given the uniform feed treatment, we are therefore able to reconstruct the feed for each seed user and quantify empirical exposure bias.

To summarize we use the tweets and retweets from 5,599 seedusers and all of the people they follow at the time of collection in 2014.

3.2 Reconstructing Feeds

To address our research questions we make use of the described empirical data. This data is important for this situation because it was collected before Twitter implemented an algorithmic recommender system for constructing feeds in 2016. Given that we know the users all experienced the same chronological feed, we can re-rank and construct new feeds to explore the exposure effects of the different feeds.

With the Twitter data we construct artificial "sessions": for each user u we assume that they browse their feed one time on any calendar day d they have any activity (either tweets or retweets). On each day d we then select all tweets and retweets that friends of u made that day and sort the tweets according to the specific feed recommendation algorithm. For analysis we only consider sessions with at least ten tweets and at least one unique friend seen.

We construct feeds according to the following strategies:

1. **Popularity.** We rank each of the tweets by their historical total number of retweets.
2. **Reverse Chronological.** We rank each of the tweets by the time they were posted (we assume the user logs in at the end of the day and observes tweets closer to the end of the day first).
3. **Random.** We rank each of the tweets randomly.
4. **Logistic Regression.** Given that the X/Twitter timeline personalization system utilizes a logistic regression model at the candidate-ranking step [35], we utilize a simplified logistic regression model with user-based features to rank tweets by the probability that user will retweet each tweet. For each user, we gather all possible tweets the user could have seen and all their retweets and then we train an individual model on that data. For users without retweets we use a similar user's model for prediction.
5. **Neural Collaborative Filtering (NCF).** We implement a dense neural network meant for personalized recommendations of tweets [18]. The model has two parallel embedding layers for user and tweet (item) inputs, which are then concatenated and passed into a dense layer with ReLU activation. The output layer predicts the likelihood of a user retweeting a tweet. The model uses a binary focal cross-entropy loss function.
6. **Wide&Deep.** Similar to the NCF model, we implement the Wide&Deep model as described in Cheng et al., 2016 [8]. We chose this model as researchers at Twitter described using a modified Wide&Deep model for ad recommendation in 2020 [17].

3.3 Simulating User Attributes

To measure the exposure bias we assign each user in the network a binary random variable $X \in \{0, 1\}$ with a fixed uniform probability $P(X = 1) = 0.10$. This variable can represent the user's ideology (e.g., liberal vs conservative), status (e.g., verified vs not), or it can represent a one-hot encoding of a belief the user shares. We choose a prevalence of 0.10 because we want to represent a minority trait of the population and observe how activity and exposure could distort it. We also run the same analyses under $P(X = 1) = 0.05$ and $P(X = 1) = 0.50$ to verify that the results are consistent.

After all accounts in the follower graph have been assigned values of the variable, we can then measure its prevalence in each user's feed. This allows the user to estimate the fraction of "activated" or "positive" (i.e., with value $x_i = 1$) friends within their network. To assess the relationship between the network structure and this random variable X we follow the attribute-swapping procedure as described in Lerman et al., 2016 [24] to vary the **degree-attribute correlation** ρ_{kx}:

$$\rho_{kx} = \frac{P(x = 1)}{\sigma_x \sigma_k}[\langle k \rangle_{x=1} - \langle k \rangle] \tag{1}$$

where x is the binary attribute, k is the degree of the node (here in-degree), σ_x, σ_k the standard deviations of the binary attribute and in-degree respectively, and $\langle k \rangle$ the average in-degree over all nodes. As ρ_{kx} increases we expect the exposure bias to increase (Table 1).

Table 1. Notation.

Notation	Description
B_{local}	Local Bias
$A(v)$	Attention function
G	Gini coefficient
ρ_{kx}	Degree-attribute correlation
M_i	No. of users with majority illusion on day i

3.4 Measures of Exposure Bias

We use the following metrics to measure exposure biases:

1. Local Perception Bias.

$$B_{\text{local}} = \mathbb{E}[q_f(X)] - \mathbb{E}[f(X)] \tag{2}$$

2. No. of users experiencing majority illusion per day.

$$M_i = |\text{fraction of positive friends seen per day} > 0.50|$$

3. Gini Coefficient G.

$$G = \frac{\sum_{i=1}^{n}(2i - n - 1)x_i}{n\sum_{i=1}^{n} x_i} \tag{3}$$

For the Gini coefficient G we assume that the tweets are distributed across all of each user's friends, i.e., that if a seed user has 50 friends and only one has tweets that are observed then G is computed over all 50 friends instead of just the one observed friend that day.

We define B_{local} as the average prevalence of the attribute among a node's friends, with each term defined as:

$$\mathbb{E}[q_f(\mathrm{X})] = \bar{d} * \mathbb{E}[f(U)A(V)|(U,V) \sim \mathrm{Uniform}(E)]$$

$\mathbb{E}[f(X)]$ is the global prevalence of the node attribute f; $\bar{d}$ represents the expected in-degree of a node; $f(U)$ the attribute value f of node U. $A(V)$ represents the attention node V pays to any particular friend.

Intuitively, this B_{local} measure is the difference between the average fraction of activated friends exposed to random users and the true prevalence of the trait (represented by $\frac{|\text{users}_{x=1}|}{|\text{users}|}$). A positive number indicates the average user should expect to see a higher fraction of friends in their feeds with $x = 1$ than the actual global prevalence.

4 Results

4.1 RQ1 Difference Between Feeds

We report the results of our analysis in Fig. 2 and Table 2. Of note is the different behavior we observe between the six conditions in Fig. 2: B_{local} for the reverse chronological feed is noticeably higher than the random and popularity feeds, comparable to the Wide&Deep feed, and seems to be lower than the logistic regression feed and NCF feeds. We note that for each of the conditions, when computing B_{local} we use the same attention function, leaving remaining differences to feed strategy. We report the statistical analysis in Table 2 where the Popularity feed is lowest according to B_{local}, Random is lowest according to G, and NCF is the lowest according to M_i. In the logistic regression plot in Fig. 2 (d) we observe that for shorter feeds the bias goes above 1.0, which may be an artifact of how we compute the bias for shorter length sessions.

4.2 RQ2 Difference Between Session Lengths

Can users compensate for exposure bias by consuming a larger share of their feeds? We model the attention users pay to their feed through the session length parameter. In the empirical data, we observe that the feed conditions behave similarly in terms of session length (Fig. 2). For B_{local} the longer sessions show less bias than the shorter sessions (Table 3). For G, we observe that the session lengths are ordered largely the same, with longer sessions being less biased than shorter ones. In contrast, the shortest feeds had lowest relative M_i compared to the longer feeds.

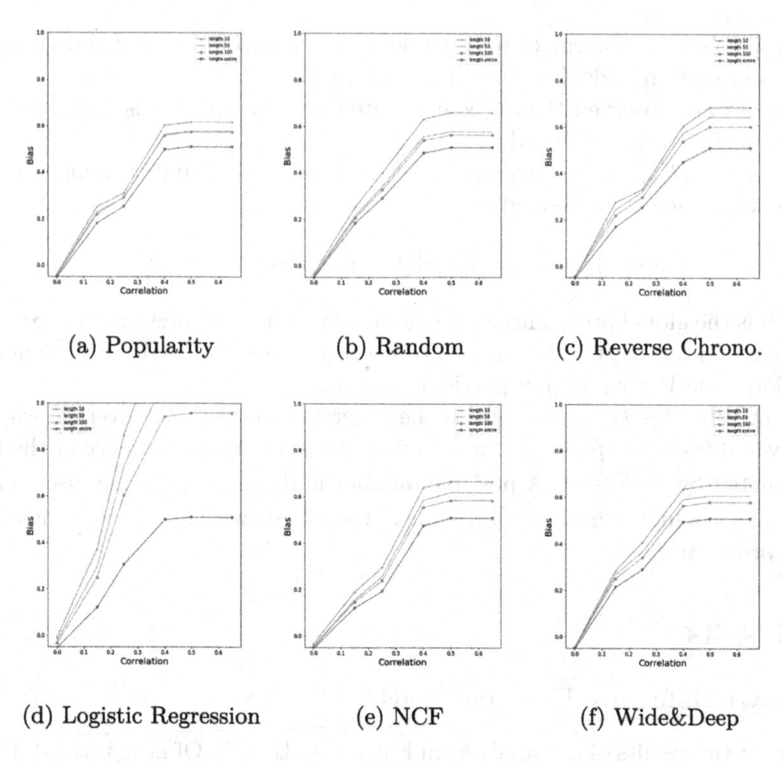

(a) Popularity (b) Random (c) Reverse Chrono.

(d) Logistic Regression (e) NCF (f) Wide&Deep

Fig. 2. Bias B_{local} versus correlation ρ_{kx}. Mean values are reported across all seed user days; bars are standard error of the mean.

Table 2. Empirical Pairwise Significance Tests. We treat the seed users in different feed conditions as paired samples and compute a paired t-test over the mean values of the various metrics. Conditions that are bolded are least biased according to the measure. * - $p < 0.05$, ** - $p < 0.01$

Feed 1	Feed 2	$P(X = 1) = 0.10$		
		B_{local}	G	M_i
Pop.	**Wide&Deep**	**−137.85******	20.99**	−0.81
Pop.	**Rand.**	**−41.36******	8.17 **	1.07
Pop.	**RevChron.**	**−162.60******	−26.44 **	5.32 **
Pop.	**NCF**	**−44.13******	8.21**	44.47 **
Pop.	**LogReg**	**−205.73******	−44.86 **	12.48 **
Wide&Deep	**Pop.**	137.85**	−20.99**	0.81
Wide&Deep	**Rand.**	−7.16**	7.47**	1.58
Wide&Deep	**RevChron.**	−204.21**	−34.99**	6.08**

continued

Table 2. continued

Feed 1	Feed 2	$P(X=1)=0.10$		
		B_{local}	G	M_i
Wide&Deep	**NCF**	−26.01**	2.49*	44.82**
Wide&Deep	**LogReg**	−213.47**	−46.03**	12.87**
Rand.	**Pop.**	41.36**	**−8.17****	−1.07
Rand.	**Wide&Deep**	7.16**	**−7.47****	−1.58
Rand.	**REvChron.**	−10.70**	**−8.95****	2.61**
Rand.	**NCF**	−25.40**	**−3.58****	43.27**
Rand.	**LogReg**	−117.02**	**−10.49****	10.49**
RevChron.	**Pop.**	162.60**	26.44**	−5.32**
RevChron.	**Wide&Deep**	204.21**	34.99**	−6.08**
RevChron.	**Rand.**	10.70**	8.95**	−2.61**
RevChron.	**NCF**	−14.37**	12.66**	43.70**
RevChron.	**LogReg**	−213.92**	−39.19**	8.75**
NCF	**Pop.**	44.13**	−8.21**	**−44.47****
NCF	**Wide&Deep**	26.01**	−2.49*	**−44.82****
NCF	**Rand.**	25.40**	3.58**	**−43.27****
NCF	**RevChron.**	14.37**	−12.66**	**−43.70****
NCF	**LogReg**	−104.33**	−20.46**	**−42.86****
LogReg	**Pop.**	205.73**	44.86**	−12.48 **
LogReg	**Wide&Deep**	213.47**	46.03 **	−12.87**
LogReg	**Rand.**	117.02**	10.49**	−10.49**
LogReg	**RevChron.**	213.92**	39.19**	−8.75**
LogReg	**NCF**	104.33**	20.46**	42.86**

5 Discussion

5.1 RQ1 Difference Between Feeds

Since the seed user follower graph and inventory is consistent across all conditions, we can treat the results as paired samples and take paired t-tests to establish difference between each pair of feeds. Table 2 shows that for B_{local} the popularity feed is less biased than other feeds, but that the order changes with the other measures. We would expect the relationship to ρ_{kx} to disappear in the random feed, making it the least biased across the metrics. This suggests we are not necessarily evaluating an appropriate random baseline. In simulated networks (not reported here), we shuffled the data generated for each seed user in order to eliminate relationship to ρ_{kx}. However doing this in the empirical data did not seem to have any significant impact.

Our findings suggest that seed users face greater exposure bias in reverse chronological feeds compared to simple retweet-based feeds, with logistic regression-based feeds showing the highest bias among all. The relatively higher Gini coefficients for chronological and logistic regression feeds suggest lower user diversity, possibly due to a temporal reliance on user activity. Interestingly the two deep learning models behave differently, suggesting the NCF is better than others at keeping the fraction of exposed friends that are positive low but higher than the true prevalence than other models.

Table 3. Empirical Session-Length Pairwise Significance Tests. We treat the seed users in different feed conditions as paired samples and compute a paired t-test over the mean values of the various metrics computed for each session length. * - p < 0.05, ** - p < 0.01

Length 1	Length 2	$P(X=1)=0.10$		
		B_{local}	G	M$_i$
10	100	245.79**	76.85**	**-30.25****
10	Full length	222.30 **	74.52**	**−28.39****
100	10	−245.79**	−76.85**	30.25**
100	Full length	195.26**	43.48**	−1.87
Full length	10	**−222.30****	**−74.52****	28.39**
Full length	100	**−195.26****	**−43.48****	1.87

5.2 RQ2 Difference Between Session Lengths

When we consider the empirical session lengths, we see a significant difference between each length within each feed condition (Table 3). Interestingly B_{local} reports higher bias for shorter sessions than longer ones; this can be explained by the number of unique friends observed in longer sessions. Longer sessions create a larger universe of edges to compute the expected value over, yielding a value closer to the population estimate. This may also create an artifact in our computation as we observe $B_{\text{local}} > 1.0$ in Fig. 2 (d), which is interesting because we observe the same effect in $P(X=1) \in \{0.05, 0.50\}$.

Given that both B_{local} and G agree on the ordering of the feeds, we interpret this as reporting that the same seed users experience significantly more exposure bias in shorter feeds than longer feeds on average. How we compute each metric (especially Gini coefficient) could be resulting in very small standard error when running the t-tests. For M_i we observe a lower relative bias, which follows from shorter feeds being sensitive to the addition and subtraction of individual users.

Our methodology effectively highlights differences between feeds. A common limitation in prior audits is the inconsistent application of treatments. For example, even when some users are subject to chronological feeds, the influence of personalized feeds used by their friends may still impact their experience. Applying uniform treatments in an archival setting, where data is generated under a uniform context, provides confidence in measuring effects between conditions.

5.3 Considerations for Practitioners

We can modulate the effective correlation ρ_{kx} in a practical manner by choosing network edges to observe to change $\langle k \rangle_{x=1} - \langle k \rangle$. However, this has ethical implications in varying how often different people get observed in expectation. This should be considered in tandem with measures of individual and group fairness to make feeds robust to biases.

Considering the number of unique friends that a user is exposed to may be a useful measure for platforms to maintain as a measure of overall fitness of the platform; it will have an impact on the measures we present in this study. More specifically, observing more unique friends seen lowers Gini coefficient and minimizes the absolute value of B_{local}.

5.4 Other Platforms

Since 2014, platforms like X have evolved to show users content from beyond their immediate follower "in-network", introducing a larger "out-of-network" user set for feed algorithms to sample from. For example, a user with friends primarily exhibiting trait $x_i = 0$ could encounter unexpected content from non-followed users with trait $x_i = 1$.

This exposure bias framework is widely applicable to other social network recommender systems, primarily through the idea of a partially observed network. If we consider repeated exposure to specific users and kinds of content as a weak tie in a social network then we can treat exposure to a user's post as someone observing an edge to that user in a partially observed network. That observed network may have a different prevalence of certain traits than what you would expect from either the follower network or total universe of possible users one could observe. This framework can apply to other platforms as follows:

1. **TikTok.** TikTok's For You Page (FYP) explicitly considers user interactions, such as videos shared and accounts followed among other factors in personalizing users' FYP. TikTok describes explicitly diversifying feeds to prevent repetitive exposure to particular users and/or types of content [33]. Exposure to different communities on TikTok (so called -Tok communities, e.g., BookTok) and certain users can be analyzed under this exposure bias framework because the exposed prevalence of certain traits within the communities can be compared to larger scale prevalence (e.g., how prevalent the traits are in the users' geographic area).
2. **Facebook.** Facebook's Feed works in a similar way as X: they gather the inventory of posts from friends, pages and groups, then calculate a relevance score and rank order each users' feed with high scoring out-of-network user posts interspersed [12]. Given the parallels to how X works, this framework readily applies to Facebook.
3. **Instagram.** Two recommender system components are worth investigating on Instagram: the Explore page and the Instagram feed. The feed works much like Facebook and X as described previously. The Explore page is explicitly

designed to show content from accounts that the user does not follow, drawing on information from followed accounts, information from posts that were interacted with and general connections on Instagram. This framework applies more readily to the feed than the Explore page.
4. **YouTube.** The YouTube front page and recommender system has been a central focus of auditing efforts for the kinds of information and news the system recommends for users to watch [28]. Signals that are used to drive recommendations include clicks, watchtime, total views, user surveys for "valued watchtime" and other interactions including shares, likes and dislikes [36]. If we consider how content creators are related to each other online (e.g., partnerships, content networks), and how content can be grouped together (e.g., political content being grouped together ideologically), we can construct a user-creator network to assess exposure.

6 Limitations and Future Work

A primary limitation for this study is the reliance on X data from 2014. The benefit of a vertical slice of data for the subgraph (with uniform feeds) is important, but it is also important to verify that the differences in recommender strategies hold in other networks as well, especially networks that are larger than the ones we analyzed here. We have simulation data of similar scale networks that suggest the results hold, but we do not report the results here.

Another limitation for this study is the assumption of binary user traits. Binary traits can be variable over time and subjective for different users. For example, person A may be perceived as being on the political left by a conservative observer B, however that same person A may be perceived as being conservative by a third liberal observer C. Allowing for dynamic non-binary traits would lead to valuable insights.

Future work could entail extending our feeds to be more complex. For instance, we could study the "out-of-network“ feed, where we would prioritize friends and allow for more users to appear in the feed as a supplement. This would get the analysis to be closer to the empirical X self-studies where they describe in-network and out-of-network tweet impressions [19,23]. Similarly, we could attempt to implement all the working components and services of the reported Twitter system, however simulating data with this could end up too contrived to be useful.

In this study, we assume a fixed social graph; future research could explore dynamic social graphs where users change their follows over time. Additionally, applying exposure bias metrics to various network substructures could reveal differences in content consumption between network cliques and highly-connected hubs for instance. We could also analyze the user heterogeneity as well as the model effects: do users with the minority trait get exposed to their network in different ways than other users?

7 Conclusion

In this study we have described exposure bias, related it to existing cognitive biases, and created metrics to quantify this bias. Our results illustrate the interconnectedness of social network structure, activity, and feed recommendation algorithm. We have shown that a mixture of bias metrics can adequately discriminate between different feed conditions. We described these metrics as being able to assess the propensity for individuals in a network to experience a distorted view of their immediate network, where a minority trait may be overrepresented and unduly more salient to the user than would be expected otherwise. Importantly we show that these feeds behave differently as the prevalence of the trait changes.

Huszár et al., 2022 shows that within an organization it is feasible to observe and record the personalized feeds for large sets of individuals [19]. Because of this we believe that both internal and external auditors should be able to use measures of exposure bias. With such measures, audits can be more closely tied to how individual users experience the system on a session-level. It would also allow for interpretable analysis to examine if different communities have disparate experiences with their personalized feeds.

Ethical Statement. All data were anonymized prior to analysis. The analysis has minimal risk to user privacy, and analysis is unlikely to involve any ethical risk.

References

1. Alipourfard, N., Nettasinghe, B., Abeliuk, A., Krishnamurthy, V., Lerman, K.: Friendship paradox biases perceptions in directed networks. Nat. Commun. **11**(1), 707 (2020)
2. Bakshy, E., Messing, S., Adamic, L.A.: Exposure to ideologically diverse news and opinion on facebook. Science **348**(6239), 1130–1132 (2015)
3. Bandy, J., Diakopoulos, N.: More accounts, fewer links: how algorithmic curation impacts media exposure in twitter timelines. Proc. ACM Hum.-Comput. Interact. **5**(CSCW1), 1–28 (2021)
4. Bartley, N., Abeliuk, A., Ferrara, E., Lerman, K.: Auditing algorithmic bias on twitter. In: 13th ACM Web Science Conference 2021, pp. 65–73 (2021)
5. Bartley, N., Burghardt, K., Lerman, K.: Evaluating content exposure bias in social networks. In: Proceedings of the International Conference on Advances in Social Networks Analysis and Mining, pp. 379–383 (2023)
6. Beattie, L., Taber, D., Cramer, H.: Challenges in translating research to practice for evaluating fairness and bias in recommendation systems. In: Proceedings of the 16th ACM Conference on Recommender Systems, pp. 528–530 (2022)
7. Chen, W., Pacheco, D., Yang, K.C., Menczer, F.: Neutral bots reveal political bias on social media. arXiv preprint arXiv:2005.08141 (2020)
8. Cheng, H.T., et al.: Wide & deep learning for recommender systems. In: Proceedings of the 1st Workshop on Deep Learning for Recommender Systems, pp. 7–10 (2016)

9. Donkers, T., Ziegler, J.: The dual echo chamber: Modeling social media polarization for interventional recommending. In: Proceedings of the 15th ACM Conference on Recommender Systems, pp. 12–22 (2021)
10. Eslami, M., Aleyasen, A., Karahalios, K., Hamilton, K., Sandvig, C.: Feedvis: a path for exploring news feed curation algorithms. In: Proceedings of the 18th ACM Conference Companion on Computer Supported Cooperative Work & Social Computing, pp. 65–68 (2015)
11. Eslami, M., et al.: First i "like" it, then i hide it: folk theories of social feeds. In: Proceedings of the 2016 CHI Conference on Human Factors in Computing Systems, pp. 2371–2382 (2016)
12. Ranking and Content. https://transparency.fb.com/features/ranking-and-content/. Accessed 01 Jan 2024
13. Galesic, M., Olsson, H., Rieskamp, J.: Social sampling explains apparent biases in judgments of social environments. Psychol. Sci. **23**(12), 1515–1523 (2012)
14. González-Bailón, S., et al.: Asymmetric ideological segregation in exposure to political news on Facebook. Science **381**(6656), 392–398 (2023)
15. González-Bailón, S., Lelkes, Y.: Do social media undermine social cohesion? a critical review. Soc. Issues Policy Rev. **17**(1), 155–180 (2023)
16. Guess, A.M., et al.: How do social media feed algorithms affect attitudes and behavior in an election campaign? Science **381**(6656), 398–404 (2023)
17. Guo, D., et al.:: Deep Bayesian bandits: exploring in online personalized recommendations. In: Fourteenth ACM Conference on Recommender Systems, pp. 456–461 (2020)
18. He, X., Liao, L., Zhang, H., Nie, L., Hu, X., Chua, T.S.: Neural collaborative filtering. In: Proceedings of the 26th International Conference on World Wide Web, pp. 173–182 (2017)
19. Huszár, F., Ktena, S.I., O'Brien, C., Belli, L., Schlaikjer, A., Hardt, M.: Algorithmic amplification of politics on twitter. Proc. Natl. Acad. Sci. **119**(1), e2025334119 (2022)
20. Juneja, P., Mitra, T.: Auditing e-commerce platforms for algorithmically curated vaccine misinformation. In: Proceedings of the 2021 Chi Conference on Human Factors in Computing Systems, pp. 1–27 (2021)
21. Kardosh, R., Sklar, A.Y., Goldstein, A., Pertzov, Y., Hassin, R.R.: Minority salience and the overestimation of individuals from minority groups in perception and memory. Proc. Natl. Acad. Sci. **119**(12), e2116884119 (2022)
22. Kooti, F., Hodas, N., Lerman, K.: Network weirdness: exploring the origins of network paradoxes. In: Proceedings of the International AAAI Conference on Web and Social Media, vol. 8, pp. 266–274 (2014)
23. Lazovich, T., et al.: Measuring disparate outcomes of content recommendation algorithms with distributional inequality metrics. Patterns **3**(8), 100568 (2022)
24. Lerman, K., Yan, X., Wu, X.Z.: The "majority illusion" in social networks. PLoS ONE **11**(2), e0147617 (2016)
25. Nikolov, D., Flammini, A., Menczer, F.: Right and left, partisanship predicts (asymmetric) vulnerability to misinformation. Harvard Kennedy School (HKS) Misinformation Review (2021)
26. Panagiotopoulos, P., Barnett, J., Bigdeli, A.Z., Sams, S.: Social media in emergency management: Twitter as a tool for communicating risks to the public. Technol. Forecast. Soc. Chang. **111**, 86–96 (2016)
27. Ramaciotti Morales, P., Cointet, J.P.: Auditing the effect of social network recommendations on polarization in geometrical ideological spaces. In: Proceedings of the 15th ACM Conference on Recommender Systems, pp. 627–632 (2021)

28. Ribeiro, M.H., Ottoni, R., West, R., Almeida, V.A., Meira Jr, W.: Auditing radicalization pathways on Youtube. In: Proceedings of the 2020 Conference on Fairness, Accountability, and Transparency, pp. 131–141 (2020)
29. Ribeiro, M.H., Veselovsky, V., West, R.: The amplification paradox in recommender systems. arXiv preprint arXiv:2302.11225 (2023)
30. Rodriguez, M.G., Gummadi, K., Schoelkopf, B.: Quantifying information overload in social media and its impact on social contagions. In: Proceedings of the International AAAI Conference on Web and Social Media, vol. 8, pp. 170–179 (2014)
31. Sandvig, C., Hamilton, K., Karahalios, K., Langbort, C.: Auditing algorithms: research methods for detecting discrimination on internet platforms. Data Discriminat.: Convert. Critic. Concerns Product. Inquiry **22**(2014), 4349–4357 (2014)
32. Sapiezynski, P., Zeng, W., E Robertson, R., Mislove, A., Wilson, C.: Quantifying the impact of user attention on fair group representation in ranked lists. In: Companion Proceedings of the 2019 World Wide Web Conference, pp. 553–562 (2019)
33. How Tiktok Recomends Videos for You. https://newsroom.tiktok.com/en-us/how-tiktok-recommends-videos-for-you. Accessed 01 Jan 2024
34. Tomlein, M., et al.: An audit of misinformation filter bubbles on Youtube: bubble bursting and recent behavior changes. In: Proceedings of the 15th ACM Conference on Recommender Systems, pp. 1–11 (2021)
35. https://github.com/twitter/the-algorithm/. Accessed 01 Jan 2024
36. On Youtube's Recommendation System. https://blog.youtube/inside-youtube/on-youtubes-recommendation-system/. Accessed 01 Jan 2024

Event Embedding Learning from Social Media Using Graph Topic Model Autoencoder

Yihong Zhang(✉) and Takahiro Hara

Multimedia Data Engineering Lab, Graduate School of Information Science and Technology, Osaka University, Osaka, Japan
yhzhang7@gmail.com, hara@ist.osaka-u.ac.jp

Abstract. Event detection from social media has been researched intensively and applied to many social problems such as disaster monitoring, rumor detection, and product sales prediction. In this paper, we deal with the underlying task of event representation. Existing works use topic modeling or graph embedding to extract events from text corpora, but have not fully utilized the available information. We propose a novel model called Graph Topic Model Autoencoder (GTMA) for improving event representation quality. The model combines non-negative matrix factorization and graph autoencoder to take advantages of the document-word matrix and the word-word co-occurrence graph. The model outputs event embeddings that can be used for topic-based event detection. We test our approach with two real-world social media datasets. Compared to several embedding learning baselines, our model generally achieves better topic quality in event detection evaluation.

Keywords: event representation · social media mining · graph topic model

1 Introduction

In recent years, social media is shown to provide rich information on societal events. Twitter, a popular social media platform, for example, currently has 330M monthly active users, and produces 350K tweets per minute[1]. These tweets cover topics of all kinds of social activities which can be helpful for many social computational tasks [19]. Researchers have been trying to develop systems that can extract events from this massive data. Generally, an event can be seen as an unusual change in activity trends [18], sometimes with geographical information [20].

Topic modeling, the technique for generating topics from a text corpus, has been widely studied in data mining and information retrieval. Latent Dirichlet Allocation (LDA) [2], one of the most well-known topic modeling techniques,

[1] https://www.bankmycell.com/blog/how-many-users-does-twitter-have.

L. M. Aiello et al. (Eds.): ASONAM 2024, LNCS 15212, pp. 54–64, 2025.
https://doi.org/10.1007/978-3-031-78538-2_4

has spanned many variations [5]. Particularly, some online variations have been deployed in social media event detection [1,13]. This technique models a document as a distribution of topics, and a topic as a distribution of words [10]. Non-negative matrix factorization (NMF) is a technique that can achieve a similar effect [15]. NMF uses the product of two low-rank matrices to approximate document-word data, and the two matrices can be seen as document-topic distribution and topic-word distribution. Because they are very similar in effect, several studies have compared the two techniques directly in various experiment settings [4]. Findings in these studies show that NMF is equivalent to and sometimes better than LDA in terms of topic quality. Because NMF is relatively new in topic modeling, and has a higher potential to be developed in a neural network framework, we choose this approach to further study in our work.

Our insight through analyzing social media data is that social media text streams can be represented as graphs. We can create a document-word graph, where the nodes are documents and words, and an edge indicates word frequency in a document. We can also create a word-word graph, where nodes are words, and an edge indicates the frequency two words co-occur in a document. The event information, which is represented by abnormal clusterings of words, is contained in these graphs. Once we represent social media data as graphs, we can use popular graph embedding techniques to encode event information in embeddings. For example, the well-known technique, TransE, and its variations, can learn graph embeddings through translation geometric [3,14]. However, it does not cope with weighted edges. Graph convolutional network (GCN) [16] is a popular graph embedding technique that can work with weighted edges. However, GCN requires label data to learn the model, and thus is not unsupervised. Another graph embedding technique, graph auto-encoder (GAE) [6], on the other hand, is unsupervised, and can be used easily in topic modeling.

Using graph embedding techniques in topic modeling is largely unexplored. In this paper, we make an initial step by proposing a model that combines topic modeling and graph embedding techniques. Our model, called Graph Topic Model Autoencoder (GTMA), combines the advantages of both document-word graph and word-word co-occurrence graph, and thus can achieve better results than existing topic model and graph embedding techniques. We use two real-world social media datasets to test our model, evaluating event detection accuracy. As the main contributions, we first propose a method to convert social media data stream into graphs, which preserves critical event information. This conversion makes way for applying advanced graph embedding techniques in event detection. Then we propose an event embedding learning model that combines topic modeling and graph embedding techniques. While relatively simple, this model takes advantages of both document-word graph and word-word co-occurrence graph, and is expected to achieve better results. Finally, we test our approach in two real-world datasets and discuss the results.

2 Related Work

We focus on introducing existing works that are technically most related to our proposed model. They are divided into two groups, namely, topic models, and graph embedding techniques. LDA [2] is a well-known topic model that produces document-topic distribution and topic-word distribution. It has many variations. For example, the bi-term topic model (BTM) [5] adds a special consideration of bi-terms to improve topic quality. The online BTM (OBTM) [5], proposed by the same authors, is an extension that stores only recent data when processing incoming new data. NMF [15] can achieve a similar effect as LDA. It also has several variations. For example, online NMF (ONMF) [8] uses sparse coding to make new latent factors converge to the past when processing new data. Evolve NMF (EvNMF) and Emerge NMF (EmNMF) are proposed together in the same work [12]. EvNMF uses a parameter to make NMF conform to a past model, while EmNMF is made to capture as many new anomalies as possible.

The representative graph embedding technique, as we mentioned, is GCN [16]. GCN also has several variations, some of which are suitable for continuous event detection. For example, Evolve GCN (EvGCN) [9] connects model parameter to past models through recurrent neural network, achieving an smooth effect. Frequency-weighted GCN (FW-GCN) [17] combines frequency changes with GCN embeddings to capture both node semantics and temporal anomalies. GCN-based methods usually perform pseudo-tasks such as link prediction when learning the model. Graph Autoencoder (GAE) [6], in contrast, reconstructs input data, and uses the difference for learning the model, and thus is fully unsupervised.

3 Preliminaries

Our method is based on two established graph embedding learning techniques, namely, non-negative matrix factorization (NMF) [12,15], and graph autoencoder (GAE) [6]. In this section, we will briefly review these two techniques.

3.1 Non-negative Matrix Factorization

NMF is a technique that is widely used in document clustering. Usually, we have an input matrix $\mathbf{X}$ that represent document-term frequency, formulated as:

$$\mathbf{X}_{ji} = tf_{ji} \cdot \log\left(\frac{n}{idf_j}\right) \tag{1}$$

where tf_{ji} is the frequency of word j in document i, and idf_j is the inverse document frequency of word j. This matrix can be also seen as a weighted graph, where values are edge weights between each pair of document and word.

Given that the number of documents is N, and the size of the vocabulary is D, NMF then tries to find two low-rank matrices $\mathbf{W}$ and $\mathbf{H}$, whose product can approximate the input matrix. To learn these two matrices, the technique tries to minimize a reconstruction loss:

$$\mathcal{L}_{NMF} = \frac{1}{2}||\mathbf{X} - \mathbf{WH}|| \quad s.t. \quad \mathbf{W}, \mathbf{H} \geq 0. \tag{2}$$

Matrix $\mathbf{W}$ of size $N \times K$ can be seen as a latent representation or embeddings of documents, while Matrix $\mathbf{H}$ of size $K \times D$ can be seen as a latent representation or embeddings of the words.

3.2 Graph Autoencoder

GAE is a technique for learning graph embeddings based on the graph data structure [6]. A graph can be represented as an adjacency matrix $\mathbf{A}$ of size $D \times D$, where D is the number of nodes. In a weighted graph, $\mathbf{A}_{ij}$ is the edge weight between node i and node j. GAE incorporates an encoder-decoder structure. The encoder outputs a latent matrix $\mathbf{Z}$ through graph convolution [7]:

$$\mathbf{Z} = \mathrm{GCONV}(\mathbf{I}, \mathbf{A}) \tag{3}$$

where the graph convolution layer GCONV is defined as

$$\mathrm{GCONV}(\mathbf{I}, \mathbf{A}) = \tilde{\mathbf{A}}\mathrm{ReLU}(\tilde{\mathbf{A}}\mathbf{I}\mathbf{W}_0)\mathbf{W}_1.$$

where $\tilde{\mathbf{A}}$ is normalized $\mathbf{A}$.

In some versions of GAE, the graph convolution can take into account node features [7]. In our work, we assume that no node feature is available. In this case, the identity matrix $\mathbf{I}$ can be used instead of the node feature matrix.

The decoder reconstruct the input from the latent matrix:

$$\hat{\mathbf{A}} = \sigma(\mathbf{Z}\mathbf{Z}^T) \tag{4}$$

where $\sigma(\cdot)$ is a logistic sigmoid activation function.

Learning the model involves minimizing both the reconstruction loss and the normalization loss:

$$\mathcal{L}_{GAE} = ||\hat{\mathbf{A}} - \mathbf{A}||^2 - \mathrm{KL}[q(\mathbf{Z}|\mathbf{I}, \mathbf{A})||p(\mathbf{Z})] \tag{5}$$

where KL is the Kullback-Leibler divergence, $q(\cdot)$ and $p(\cdot)$ are the probabilities based on a Gaussian distribution. We note that in the original paper of GAE [6], the authors proposed a more complex version called variational GAE (VGAE), which allows a probabilistic definition of the latent matrix $\mathbf{Z}$. After experimenting, however, we find that VGAE tends to generate less stable embeddings. So in this paper, we use the basic version of GAE.

4 Graph Topic Model Autoencoder for Event Embedding Learning

We propose a method to covert raw social media data stream into event embeddings. Because we follow a graph-based approach, we need to first convert social media data into graphs. Then we generate embeddings using a novel graph-based technique. In this section, we will present our method in detail.

4.1 Generating Event Graph from Social Media

We first convert raw social media data stream into graphs. The raw social media data stream is defined as the following. We first set a time unit length s. As we collect data, we obtain data in a number of time units, $\mathbf{C} = \{C_1, ..., C_t\}$. Each C_t is a collection of short texts, $C_t = \{c_1, ...c_{N_t}\}$, collected in the time period $[t, t+s]$. The number of short texts, N_t, is variable in different time units. The vocabulary size, D, is assumed to be fixed for all time units.

We can construct two graphs from C_t. The first is document-word graph. We use the same approach as the NMF to construct the graph, which we described in Eq. (1). This graph is denoted as G^{DW}, where each G^{DW}_{ji} is the tf-idf value of word j in document i.

The document-word graph comes naturally with a collection of short text documents. However, this graph may generate some noises by confining words to documents. For example, one account may spam some similar text in one time period. The document-word graph may capture this anomaly, but it is really not related to overall events in the period. In contrast, the word-word co-occurrence graph captures overall events by removing the confinement of documents. The word-word co-occurrence graph is simple to construct. This graph, denoted as G^{WW}, is a $D \times D$ matrix, where each value G^{WW}_{ij} is the number of documents word i and word j co-occur:

$$G^{WW}_{ij} = |C_{co}|, \forall (w_i, w_j) \in C_{co}, C_{co} \subseteq C_t \tag{6}$$

Above two graphs G^{DW} and G^{WW} can capture anomalies in word frequency and word co-occurrence in the time unit, thus the event information. After constructing these two graphs, we use a unified graph embedding technique to further refine the information present in the graphs. As a topic model approach, we manually define the number of topics as K.

4.2 Graph Topic Model Autoencoder

We propose a model called Graph Topic Model Autoencoder (GTMA) that takes advantages of both the document-word graph and word-word co-occurrence graph. Our model combines the approaches of NMF and GAE. Both approaches are unsupervised approaches based on data reconstruction, so our model is also unsupervised. Both methods reconstruct the graph by processing the original graph with latent factors. We let the NMF model process G^{DW}, and the reconstruction is done by calculating $\mathbf{WH}$, the product of the document latent factor and the word latent factor. The word latent factor $\mathbf{H}$ is usually used for topic modeling and event detection. We then let the GAE model process G^{WW}, reconstructing the graph using Eq. (4). The latent factor $\mathbf{Z}$ in the equation is generated using the encoder defined in Eq. (3). We find that this latent factor $\mathbf{Z}$, of the size of $K \times D$, can also be used in topic modeling and event detection.

We combine the latent factor from two models to form the final event embedding of our model. Like existing studies on topic modeling, the latent factors are processed by softmax, so that the value in each topic is summed to 1. Formally, each row of our event embedding matrix $\mathbf{E}$ is calculated as:

$$\mathbf{E}_{[i,:]} = \text{softmax}(\mathbf{H}_{[i,:]}) + \text{softmax}(\mathbf{Z}_{[i,:]}) \tag{7}$$

This embedding matrix $\mathbf{E}$ is our final output, which can then be used for event detection and other downstream tasks.

4.3 Temporal Smoothing

Our model described in the above section captures events in each time unit, treating text messages in each time unit as an independent corpus. This model has some disadvantages. First, events are often continuous. Treating each time unit as independent cannot capture the evolution of events. Second, treating each time unit as independent makes the model too sensitive to the anomaly in the time unit. Particularly, the model will be vulnerable to spams in the time unit. As such, following existing approaches, we perform temporal smoothing on the model output. We use a loss function for minimizing the difference between current event embedding and event embedding of the past time unit:

$$\mathcal{L}_{smooth} = \gamma||\mathbf{E}_t - \mathbf{E}_{(t-1)}||_2 \tag{8}$$

where $||\cdot||_2$ is the L2 norm, and γ is a scaling factor. In this way, we enforce current event embedding to conform to past embedding, capturing event evolution and removing noises.

4.4 Model Learning

We follow the standard machine learning approach to learn the parameters in the model through gradient descent. Particularly, we calculate the final loss value by combining three loss functions presented in the above section:

$$\mathcal{L} = \mathcal{L}_{NMF} + \mathcal{L}_{GAE} + \mathcal{L}_{smooth} \tag{9}$$

4.5 Event Detection

The event embedding matrix $\mathbf{E}$ of the size $K \times D$ is considered containing information of K events. Each of the K events is represented by a number of topic words. The topic words can be obtained by ranking D words by their corresponding value in each row of $\mathbf{E}$. These top l topic words are considered most relevant for the topic and can be used for explaining and characterizing events. In the experiment, we use these top l words to represent the detected events, and compare them with ground truth labels, which are also some descriptive words representing the events.

5 Event Detection Evaluation

We conduct experimental evaluations to test the effectiveness of our method in the task of event detection. In this section, we will present the experiment setup and baseline methods, and discuss the results.

5.1 Datasets

In our experimental evaluation we use two real-world social media datasets. The first is called Reseed [11] dataset, which is currently publicly available[2]. The dataset contains about 437k Flickr posts, which are collected through Flickr API by filtering posts that have a social event tag. The posts are associated with textual titles, textual tags, and photos. Each post has a post date and a photo taken date, which are concentrated around 2006 to 2012. The second dataset is called Politics, which is collected by our own effort. First, we collect a list of Japanese politician Twitter accounts[3]. Next, we collect the follower accounts of these politicians. Then we collect from these follower accounts tweets that dated between Jun and September 2017. In total, this dataset contains about 2,464k Japanese tweets from 33,443 accounts.

We clean both datasets by removing less frequent words. For the Reseed dataset, we remove words with a frequency less than 100, resulting in a vocabulary of 2,438. For the Politics dataset, we remove words with a frequency less than 200, resulting in a vocabulary size of 6,955. A summary of these statistics about the dataset is shown in Table 1.

Table 1. Dataset Statistics

dataset	Reseed	Politics
period	2006.01–2010.07	2017.06–2017.09
time unit length	14 days	1 day
number of units	120	120
number of posts	437k	2,464k
vocabulary size	2,438	6,955

For evaluating event detection accuracy, we prepare corresponding ground truth data. The Reseed dataset comes with a ground truth file that contains a list of events that fall in the same period as the posts. The list is constructed from the web service Upcoming, which provides information on musical events. The events are associated with textual descriptions and timestamps. We tokenize the textual descriptions using a standard tokenizer and assign them to time units

[2] https://qualinet.github.io/databases/image/reseed_social_event_detection_dataset/.
[3] An example list is provided by the website Meyou with the url https://meyou.jp/group/category/politician/.

as the ground truth label. For the Politics dataset, we obtain the ground truth from the pre-trained large language model, GPT-4[4]. GPT-4 is trained on the entire Web data up to 2021. It cannot generate answers for information related to a specific date, but is relatively accurate on a monthly level. We query GPT-4 with this prompt, "What are the top 10 Japanese political events that happened in *month* 2017?", where $month = \{June|July|August|September\}$, and use its answer, textual descriptions of top 10 events, as the ground truth label.

5.2 Evaluation Metrics

We measure the accuracy of event detection by comparing the output of various embedding approaches with the ground truth. As we mentioned earlier, the events are predicted by ranking the words according to their values in a latent dimension. So for each time unit t, we predict top l words in all K latent dimensions as P_t. The ground truth label is a collection of words L_t. As such, the true positive count of prediction is $TP = \sum_{t \in T} |P_t \cap L_t|$. Then we measure precision, recall, and the F-value, where precision is $\frac{TP}{\sum_{t \in T} |P_t|}$, recall is $\frac{TP}{\sum_{t \in T} |L_t|}$, and F-value is $2 \cdot \frac{Precision \times Recall}{Precision + Recall}$. For experiments with the Politics dataset, the ground truth is of a month, and the prediction is of a day. In this case, when measuring precision, we use a day as the unit, and repeat the ground truth for each day of the month. When measuring recall, we use a month as the unit, and aggregate the daily predictions to a month.

5.3 Baseline Methods and Implementation

We compare our model with seven baseline methods, some are topic model methods, and some are graph embedding methods. These include LDA [2], OBTM [5], ONMF [8], EvNMF [12], EmNMF [12], GCN [16], and EvGCN [9], all of which we have briefly introduced in the Related Work section.

We implement all models except OBTM as neural networks using Python and TensorFlow. For OBTM, we use the code provided by the authors[5]. We set the dimension of the latent factor, i.e., the number of topics, to 10 for all models. For the γ in GTMA, we set it to 1000, which seems to provide optimal results. We will make the code for our model available soon.

5.4 Results and Discussion

The accuracy results for the Reseed and Politics datasets are shown in Tables 2 and 3, respectively. The best result among compared methods is highlighted in bold font.

For the Reseed dataset, we see that the best method is GTMA. When l is 20, the accuracy is improved by 3.7% compared to the best baseline. For

[4] https://openai.com/research/gpt-4.
[5] https://github.com/xiaohuiyan/OnlineBTM.

Table 2. Event detection accuracy results on the Reseed dataset

	P@20	P@50	R@20	R@50	F@20	F@50
LDA	0.129	0.107	0.149	0.308	0.138	0.158
OBTM	0.169	0.146	0.195	0.421	0.181	0.217
ONMF	0.156	0.147	0.180	0.423	0.167	0.218
EvNMF	0.158	0.157	0.182	0.453	0.169	0.233
EmNMF	0.158	0.158	0.182	0.457	0.169	0.235
GCN	0.146	0.136	0.168	0.392	0.156	0.202
EvGCN	0.152	0.121	0.176	0.350	0.163	0.180
GTMA	**0.175**	**0.160**	**0.202**	**0.462**	**0.188**	**0.238**

Table 3. Event detection accuracy results on the Politics dataset

	P@20	P@50	R@20	R@50	F@20	F@50
LDA	0.110	0.081	0.469	0.589	0.178	0.143
OBTM	0.158	0.109	0.214	0.340	0.182	0.165
ONMF	0.128	0.102	0.226	0.334	0.163	0.157
EvNMF	0.117	0.094	0.357	0.563	0.177	0.161
EmNMF	0.113	0.091	**0.491**	**0.674**	0.183	0.161
GCN	0.127	0.103	0.454	0.606	0.199	0.176
EvGCN	0.138	0.102	0.451	0.617	0.211	0.175
GTMA	**0.233**	**0.177**	0.300	0.469	**0.262**	**0.257**

the Politics dataset, we see that some methods tend to have higher precision while others tend to have higher recall. To have higher precision, it is better to predict critical and similar topic words, while to have a higher recall, it is better to predict words of higher variety. From the results, we see that OBTM, ONMF, and GTMA have high precision and low recall, while LDA, EmNMF, GCN, and EvGCN have high recall and low precision, separated by the cause we described. However, even though GTMA achieves higher precision by predicting fewer words, its recall is nonetheless relatively high compared to similar methods. As a result, it achieves the best F-value. This shows that GTMA has the best balance between picking a large number of words and pickingcritical topic words.

6 Conclusion

Event detection from social media is an important and challenging problem. In existing works, topic models and graph embedding methods have been used for learning event embedding from social media. In this paper, we propose a model to combine the advantages of topic models and graph embeddings for better event embedding learning. Our model consists of an NMF module, a GAE module, and a temporal smoothing module. Experimental evaluation with two real-world social media datasets shows that each module contributes positively to the model performance. We also compare our model with a number of existing approaches and show that our model has a superior performance. In the future, we plan to further develop our model to include a variable vocabulary.

References

1. AlSumait, L., Barbará, D., Domeniconi, C.: On-line lda: adaptive topic models for mining text streams with applications to topic detection and tracking. In: 2008 Eighth IEEE International Conference on Data Mining, pp. 3–12. IEEE (2008)
2. Blei, D.M., Ng, A.Y., Jordan, M.I.: Latent dirichlet allocation. J. Mach. Learn. Res. **3**, 993–1022 (2003)
3. Bordes, A., Usunier, N., Garcia-Duran, A., Weston, J., Yakhnenko, O.: Translating embeddings for modeling multi-relational data. Adv. Neural Inf. Process. Syst. **26** (2013)
4. Chen, Y., Zhang, H., Liu, R., Ye, Z., Lin, J.: Experimental explorations on short text topic mining between LDA and NMF based schemes. Knowl.-Based Syst. **163**, 1–13 (2019)
5. Cheng, X., Yan, X., Lan, Y., Guo, J.: BTM: topic modeling over short texts. IEEE Trans. Knowl. Data Eng. **26**(12), 2928–2941 (2014)
6. Kipf, T.N., Welling, M.: Variational graph auto-encoders. In: NIPS Workshop on Bayesian Deep Learning (2016)
7. Kipf, T.N., Welling, M.: Semi-supervised classification with graph convolutional networks. In: Proceedings of the Fifth International Conference on Learning Representations ICLR (2017)
8. Mairal, J., Bach, F., Ponce, J., Sapiro, G.: Online learning for matrix factorization and sparse coding. J. Mach. Learn. Res. **11**(1) (2010)
9. Pareja, A., et al.: Evolvegcn: evolving graph convolutional networks for dynamic graphs. Proc. AAAI Conf. Artif. Intell. **34**, 5363–5370 (2020)
10. Porteous, I., Newman, D., Ihler, A., Asuncion, A., Smyth, P., Welling, M.: Fast collapsed gibbs sampling for latent dirichlet allocation. In: Proceedings of the 14th ACM SIGKDD International Conference on Knowledge Discovery and Data Mining, pp. 569–577. ACM (2008)
11. Reuter, T., Papadopoulos, S., Mezaris, V., Cimiano, P.: Reseed: social event detection dataset. In: Proceedings of the 5th ACM Multimedia Systems Conference, pp. 35–40 (2014)
12. Saha, A., Sindhwani, V.: Learning evolving and emerging topics in social media: a dynamic NMF approach with temporal regularization. In: Proceedings of the Fifth ACM International Conference on Web Search and Data Mining, pp. 693–702 (2012)
13. Wang, Y., Agichtein, E., Benzi M.: Tm-lda: efficient online modeling of latent topic transitions in social media. In: Proceedings of the 18th ACM SIGKDD International Conference on Knowledge Discovery and Data Mining, pp. 123–131 (2012)
14. Wang, Z., Zhang, J., Feng, J., Chen, Z.: Knowledge graph embedding by translating on hyperplanes. In: Proceedings of the AAAI Conference on Artificial Intelligence, vol. 28 (2014)

15. Xu, W., Liu, X., Gong, Y.: Document clustering based on non-negative matrix factorization. In: Proceedings of the 26th Annual International ACM SIGIR Conference on Research and Development in Information Retrieval, pp. 267–273. ACM (2003)
16. Zhang, S., Tong, H., Xu, J., Maciejewski, R.: Graph convolutional networks: a comprehensive review. Comput. Soc. Netw. **6**(1), 1–23 (2019)
17. Zhang, Y., Fang, X.S., Hara, T.: Evolving social media background representation with frequency weights and co-occurrence graphs. ACM Trans. Knowl. Discov. Data **17**(7), 1–17 (2023)
18. Zhang, Y., Shirakawa, M., Hara, T.: Generalized durative event detection on social media. J. Intell. Inf. Syst. 1–23 (2022)
19. Zhang, Y., Shirakawa, M., Wang, Y., Li, Z., Hara, T.: Twitter-aided decision making: a review of recent developments. Appl. Intell. **52**(12), 13839–13854 (2022)
20. Zhang, Y., Szabo, C., Sheng, Q.Z.: Sense and focus: towards effective location inference and event detection on twitter. In: Proceedings of the 16th International Conference on Web Information Systems Engineering Part I, pp. 463–477 (2015)

The Impact of Featuring Comments in Online Discussions

Cedric Waterschoot[1](✉), Ernst van den Hemel[1], and Antal van den Bosch[2]

[1] KNAW Meertens Instituut, Amsterdam, The Netherlands
{cedric.waterschoot,ernst.van.den.hemel2}@meertens.knaw.nl
[2] Institute for Language Sciences, Utrecht University, Utrecht, The Netherlands
a.p.j.vandenbosch@uu.nl

Abstract. A widespread moderation strategy by online news platforms is to feature what the platform deems high quality comments, usually called editor picks or featured comments. In this paper, we compare online discussions of news articles in which certain comments are featured, versus discussions in which no comments are featured. We measure the impact of featuring comments on the discussion, by estimating and comparing the quality of discussions from the perspective of the user base and the platform itself. Our analysis shows that the impact on discussion quality is limited. However, we do observe an increase in discussion activity after the first comments are featured by moderators, suggesting that the moderation strategy might be used to increase user engagement and to postpone the natural decline in user activity over time.

Keywords: Content moderation · online discussions · NLP

1 Introduction

How to foster constructive debate in the comment section? This question is of increasing concern for online media outlets. For at least two decades, online news platforms have been struggling to curtail dark participation and trolling in comment spaces [13]. In recent years, however, the moderator received an increasingly wider range of tasks besides merely deleting undesired user content [17]. The moderator is now also tasked with recognizing and promoting *good* comments. Featuring quality comments as a norm-setting strategy has become widespread among large online outlets such as, for example, the New York Times and the Guardian [1,16]. Presented as examples of constructive discussion among users, certain comments are pinned to a highly visible position within the comment interface. However, much remains unclear regarding the actual impact and implications of this moderation strategy on the discussion. What happens in the comment space when moderators start promoting certain comments and commenters?

In this paper, we analyze discussions in which moderators performed the moderation strategy of featuring high quality comments, by comparing them

L. M. Aiello et al. (Eds.): ASONAM 2024, LNCS 15212, pp. 65–74, 2025.
https://doi.org/10.1007/978-3-031-78538-2_5

to discussions in which no comments were featured. By further splitting online discussions, either with or without featured comments, in before and after subgroups based on the featuring time of comments in the data, we are able to pinpoint differences in discussion quality and activity between the control and the featured content discussions. Overall, discussion quality was unaffected. Quality as estimated from the editorial perspective was not impacted by the presence of featured comments. Due to the fact that the control discussions' activity dwindled down faster over time, we end our analysis by hypothesizing whether featuring comments can be used to extend activity on the discussion platform. Indeed, we observe more engagement in terms of posts and involved users in discussions with featured comments.

2 Background

The task of content moderation is defined as the screening of user-generated content to assess the appropriateness for the given platform [3,14]. The practice has been evolving to address the needs of a growing, contemporary online community. This development expanded the task set of the moderator, who needs to swiftly make interpretative moderation choices [11]. Practically speaking, the moderation task has been described as a gatekeeping role [20]. This function is twofold. First, moderators ought to keep the comment space clean of unwanted content [11]. Described under the umbrella of dark participation, this content can take the form of trolling, cyberbullying or even organized misinformation campaigns [7,8,13]. The moderator is charged with deleting or reducing the visibility of such content [4]. Additionally, coping with these negative influences required the platforms to expand the online content moderation practice [19]. This expansion, among other things, led to the promotion of *good* content [1,20]. The bulk of the literature on content moderation focuses on the bad and unwanted comments and actors within the comment space. Relatively little is known about the active promotion of good commenting behaviour, even though the practice by now is widespread among online news platforms.

As a part of the modern comment section, platforms have been highlighting quality comments on their discussion page [12]. In a practical sense, it takes the form of NYT Picks at the New York Times [1], Guardian Picks at The Guardian [15] and featured comments at Dutch news platform NU.nl [10], for instance. Roughly speaking, the platforms define such quality comments as "substantiated", "most interesting and thoughtful" or "presenting a range of perspectives" [2,10]. In general, featuring what they deem high-quality content is an attempt by platforms at norm-setting [16]. Dutch online platform NU.nl specifically states that these comments serve as examples for other users [10].

Even though most research on content moderation focuses on unwanted comments, some have specifically looked at aspects of featured comments. NYT picks in particular have been used as examples of constructive comments in classification tasks [5]. Wang and Diakopoulos use a classifier trained on NYT picks to assign quality scores to other comments, concluding that users who receive a

NYT pick subsequently write higher quality comments, an effect that diminished over time [16]. Yahoo News comment threads have been used for the annotation of good content, more specifically in terms of "ERICs: Engaging, Respectful, and/or Informative Conversations" [9]. The authors focuses on the thread level rather than on the comment and did not use an editorial standard as their labelling, as is the case with NYT picks [9]. Additionally, research has annotated constructive comments as containing specific evidence or solutions, as well as personal anecdotes or stimulating dialogue [6]. As part of their visual CommentIQ interface, prior research classifies and ranks comments using comment and user history criteria. These criteria included readability scores and the number of likes a comment had received and relevance scores introduced in earlier research by Diakopoulos [2,12]. The visual interface allows for different plots and ranking possibilities and uses NYT picks [12]. With the similar goal of supporting the moderators to pick featured comments, Waterschoot and Van den Bosch trained classifiers to rank comments based on the probability that moderators picked them as featured [18]. Using data from the Dutch news platform NU.nl, the authors supplemented comment and user information with text representation [18]. The models were tested on the discussion level on unseen articles from the platform and evaluated by the NUjij moderators themselves, yielding positive results in regard to the ranking of comments.

In sum, while literature on online content moderation is mostly aimed at toxic or other unwanted content in the comment space, moderation strategies aimed at promoting good user-generated content are widespread. While previous work did look at practical support for the moderator in picking content and the effect of highlighted comments on the user and replies to the comment, an analysis on the discussion level comparing discussions with featured content to those without has not yet been performed. The current study aims to address the potential impact of highlighted quality comments on the discussion by comparing them to a control set of discussions in which this strategy was not performed. Furthermore, we broaden the concept of discussion quality by including a user perspective as well, aside from the editorial definition used in previous work.

3 Methodology

We use a 2023 Dutch language dataset from the comment platform NUjij[1], part of the Dutch online newspaper NU.nl[2]. The platform allows users to comment on the news articles published by the outlet. Aside from the set of articles in which moderators picked featured comments ($n = 143$; 86,157 comments, on average 602 comments per article, 1,235 featured comments), we also obtained a control set of articles in which no featured content was chosen ($n = 66$; 32,862 comments, on average 498 comments per article). Articles in both groups have publication times spread throughout the day. Included in the data are comments that were rejected by the moderators. The articles cover a range of topics such

[1] Data available at github.com/cwaterschoot/Featured_Comments_Impact.

[2] https://nu.nl.

as climate change, the local elections, the nitrogen issue in the Netherlands and the war in Ukraine. Comments are not featured based on the popularity of the news topic. Both the featured group of discussions as well as the control group includes articles covering these topics. In the case of featured comments, a timestamp indicates the exact time that moderators highlighted the comment.

In order to assess whether the presence of featured content had any impact, each discussion is split in two subgroups: (1) comments before featured content was chosen and (2) comments posted after. For the discussions in which comments were featured by moderators (group Featured), the cut-off was made at the featuring time of the first featured comment per discussion. In the case of control discussions (group Control), the split was made at the median time of these first featured comments relative to the publication time of the article (123 minutes). We decided for the median as opposed to the mean (231 minutes) due to the impact of articles published late at night, for which the comment sections only opened up in the morning, leading to outliers.

Table 1. Logistic regression: Characteristics before cut-off

Variable	Coeff (std er.)	p-value
Respect count	−0.055 (0.039)	0.165
Featured candidates	0.038 (0.030)	0.209
Flagged comments	0.034 (0.035)	0.338
Rejection rate	−0.006 (0.015)	0.678
User count	0.007 (0.005)	0.158
Post count	−0.003 (0.002)	0.172

Dependent variable: group (control before/ featured before)

To test the validity of the comparison between the featured and control group, we constructed a logistic regression model based on the before data (Table 1). As dependent variable, we included the group identifier, either control or featured. Thus, the model tests whether it is capable to predict if a before discussions belongs to the featured or control group. If this would be possible, the groups show different discussion characteristics. Independent variables are listed in Table 1. These variables were averaged across all comments before the cut-off, i.e. for the control group the 123 minute mark and for the featured group comments posted before the first featuring timestamp in the discussion. Significant effects of these discussion characteristics imply that we cannot conclude that the discussion groups showed similar discussion characteristics before the moderation strategy was performed. This result would suggest the invalidity of our between-group testing. However, the included discussion characteristics were not statistically significant between the two groups (Table 1). All comparisons between the after subgroups were made using Mann Whitney U tests with Bonferonni correction to correct for Type I error.

3.1 Measuring Influence: Quality and Activity

We aim to analyze whether discussions in which featured comments are highlighted contain more quality comments compared to discussions without the moderation strategy. We operationalized the concept of *discussion quality* based on two categories, each further broken down into two perspectives. The two categories of discussion quality are (1) the absence of bad content and (2) the presence of quality comments. Each category is analyzed from both the user and editorial (platform) perspective. We contrasted these markers of discussion quality between the before and the after subgroups for the control and featured discussions.

The absence of bad quality from the user perspective is tested by averaging the percentage of comments that were flagged. A higher rate of flagged comments could be seen as an indication that the user base decided the discussion contained less quality comments. The editorial perspective in this category is operationalized through the rejection rate. This variable captures the percentage of comments in each discussion that moderators decided to delete. Discussion quality would increase if the need to delete unwanted content decreases.

The second category related to discussion quality aims to capture the opposite, i.e. the presence of high quality comments. The user perspective is defined through the average number of likes comments received. For both the before and after subgroups within the control and featured discussions, we calculated the average number of likes. We assessed discussion quality from the editorial perspective by following the procedure outlined in by previous research [16]. Using a different NUjij dataset containing comments from 2020, we replicated the model for scoring unseen comments in terms of comment quality, operationalized as the class probability for being featured [16,18]. The variables used for training the model are described in previous research and include both comment and user features [18]. For training and testing, we split the 2020 dataset into an 80/20 split, resulting in 6,679 featured comments in the training and 1,661 featured comments in the test set. A random forest model was trained on a balanced set containing the 6,679 featured comments alongside an equal number of random non-featured comments. The final model, calculated on unseen test set, achieved an F1-score of 0.86, a similar result as reported in previous work [16]. Each comment from the current dataset received a probability of being featured-worthy, a proxy for the editorial view of comment quality [16]. We calculated the averaged percentage of featured-worthy candidates (those with a class probability above 0.5).

The final point of focus relates to the evolution of activity within discussions on the platform. We assessed how the discussions evolved in the first 10 hours by counting the average number of unique users and mean total comments, comparing the before and after subgroups for both the control and featured groups. For this analysis, we only included the accepted comments in the discussion, leaving out the comments that were deleted by a moderator. We are particularly interested in analyzing whether the featured content caused a different evolution in discussion growth.

4 Results

Discussion quality was studied comparing the before/after groups and by contrasting the control group with the discussions in which featured content was chosen. We divided quality into two categories: (1) absence of bad content and, (2) presence of high-quality comments. The discussion quality framework resulted in four variables calculated for the before and after subgroups for both the control and feature sets (Table 2).

Table 2. Discussion Quality: differences between the before and after subgroups

Category	Perspective	Variable		Before	After	Δ
Absence of	User	Flagged comments	Control	9.08%	10.07%	+0.99pp
bad content			Feat.	9.74%	9.7%	−0.04pp
	Editorial	Rejection rate	Control	23.25%	22.20%	−1.05pp
			Feat.	23.54%	20.91%	−2.63pp
Presence quality	User	Respect Count	Control	5.69	3.61	−2.08
comments			Feat.	5.65	3.43	−2.22
	Editorial	Featured candidates	Control	14.04%	8.90%	−5.14pp
			Feat.	14.39%	8.30%	−6.09pp

The user perspective on the absence of bad quality content is expressed through the average percentage of user flagged comments. In the control group, we found that 9.08% before the 123 minute mark and 10.07% after the cut-off were flagged by at least one user (+0.99pp). For the featured group, we calculated a decrease from 9.74% before to 9.70% (-0.04pp). The results indicate that, while featuring content did not decrease the flagging of bad comments by users, it could prevent more flagged comments later on in the discussion. However, the difference between the average number of flagged comments in the after discussions is not significant ($U = 4830.50, p = 0.42$). The editorial perspective on absence of bad content was calculated by the average percentage of comments deleted by moderators. Both the control group (−1.05pp) and the featured group (−2.63pp) showed a decline (Table 2). Comparing the after groups of both the featured on control group, we did not find a significant difference ($U = 4021.5, p = 0.38$). This result suggest that the presence of featured comments did not reduce the need for moderators to reject incoming comments.

The second category of discussion quality aimed to capture the presence of quality content. The user perspective of this category was operationalized by calculating the average number of respect points that comments received before and after the cut-off. Over time, the average number of likes declined, potentially due to the fact that comments posted late in the discussion are read less often. We specifically looked at whether this decline would be less steep in the featured discussion. In the featured group, the average number of likes declined by 2.22

respect points. In the control group, the average declined by 2.08 respect points. This difference between the control and featured group after the cut-off is not significant ($U = 4420.0, p = 0.8$).

The final marker for discussion quality was the percentage of featured candidates (class probability > 0.5). The average percentage of candidates before the cut-off showed no difference comparing the featured group (14.39%) and the control group (14.04%). As the discussion continued, the average share of featured candidates comments decreased in both featured ($-6.09pp$) and control ($-5.08pp$) groups (Table 2). This difference in average number of featured candidates in the after subgroups were not significant ($U = 4093.5, p = 0.49$). The average percentage of featured candidates in the featured and control groups after the cut-off were 8.30 and 8.90, respectively (Table 2).

Finally, we analyzed discussion activity before and after the cut-off based on two factors: the set of users commenting and, the average number of comments in the discussions. This analysis omitted those that were rejected by moderators.

Unlike the quality markers discussed earlier, the discussion activity progressed differently within the featured group as opposed to the control discussions. The speed of discussion activity slowed down over time, which is expected due to the fact that users move on to more recent articles. However, the average discussion activity in the control group dwindled down much quicker; the featured discussions continued on and slowed down at a point later on (Fig. 1).

Before the cut-off, an average of 112 users commented before the 123 minute mark in the control discussions, while 115 users participated before the first comment was featured in the featured group. In the case of the after groups, however, a significant difference was found, indicating that more unique users commented in the featured discussions after content was featured ($U = 3205.5, p < 0.001$). The same result was found for the post count. The average post count before the 123 minute mark in the control discussions was 223, while in the featured group this was 213 comments. After the cut-off, the average control discussion went on for 126 more comments, while the mean post count in the featured group was 207. This difference in after groups is statistically significant ($U = 3161.5, p < 0.001$).

5 Discussion

In the following sections, we discuss the apparent inability to influence discussion quality with featured comments. We end the discussion by outlining several open questions for future work and the limitations of the current study.

First, featuring content did not lower the necessity to reject incoming comments, implying that the highlighted examples did not deter people from posting uncivil or off-topic content. While the rejection rate did decrease over time, control discussions without featured comments evolved in the same manner. From the user perspective, the control group experienced an increase in user flagging in the after group, while this increase was not found in the group with featured comments. However, this difference was not significant.

Second, we found no difference between the two groups in regard to the user perspective on quality, represented in the study by the number of likes

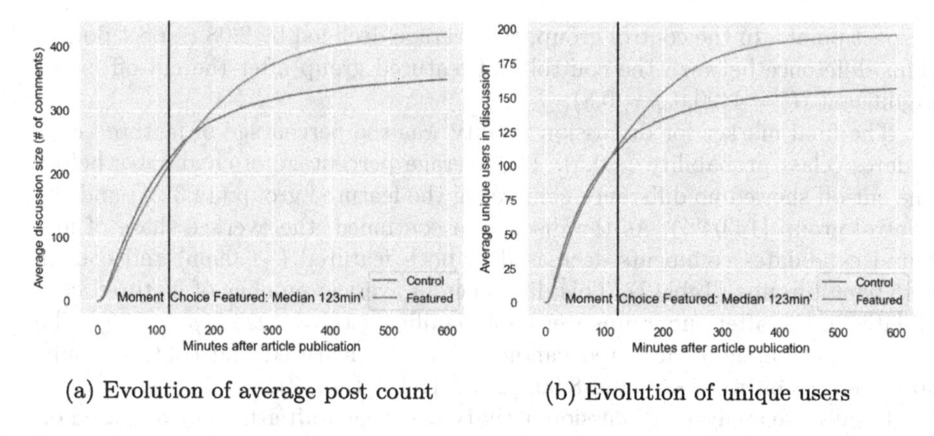

(a) Evolution of average post count

(b) Evolution of unique users

Fig. 1. Growth of discussion activity in featured and control group (first 600 min)

comments received. Over time, the average number of likes a comment received decreased significantly with or without featured comments. The final marker of discussion quality entailed featured-worthy comments. If such contributions would successfully serve as examples to users of the editorial standard of quality, one would expect the number of comments qualifying as featured-worthy to be significantly higher in the after group of featured discussion compared to the after group comprising of control discussions. However, we did not find such a significant difference. In both groups, the number of featured-worthy comments decreased in the after subgroup. Overall, these results indicate that discussion quality decreased over time. Even though the practice of featuring user comments is widespread among online news platforms, the inability to influence discussion quality in all but one of the variables (and not significantly so) indicates that users do not use these comments as examples. In particular the lack of change in the editorial perspective showcased that the platform does not succeed in shaping the discussion to what they themselves deem good discussion.

We hypothesize that featured content could be used to postpone the natural decrease in discussion activity. Further studies should focus on this particular point, eliminating other factors through, among other things, A/B testing. Additionally, cross-platform analysis is necessary to assess the general impact of the moderation strategy. Platforms like, for example, the New York Times and the Guardian highlight certain comments as well. Such an analysis can compare platforms to test whether moderators at different outlets feature similar content.

Research aimed at analyzing the effects of online moderation has the inherent constraint of the availability of data and metadata. To replicate the current study, researchers require not only the comments that were published at the time, but also information about those which were rejected or later deleted by moderators. The latter is not publicly available, typically, cannot be published, and requires cooperation with online news platforms to be obtained, or working from within the platform's organization. Furthermore, individual comments require metadata indicating to which article they were posted, such that researchers can separate different discussions. Timestamps indicating when comments were featured are also not publicly available, typically, as well as information as to how many times a comment might have been flagged by other users. All in all, without this crucial information, the potential effect of these moderation strategies cannot be adequately analyzed. Cooperation with the platforms is needed to obtain such unpublished information.

6 Conclusion

To sum up, we did not find evidence indicating an impact on discussion quality as a result of the featured content, especially from the editorial perspective. The rejection percentage was unaffected, as well as the number of featured-worthy comments. Both aspects of discussion quality decreased in both the control and featured groups in similar fashion. A similar decrease was found in the average respect points comments had received, the user perspective of presence of quality comments. What did change, however, was the average number of flagged comments by users. As opposed to the other quality variables in the framework, the average number of flagged comments increased in the after group of the control discussion, while this increase was not found in the featured group. However, this difference was not statistically significant between after groups.

Finally, we did find differences in discussion activity between the featured and control group. The results show that discussion activity declined slower in the featured group. It would seem that this moderation strategy can be used to postpone the decline in user activity. However, future research is necessary to eliminate other factors potentially influencing discussion growth.

Acknowledgments. This study is financed by project number 410.19.006 of the research program 'Digital Society - The Informed Citizen' which is financed by the Dutch Research Council (NWO).

Disclosure of Interests. The authors have no competing interests to declare that are relevant to the content of this article.

References

1. Diakopoulos, N.: Picking the NYT picks: editorial criteria and automation in the curation of online news comments. ISOJ **5**(1), 147–166 (2015)
2. Diakopoulos, N.: The editor's eye: curation and comment relevance on the New York Times. In: Proceedings of the 2015 ACM International Conference on Computer-Supported Cooperative Work and Social Computing (CSCW 2015), pp. 1153–1157 (2015). https://doi.org/10.1145/2675133.2675160
3. Gillespie, T.: Custodians of the Internet: Platforms, Content Moderation, and the Hidden Decisions that Shape Social Media. Yale University Press (2018)
4. Gillespie, T.: Do not recommend? reduction as a form of content moderation. Soc. Media Soc. **8**(3), 1–13 (2022). https://doi.org/10.1177/20563051221117552

5. Kolhatkar, V., Taboada, M.: Constructive language in news comments. In: Proceedings of the Annual Meeting of the Association for Computational Linguistics, pp. 11–17 (2017). https://doi.org/10.18653/v1/w17-3002
6. Kolhatkar, V., Thain, N., Sorensen, J., Dixon, L., Taboada, M.: Classifying constructive comments. First Monday **28**(4), 1–16 (2023). https://doi.org/10.5210/fm.v28i4.13163
7. Lewandowsky, S., Ecker, U.K., Cook, J.: Beyond misinformation: understanding and coping with the "Post-Truth" era. J. Appl. Res. Mem. Cogn. **6**(4), 353–369 (2017). https://doi.org/10.1016/j.jarmac.2017.07.008
8. van der Linden, S., Leiserowitz, A., Rosenthal, S., Maibach, E.: Inoculating the public against misinformation about climate change. Global Chall. **1**(2), 1600008 (2017). https://doi.org/10.1002/gch2.201600008
9. Napoles, C., Tetreault, J., Rosato, E., Provenzale, B., Pappu, A.: Finding good conversations online: the yahoo news annotated comments corpus. In: 11th Linguistic Annotation Workshop, Proceedings of the Workshop (LAW 2017), pp. 13–23 (2017). https://doi.org/10.18653/v1/w17-0802
10. NUJij. NUjij - Veelgestelde vragen (2018). https://www.nu.nl/nujij/5215910/nujij-veelgestelde-vragen.html
11. Paasch-Colberg, S., Strippel, C.: "The Boundaries are Blurry...": How Comment Moderators in Germany See and Respond to Hate Comments. Journalism Stud. **23**(2), 224–244 (2022).https://doi.org/10.1080/1461670X.2021.2017793
12. Park, D., Sachar, S., Diakopoulos, N., Elmqvist, N.: Supporting comment moderators in identifying high quality online news comments. In: Proceedings of the Conference on Human Factors in Computing Systems, pp. 1114–1125 (2016). https://doi.org/10.1145/2858036.2858389
13. Quandt, T.: Dark participation. Media Commun. **6**(4), 36–48 (2018). https://doi.org/10.17645/mac.v6i4.1519
14. Roberts, S.T.: Content moderation. In: Encyclopedia of Big Data, pp. 1–4 (2017). https://doi.org/10.1201/9781003293125-7
15. The Guardian. Frequently Asked Questions about Community on the Guardian Website (2009). https://www.theguardian.com/community-faqs
16. Wang, Y., Diakopoulos, N.: Highlighting high-quality content as a moderation strategyâĂŕ: the role of new york times picks in comment quality. ACM Trans. Soc. Comput. **4**(4), 1–24 (2022)
17. Waterschoot, C.: Governing the 'Third half of the internet': the dynamics of human and AI-assisted content moderation. In: van Dijck, J., van Es, K., Helmond, A., van der Vlist, F. (eds.) Governing the Digital Society: Platforms, Artificial Intelligence, and Public Values. Amsterdam University Press (2024), forthcoming
18. Waterschoot, C., van den Bosch, A.: A time-robust group recommender for featured comments on news platforms. Front. Big Data **7** (2024). https://doi.org/10.3389/fdata.2024.1399739
19. Wintterlin, F., Schatto-Eckrodt, T., Frischlich, L., Boberg, S., Quandt, T.: How to cope with dark participation: moderation practices in German newsrooms. Digit. J. **8**(7), 904–924 (2020). https://doi.org/10.1080/21670811.2020.1797519
20. Wolfgang, J.D.: Cleaning up the "Fetid Swamp": examining how journalists construct policies and practices for moderating comments. Digital Journalism **6**(1), 21–40 (2018). https://doi.org/10.1080/21670811.2017.1343090

Multicriteria Recommendation System by Leveraging Predefined, Implicit, and Undefined Criteria

Emrul Hasan[1,2](✉) and Chen Ding[2]

[1] Department of Computer Science, Toronto Metropolitan University, Toronto, Canada
e1hasan@torontomu.ca

[2] Vector Institute, Toronto, Canada
cding@torontomu.ca

Abstract. In this study, we propose a novel multi-criteria recommendation model that utilizes predefined, implicit, and undefined criteria. We use a semantic similarity-based sentence clustering method to identify the predefined and implicit criteria and a sentiment analyzer to estimate their ratings. Semantic similarity between each sentence in the review and the predefined criteria are calculated, and then the sentence is assigned to the most similar criteria. The criteria that are extracted from the review and are aligned with predefined criteria are referred to as implicit criteria. A sentence is considered as expressing opinions on an undefined criterion if the similarity score between this sentence and all the predefined criteria is lower than a predefined threshold. Ratings are computed for each extracted implicit criterion and the undefined criterion based on the review content. Finally, we use all three types of criteria and an aggregation model to make the final rating prediction for the recommendation system. Our proposed method demonstrates the superiority compared to several baselines on TripAdvisor and Beer Advocate datasets.

Keywords: Criteria rating · implicit criteria · undefined criteria · predefined criteria

1 Introduction

Due to the rapid growth of online data, recommendation system has become a ubiquitous tool for e-commerce and content providers such as Amazon, Netflix, YouTube, etc. When users choose a product or service, they typically consider one or multiple attributes rather than relying solely on overall ratings. In the context of hotels, a user might prioritize factors such as location and service, or focus solely on price. The overall rating could be high due to satisfaction with other criteria, leading to a situation where the user may not receive appropriate recommendations. To address this issue, the researchers turned their attention

L. M. Aiello et al. (Eds.): ASONAM 2024, LNCS 15212, pp. 75–84, 2025.
https://doi.org/10.1007/978-3-031-78538-2_6

to multi-criteria recommendation systems where multiple criteria ratings are utilized to learn user preferences [1–3]. Another direction is to build aspect-based systems where user preferences and item attributes are learned based on different aspects that users discussed in their reviews [3–5,8].

While aspect-based [6] and multi-criteria rating-based [7] recommendations rely on a single data source (review or criteria) for user preference learning, we argue that a more accurate representation of user preferences can be learned when we consider both the criteria ratings and the sentiment expressed through reviews. It is observed that in addition to pre-defined criteria (domain-specific criteria that are defined by the business), users express their opinions on different aspects more freely through reviews.

To this end, we propose a novel multi-criteria recommendation model that utilizes both the predefined criteria and the criteria (aspects) extracted from the review. The predefined criteria may also be discussed in the review, however, the ratings (sentiment scores) associated with them may not be the same as the ratings users explicitly give to the predefined criteria. So, to differentiate the two, we call the ratings based on the review as implicit criteria ratings. If the aspects extracted from the review are not available in the predefined criteria list, they are named undefined criteria.

We assume that each sentence or sub-sentence in the review contains the user's opinion on one criterion. Some criteria are discussed in multiple sentences. To extract the criteria from the review, we split the review into sentences and clustered them by comparing the semantic similarity between predefined criteria and the split sentences. The sentences that have high similarity degrees with predefined criteria are clustered into different homogeneous groups, each of which corresponds to a predefined criterion. The following are the major contributions of our research:

1. We propose a novel multi-criteria recommendation model by leveraging both reviews and the predefined criteria. We show that the user's overall preference for an item not only depends on the predefined criteria ratings but also on the ratings extracted from reviews.
2. We leverage all three types of criteria ratings to learn the overall rating e.g. predefined, implicit, and undefined. To the best of our knowledge, our cluster-based aspect extraction method is new, and the process of aligning the criteria extracted from reviews with the predefined criteria to get undefined criteria ratings and include them in the overall rating prediction is also considered novel.

2 Related Work

Nassar et al. [1] introduce an aggregation-function-based recommendation consisting of two parts, criteria ratings prediction using a deep collaborative filtering model by leveraging DNN and MF, and learning an aggregation function using a DNN. Hong et al. [3] develop a multi-criteria tensor model for tourism recommender systems. They explore user preferences by incorporating cultural factors into the evaluation of multiple criteria ratings. Wang et al. [9] introduce a recommendation system with a Hybrid Deep Tensor Decomposition model, incorporating multiple criteria. It is observed that the consideration of multiple factors has an impact on recommendation performance. Hong et al. [10] introduce ClustPTF (Cluster-based Parallel Tensor Factorization), which employs sentiment analysis to mitigate data sparsity within the tensor model [11]. Next, K-means clustering is used to cluster similar user preferences, thereby improving recommendation diversity. Shambour [2] introduces a deep learning-based multi-criteria recommendation system. Although all the above-mentioned multi-criteria recommendation models have proven to be successful, they use predefined criteria while ignoring the user sentiment expressed through reviews.

On the other hand, aspect-based approaches concentrate on learning the aspect representations from reviews and utilizing them to predict overall preferences. ANR [4] introduces an Aspect-based Neural Recommendation model incorporating an attention mechanism to learn user and item representations simultaneously. Similarly, Li et al. [5] develop an aspect-based fashion recommendation system that extracts both local and global features to estimate overall ratings. Since relying on explicit criteria ratings, aspect-based recommendation systems face the challenge that inferred ratings from reviews may not always be accurate. Again, aspect-based recommendation considers only review for preference learning, ignoring the explicit multiple criteria ratings.

In a more recent development, Hasan et al. [12] propose a multi-criteria recommendation model that learns aspect representation from the review and then combines the aspect ratings with the explicit criteria ratings for the overall rating prediction. The most similar research to our proposed model are the ones by Nassar et al. [1] and by Hasan et al. [12]. The major difference between our work and the former is that they solely rely on criteria ratings without considering textual reviews. Although the latter considers both reviews and criteria, they ignore the undefined criteria. In contrast, our model considers predefined, implicit, and undefined criteria for a more comprehensive analysis.

3 Proposed Model

3.1 Preliminaries

Consider a user $u \in U$ provided feedback to an item $i \in I$ with three different modes: review $x \in X$, criteria rating $r_K \in R$, and overall rating $r_0 \in R$ where K the number of predefined criteria. Our first aim is to extract the implicit and undefined criteria and corresponding ratings from the review. In the second step, our goal is to predict the criteria ratings using Neural collaborative filtering. In the third step, we aim to learn an aggregation function using all criteria (undefined, implicit, and predefined) and the overall rating. Finally, the overall rating for the unknown item is predicted using predicted criteria and the learned aggregation function. The architecture of the model is shown in Fig. 1.

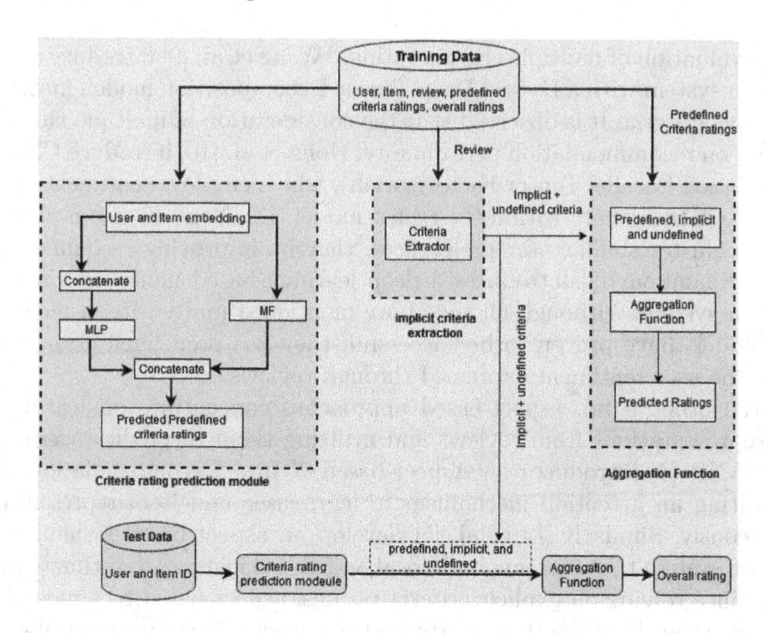

Fig. 1. The detailed architecture of the proposed model.

3.2 Criteria Extraction from the Review

In this step of the process, we split the review into sentences on punctuation symbol ".", and words such as "but", "while", "however", "nonetheless". Here we assume that when using these conjunctions users start to speak on different aspects. Next, we compute the semantic similarity between each sentence and the predefined criteria using cosine similarity measure. We utilize the pre-trained sentence transformer (e.g. SBERT [15]) to generate embeddings for both sentences and the predefined criteria. Sentences are assigned to one of the predefined criteria if the similarity score with that criterion is bigger than a threshold value. Sentences whose cosine similarity scores with all the predefined criteria are less than this threshold value are assigned to one cluster, corresponding to the undefined criteria. We consider these sentences to express users' opinions on aspects that are not covered by predefined criteria. Once the clustering is done, all the sentences in the cluster for each criterion are concatenated, and finally, a sentiment analyzer is used to estimate the rating for each of the clusters.

3.3 Criteria Rating Prediction

Inspired by Nassar et al. [1], we employ Neural Multi-criteria Collaborative Filtering for criteria rating prediction. Since the IDs are categorical features, motivated by He et al. [13], the user ID and item ID are vectorized by leveraging the embedding technique and initialized with random values. Consider the embedding for user u and item i are e_u and e_i respectively. The MF and MLP are fused to learn the user-item interaction.

$$\hat{y}_{ui} = \sigma\left(W^T(e_u \odot e_i) + h^T \begin{bmatrix} e_u \\ e_i \end{bmatrix} + b\right) \tag{1}$$

Rectified Linear Unit (ReLU) [14] and sigmoid σ are the activation function for hidden and final layers respectively.

3.4 Learning Aggregation Function

We employ a DNN-based aggregation function to capture the relationship between the criteria and the overall ratings. Mathematically, overall rating is:

$$r_0 = f(r_1, r_2, ..., r_K) \tag{2}$$

where k is the number of criteria, and r_0 is the overall rating. The final input vector is then passed through several dense layers followed by a *ReLU* layer.

3.5 Overall Rating Prediction

The predicted ratings are used as input to the learned function $f(r)$ to predict the final output and are defined as

$$\hat{r}_o = f(\hat{r}_1, \hat{r}_2, ..., \hat{r}_K) \tag{3}$$

where $\hat{r}_o$ is the predicted final overall rating and $\hat{r}_1, \hat{r}_2, ..., \hat{r}_K$ are the predicted criteria ratings. We exploit the regression with squared loss as the objective function

$$J = \sum (\hat{r}_0 - r_0)^2 + \lambda_\theta \|\Theta\|^2 \tag{4}$$

r_0 and $\hat{r}_0$ are true and the predicted ratings respectively, and $\lambda_\theta \|\Theta\|^2$ is used as regularization to prevent the model from overfitting.

4 Experiments

Datasets: We use two publicly available multicriteria datasets: Tripadvisor[1] and BeerAdvocate[2]. The reason we choose these datasets is that they come with three forms of feedback, e.g. textual review, criteria ratings, and overall rating. Tripadvisor.com is a famous hotel and tourism management platform where users can book hotels, flights, and tourist attractions. The TripAdvisor dataset was

[1] https://www.cs.virginia.edu/~hw5x/Data/LARA/TripAdvisor/.
[2] https://cseweb.ucsd.edu/~jmcauley/datasets.html#multi_aspect.

gathered in 8 years, from 2004 to 2012. The scores for both individual criteria and overall ratings span a scale from −1 to 5.

Baselines: Our comparison is restricted to papers that have openly shared their code. **MRRRec** [12] is a multi-criteria recommendation model that leverages both the criteria ratings and the review to learn the user preferences, demonstrating the performance improvement over several baselines. **ANR** [4] leverages user and item reviews to learn user preferences and item features, increasing performance and explainability. **Deep Cooperative Neural Network (DeepCoNN)** [13] is a state-of-the-art recommendation model that incorporates reviews for predicting ratings. In contrast, **Multi-criteria RS by Nassar et al.** [17] represents a multi-criteria recommendation system based on deep learning. The model predicts overall criteria ratings through a two-step process. **NARRE** [18] employs attention mechanism to capture the representation of the user and item while predicting the overall ratings for the user. **DAML** [19] introduces a dual attention mutual learning method for recommendation by leveraging ratings and reviews. Review features are jointly learned by local and mutual attention, which can increase the interpretability. **MPCN** [13] tries to improve the performance over the state-of-the-art D-ATTN [20] and DeepCoNN model [21].

We evaluate the performance of our model with two error metrics: MSE (Means Squared Error) and MAE (Mean Absolute Error). The reason behind choosing these two metrics is due to their widespread adoption in state-of-the-art rating prediction recommendation models [4,19,22,23]. We further extend our evaluation to precision, recall, and F1 accuracy measures. To compute the precision, recall, and F1, we formulate the problem as a binary classification problem by converting the ratings to 0 and 1 based on a threshold score. In our case, we set this threshold to 3.5. A rating above 3.5 is converted to 1; if it is less than 3.5, it is converted to 0.

4.1 Experimental Settings

We partitioned the datasets into three subsets: training, validation, and test sets, using an 80, 10, 10 ratio. We implemented our model and the baselines using the Pytorch library and T4 GPU from Google Colab Pro. For a fair comparison, we kept the hyperparameters the same as the baseline papers. In our experiments, for clustering tasks, we used different similarity threshold scores, ranging from 0.10 to 0.40 with an increment of 0.05.

5 Results

The model obtains 3.08% and 2.02% lower MSE and MAE respectively compared to the best-performing baseline, i.e., MRRRec, on TripAdvisor dataset (shown in Table 1). For the Beer Advocate dataset, the model achieves a 3.65% and 2.97% lower MSE and MAE respectively. The existing state-of-the-art recommendation models use only MSE and MAE for performance evaluation [4,19,24,25] because they use their models for rating prediction tasks. We further extend our evaluations to precision, recall, and F1 scores to understand the recommendation accuracy. Table 2 depicts the accuracy measures of our proposed model and compares them with the baselines. On the Tripadvisor dataset, our model achieves higher accuracy compared to all the baselines in terms of precision, recall, and F1 measures. In comparison to the best-performing baseline, i.e., MRRRec, precision, recall, and F1 are increased by 0.69%, 0.35%, and 0.93% respectively. For the BeerAdvocate dataset, recall and F1 scores are increased by 1.7% and 1.95% respectively. However, precision sees a decrease of 0.25%, indicating a slightly lower performance compared to the ANR model.

Table 1. Performance comparison in terms of MSE and MAE

Performance Comparison				
Model	TripAdvisor		BeerAdvocate	
	MSE	MAE	MSE	MAE
Proposed	**0.3912**	**0.3721**	**0.2921**	**0.3024**
MRRRec	0.4220	0.3923	0.3286	0.3762
NARRE	0.5798	0.5167	0.5165	0.4602
DAML	0.6198	0.5722	0.4918	0.5013
MPCN	0.5526	0.5282	0.4526	0.5282
ANR	0.4431	0.4110	0.4452	0.3921
DeepCoNN	0.6002	0.5535	0.5006	0.4654
Nassar et al. [1]	0.4670	0.3945	0.3979	0.3321

Table 2. Performance comparison in terms of precision, recall, and F1

Performance Comparison						
Model	TripAdvisor			BeerAdvocate		
	Precision	Recall	F1	Precision	Recall	F1
Proposed	**0.9511**	**0.9596**	**0.9553**	0.9095	**0.9421**	**0.9255**
MRRRec	0.9301	0.9490	0.94901	0.8938	0.9110	0.9023
NARRE	0.8521	0.9116	0.8808	0.8935	0.9257	0.9093
DAML	0.9010	0.8793	0.8793	0.8968	0.8813	0.8889
MPCN	0.8323	0.9021	0.9021	0.8776	0.9317	0.9038
ANR	0.9092	0.9521	0.9302	0.9121	0.9001	0.9060
DeepCoNN	0.8725	0.9295	0.9000	0.8582	0.9176	0.8869
Nassar et al. [1]	0.8911	0.9201	0.9053	0.7454	0.9247	0.8254

Table 3. Performance with different scenarios of criteria ratings

Criteria	MSE	MAE
implicit+undefined+predefined	0.2053	0.3226
implicit+predefined	0.2119	0.3251
undefined+predefined	0.2068	0.3226
undefined+implicit	0.7854	0.6942
Only predefined	0.2121	0.3237

5.1 Ablation Study

We analyze the rating prediction performance for five different scenarios: 1) implicit criteria, undefined criteria, and predefined criteria, 2) implicit and predefined criteria, 3) undefined and predefined criteria, 4) undefined criteria and implicit criteria, 5) predefined criteria. The focus in this set of experiments is to study the impact of different criteria ratings on overall ratings. Once we identify the scenario that gives the best performance on predicting overall ratings, we use that combination of criteria in our final model. The result is shown in Table 3. It is important to note that we conduct the ablation study on TripAdvisor datasets only, because TripAdvisor datasets are the most widely used datasets in multi-criteria rating research [3,17,24]. For each scenario, we predict criteria ratings and learn the aggregation function for overall rating prediction. Among the five cases, we obtain the best performance when we use all the criteria ratings, e.g. implicit, undefined, and predefined criteria ratings shown in Table 3. For all criteria, MSE and MAE are decreased by 0.678% and 0.11% respectively. The worst performance is noted from the undefined and implicit criteria, in which we obtain the MSE value of 0.7854 and the MAE value of 0.6942, respectively. This suggests that relying solely on undefined and implicit criteria extracted from the review is insufficient for capturing the user's overall preferences. Likewise, relying solely on predefined criteria is inadequate, as indicated by MSE and MAE values of 0.2121 and 0.3237, respectively. Therefore, there is a need to combine predefined, undefined, and implicit criteria to learn user preferences.

5.2 Discussion

It is challenging to cluster free-form reviews because of the length and structure of the content. The reason we chose the sentence clustering method using the semantic search technique is that other clustering algorithms (e.g. k-means) require setting a predefined number of clusters which is infeasible due to the variable review lengths. We tried to find a similarity score between sentences and criteria at which model gives the best clusters. We obtained the best performance at a similarity threshold value of 0.30. Each review contains multiple sentences and it is computationally expensive and time-consuming.

Existing approaches utilize either multiple criteria or reviews to capture the relationship between features (e.g. aspect, and criteria) and ratings. We leverage a sentiment analyzer to estimate the criteria ratings from the review. The ratings on predefined criteria and implicit criteria are different. From our results, it is evident that the criteria ratings extracted from reviews have an impact on user preference learning. The results also reveal that it is important to consider users' sentiments on criteria in addition to predefined criteria.

6 Conclusion

We introduce a novel multicriteria recommendation system that utilizes implicit, undefined, and predefined criteria ratings. We show that users' preferences for an item not only depend on explicit ratings given on predefined criteria but also on the implicit and undefined criteria that users express in the review. We employ an aggregation-based method to develop a multi-criteria recommendation system. Predefined, implicit, and undefined criteria are utilized to learn the aggregation function and predict the overall ratings. We show that our proposed method outperforms the existing baselines in terms of both error and accuracy measures on the TripAdvisor dataset and Beer Advocate dataset.

Acknowledgments. This work is partially sponsored by Natural Science and Engineering Research Council of Canada (grant 2020-04760). The authors would like to thank our reviewers for their tremendous efforts in review and the constructive feedback.

References

1. Nassar, N., Jafar, A., Rahhal, Y.: Multi-criteria collaborative filtering recommender by fusing deep neural network and matrix factorization. J. Big Data **7**(1), 1–12 (2020)
2. Shambour, Q.: A deep learning based algorithm for multi-criteria recommender systems. Knowl.-Based Syst. **211**, 106545 (2021)
3. Hong, M., Jung, J.J.: Multi-criteria tensor model for tourism recommender systems. Expert Syst. Appl. **170**, 114537 (2021)
4. Chin, J.Y., Zhao, K., Joty, S., Cong, G.: ANR: aspect-based neural recommender. In: Proceedings of the 27th ACM International Conference on Information and Knowledge Management, pp. 147–156 (2018)
5. Li, W., Xu, B.: Aspect-based recommendation model for fashion merchandising. In: Martínez-López, F.J., López López, D. (eds.) DMEC 2021. SPBE, pp. 243–250. Springer, Cham (2021). https://doi.org/10.1007/978-3-030-76520-0_25
6. Hasan, E., Rahman, M., Ding, C., Huang, J.X., Raza, S.: Based recommender systems: a survey of approaches, challenges and future perspectives. arXiv preprint arXiv:2405.05562
7. Hasan, E., Ding, C.:. Aspect-aware multi-criteria recommendation model with aspect representation learning. In 2023 IEEE International Conference on Web Intelligence and Intelligent Agent Technology (WI-IAT), pp. 268–272. IEEE (2023)
8. Wu, C., Wu, F., Liu, J., Huang, Y., Xie, X.: ARP: aspect-aware neural review rating prediction. In: Proceedings of the 28th ACM International Conference on Information and Knowledge Management, pp. 2169–2172 (2019)

9. Wang, W., Liu, D., Liu, X., Pan, L.: Fuzzy overlapping community detection based on local random walk and multidimensional scaling. Phys. A **392**, 6578–6586 (2013)
10. Hong, M., Jung, J.J.: ClustPTF: clustering-based parallel tensor factorization for the diverse multi-criteria recommendation. Electron. Commer. Res. Appl. **47**, 101041 (2021)
11. Kim, Y.-D., Choi, S.: Nonnegative tucker decomposition. In: IEEE Conference on Computer Vision and Pattern Recognition. pp. 1–8. IEEE (2007)
12. Emrul, H., Chen, D., Alfredo, C.: Multi-criteria rating and review based recommendation model. In: Proceedings of the 33 IEEE International Conference on Big Data (2022)
13. He, X., Liao, L., Zhang, H., Nie, L., Hu, X., Chua, T.-S.: Neural collaborative filtering. In: Proceedings of the 26th International Conference on World Wide Web, pp. 173–182 (2017)
14. Nair, V., Hinton, G.E.: Rectified linear units improve restricted Boltzmann machines. In: ICML (2010)
15. Reimers, N., Gurevych, I.: Sentence-BERT: sentence embeddings using Siamese BERT-networks. arXiv preprint arXiv:1908.10084 (2019)
16. Ioffe, S., Szegedy, C.: Batch normalization: accelerating deep network training by reducing internal covariate shift. In: International Conference on Machine Learning, pp. 448–456. PMLR (2015)
17. Nassar, N., Jafar, A., Rahhal, Y.: A novel deep multi-criteria collaborative filtering model for recommendation system. Knowl.-Based Syst. **187**, 104811 (2020)
18. Chen, C., Zhang, M., Liu, Y., Ma, S.: Neural attentional rating regression with review-level explanations (2018)
19. Liu, D., Li, J., Du, B., Chang, J., Gao, R.: DAML: dual attention mutual learning between ratings and reviews for item recommendation. In: Proceedings of the 25th ACM SIGKDD International Conference on Knowledge Discovery and Data Mining, pp. 344–352 (2019)
20. Seo, S., Huang, J., Yang, H., Liu, Y.: Interpretable convolutional neural networks with dual local and global attention for review rating prediction. In: Proceedings of the Eleventh ACM Conference on recommender Systems, pp. 297–305 (2017)
21. Zheng, L., Noroozi, V., Yu, P.S.: Joint deep modeling of users and items using reviews for recommendation. In: Proceedings of the Tenth ACM International Conference on Web Search and Data Mining, pp. 425–434 (2017)
22. Deng, D., Jing, L., Yu, J., Sun, S., Zhou, H.: Neural gaussian mixture model for review-based rating prediction. In: Proceedings of the 12th ACM Conference on Recommender Systems, pp. 113–121 (2018)
23. Rafailidis, D., Crestani, F.: Adversarial training for review-based recommendations. In: Proceedings of the 42nd International ACM SIGIR Conference on Research and Development in Information Retrieval, pp. 1057–1060 (2019)
24. Nassar, N., Jafar, A., Rahhal, Y.: Multi-criteria collaborative filtering recommender by fusing deep neural network and matrix factorization. J Big Data **7**, 34 (2020)
25. Shi, C., et al.: Deep collaborative filtering with multi-aspect information in heterogeneous networks. IEEE Trans. Knowl. Data Eng. **33**(4), 1413–1425 (2019)

Non-binary Gender Expression in Online Interactions

Rebecca Dorn(✉), Negar Mokhberian, Julie Jiang, Jeremy Abramson, Fred Morstatter, and Kristina Lerman

Information Science Institute, University of Southern California, Marina del Rey, CA, USA
{rdorn,nmokhber,yioujian}@usc.edu, {abramson,fredmors,lerman}@isi.edu

Abstract. *Trigger Warning: Profane Language, Slurs.*

The presence of openly non-binary gender individuals on social networks is growing. However, the relationship between gender, activity, and language in online interactions has not been extensively explored. Lack of understanding surrounding this interaction can result in the disparate treatment of non-binary gender individuals on online platforms. We investigate patterns of gender-based behavior identity on Twitter, focusing on gender expression as represented by users' expression of pronouns from eight different pronoun groups. We find that non-binary gender groups tend to receive substantially less attention in the form of likes and followers compared to binary groups. Additionally, non-binary users send and receive tweets with higher toxicity scores than other groups. This study identifies differences in the language and online activity of users with non-binary gender identity, and highlights a need for further evaluation of potential disparate treatment by algorithms used by online platforms.

Keywords: Big Data applications · Social computing · Text analysis

1 Introduction

An individual's identity, defined along multiple dimensions such as age, gender, and race, influences self expression and social connection [7]. As social interactions continue to migrate online, social media platforms play an increasingly critical role in identity formation [15]. While gender has been traditionally conceptualized in Western society as binary—specifically 'male' vs 'female'—two recent developments have transformed how we think about gender. First, there is growing recognition of gender as a cultural construct, distinct from the biologically-based sex [1]; second, there is growing awareness that gender forms a spectrum, rather than a binary identity [17].

We study individual identity on X (formerly Twitter), and how it mediates online expression and interactions. We focus on gender, one of the core dimensions of individual identity. As a proxy of gender expression, we study pronouns

L. M. Aiello et al. (Eds.): ASONAM 2024, LNCS 15212, pp. 85–95, 2025.
https://doi.org/10.1007/978-3-031-78538-2_7

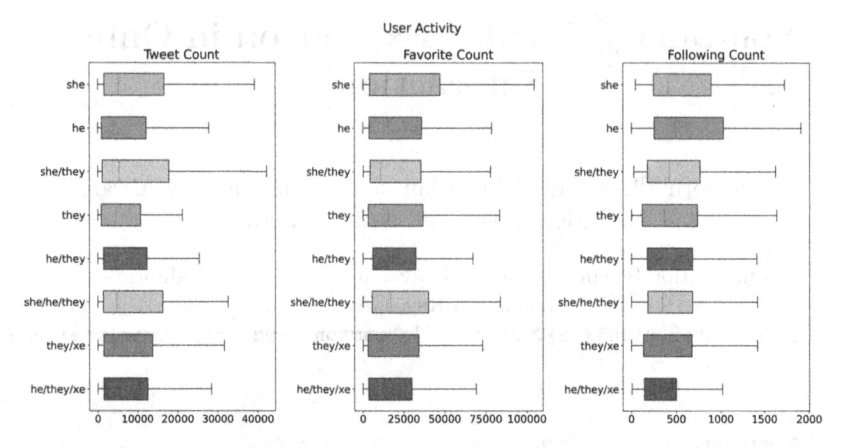

Fig. 1. User activity by pronoun group. Number of followers and favorites is slightly lower for non-binary pronoun groups. Pronoun groups are ordered from top to bottom by representation in the `Seed Set` of tweets.

users choose to display in their online profile or biography. These pronouns range from the traditional binary gender categories, such as 'she/her' and 'he/him', to non-binary and gender nonconforming [12] categories 'they/them', 'she/ze', 'she/they/xe', etc.

Creating safe online spaces for gender minorities is crucial, as individuals often turn to online communities for support [16] in the face of discrimination and social isolation they experience in real life [25]. In this work, we collect and study `NB-TwitCorpus3M`: a dataset of 3 million tweets annotated with author's self-provided pronouns, majority with non-binary pronouns. We analyze toxicity in tweets and replies to understand the role offensive content plays in online dynamics in gender minority communities.

To explore differences in language and activity of groups we investigate the following research questions:

- **RQ1**: How do users in different pronoun groups vary in their level of online activity and the online attention they receive from others? Are there systematic differences across the spectrum of gender identity?
- **RQ2**: Which user pronoun groups convey and experience more toxicity on Twitter?

We measure *activity* with number of tweets a user interacts with, and *attention* through external engagement with a user's messages. We find that non-binary groups have lower rates of activity online, and messages from non-binary groups get less attention through likes. We find that non-binary users tend to receive more toxic replies to their tweets. Surprisingly, we find that non-binary groups also post more toxic messages. These findings signal an important avenue for future work, as the consequences of gender variant communities having higher toxicity scores can lead to increased social exclusion.

2 Related Works

Gender as Identity. Gender is one of the earliest forms of social stratification and is a core dimension of identity. Inspired by second-wave feminism [23] people have started to draw a distinction between sex (biologically-produced) and gender (culturally-produced) identity. This has helped resolve the tension between the traditional conceptualization of gender in Western society and science as binary (i.e., 'male' and 'female'), and historical and societal evidence of the presence of non-binary individuals [19]. In English, third-person singular pronouns are used to express some form of gender identity [21]. The traditional two-sex naming system uses pronouns 'he' and 'she' to convey gender-binary identity. Individuals falling outside of the two-sex system (such as folks identifying with neither, both or a fluctuating set of binary gender identities) have begun to adopt pronouns such as 'they', 'xe' and 'ze' to convey their non-socially normative gender identity [26].

LGBTQ+ Online Interactions. Transgender adolescents participate in virtual communities for emotional and information support [24] and to explore their gender identity and the process of coming out [11]. In fact, queer adolescents demonstrate a tendency to participate in online LGBTQ+ communities over in-person communities [16]. With online communities playing such a large role in so many LGTBQ+ peoples lives, promoting safe queer online spaces is crucial for promoting long and happy LGBTQ+ lives. However, since 2018, there has been a stark rise in the visibility of LGBTQ+ hate speech[1] that has been reflected in social media. For example, despite its purported hate speech detection methods, X is wrought with transphobia [14], LGBTQ+ friendly on the platform 'Gab' have been overtaken with queer-phobic fetishization in the name of free speech [4] and Facebook comments against the LGBTQ+ community have a pattern of denying any gender conception outside of a two-sex gender system [2]. The need for understanding and mitigating discrimination towards queer people online is imminent.

Gender Behavior on X. X biographies measure expressions of personal identity and cultural trends. Longitudinal Online Profile Sampling (LOPS) measures identity formation through the evolution of a user's X (then Twitter) biography, finding that the tokens with the highest prevalence within biographies over 5 years were *he*, *him*, *she* and *her* [20]. The LOPS studies relied upon the notion of *personally expressed identity* where individuals declare *their own* attributes. Previous work assessing X activity across a spectrum of gender relied upon Census data to infer gender, a practice which excludes gender variance [3]. In this work, we allow users to conceptualize their own gender through their X biographies.

[1] https://www.adl.org/resources/report/online-hate-and-harassment-american-experience-2023.

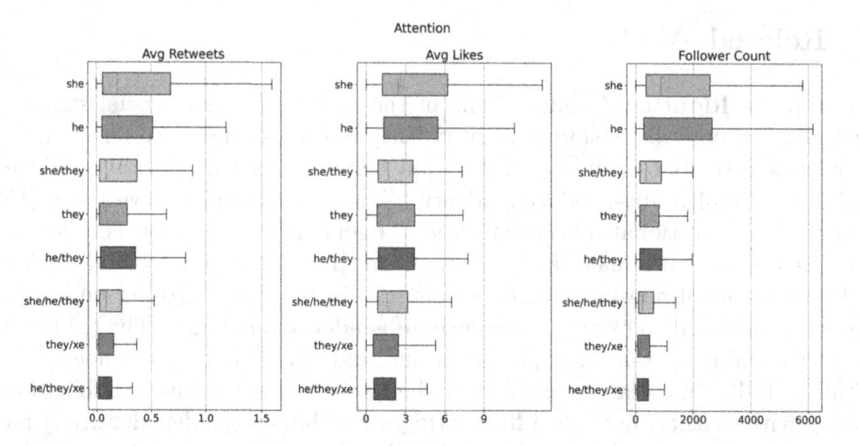

Fig. 2. Attention. Users with more representation receive more attention in retweet and like averages, as well as followers.

3 Methodology

3.1 Gender Spectrum

In this work, we will use the term **non-binary** to describe anyone using pronouns outside of 'she/her/hers' or 'he/him/his'. While this conceptualization of non-binary does not include many gender-queer populations (such as transgender people with a binary gender and non-binary individuals without pronouns in their biography) we find that this conceptualization is interoperable for downstream applications while maximizing user ability to self-describe their gender.

3.2 Data

Rather than collecting a key-word based sample, representation of a tweet sample can be increased by sampling users from a seed data set [9]. Users are first collected from a collection of over 2 billion tweets related to the Covid-19 pandemic collected between January 21, 2020, and November 5, 2021 [5]. From here on out we will call this dataset `Seed Data`. As most platform-engaged Twitter users tweeted about Covid-19 at some point, this generates a sample of active Twitter users. This dataset includes tweets from 2,066,165 users with specified pronouns in their Twitter profiles or biographies. The presence of pronouns is determined by whether a user has specified any combination of {he, him, his, she, her, hers, they, them, theirs, their, xe, xem, ze, zem} separated by forward slashes or commas, with any or no white space in their profile descriptions [13]. Profile descriptions of the users are recorded at the time of the first tweet collected for the sample.

We group the pronouns into five different series: she/her/hers, he/him/his, they/them/theirs, xe/xem and ze/zem. We encode the combinations of these pronouns series via a 5-digit dummy variable that is malleable to a range of gender representations and computationally efficient (e.g. 'she/they' represented as 10100). We encode the pronouns of all ~2 million users into this 5-digit schema. We identify all pronoun groups with at least 1,000 members, and randomly sample up to 600 valid users from each group. This yields eight pronoun groups with at least 375 users. Table 1 reports groups in decreasing order of their size within `Seed Data`. For each user in our sample, we collect at most 1,000 of their most recent tweets posted before September 30, 2022. Tweets are retrieved using the API's `user_timeline` call. Table 1 reports the total tweets in our sample authored by each pronoun group. The resulting collection of tweets is the `NB-TwitCorpus3M` dataset.

Table 1. Pronoun Group composition. 'Original Users' denotes the number of users within each pronoun group of `Seed Data`, 'Sample Users' shows user number in our new dataset, and 'Tweets' shows the number of tweets collected for each pronoun group.

Group	Original Users	Sample Users	Tweets
She	1,194,565	508	464,262
He	461,264	559	503,780
She/They	158,025	508	463,599
They	132,374	560	506,064
He/They	77,951	514	469,328
She/He/They	20,882	557	611,227
They/Xe	1,312	468	462,775
He/They/Xe	1,015	377	387,722
Total	**2,047,388**	**4,051**	**3,868,757**

3.3 Toxicity Inference

To measure toxicity we use the *Detoxify* model [10], a RoBERTa model trained on open source data emphasizing toxicity towards specific identities [6]. This model has an AUC score of 92.11 on the Kaggle dataset. The model outputs a continuous value between zero and one that captures the toxicity of language in the tweet. Values close to one are associated with high toxicity.

4 Results

4.1 RQ1: Activity and Attention

LGBTQ+ data is notoriously sparse and often low-quality [22]. Leveraging online engagement provides a novel opportunity to improve the representation and quality of queer-related data. We focus on measuring user activity, which encompasses a user's outward engagement on Twitter, including the number of original tweets, likes, and accounts followed.

Our analysis reveals lower levels of activity among non-binary users compared to binary users. Figure 1 illustrates the distribution of activity for each pronoun group, with outliers excluded to highlight group differences. Visually, non-binary pronoun groups exhibit slightly lower levels of outward engagement on Twitter compared to binary pronoun groups.

We investigate the interaction between pronoun group representation in the `Seed Set` and group-level online activity, as representation in the user pool may portray overall representation on the X platform. We find a negative correlation between the number of tweets sent out by users and the size of the group in the `Seed Set` (Spearman's $\rho = -0.272$, $p < 0.01$). Similarly, the favorite count (the number of tweets liked by the user) and the following count (the number of other users someone follows) show negative correlations with group size (Spearman's $\rho = -0.216$, $p < 0.01$ and Spearman's $\rho = -0.352$, $p < 0.01$, respectively). Further, we observe that the pronoun group with the highest median following count is *he* (494), while the lowest is *she/he/they* (346). Overall, our findings suggest that user activity, as measured by likes, retweets, and following count, tends to be lower for minority non-binary groups compared to groups with higher representation. This raises concerns regarding the potential bias in randomly sampled Twitter data, which may disproportionately represent binary users over non-binary users, thus posing challenges for achieving adequate representation in downstream analyses.

The allocation of attention on social media platforms carry significant political, economic, and social ramifications [18]. In this study, we operationalize *attention* as the level of inward engagement and prominence that users attain on Twitter. We quantify attention using metrics such as the number of followers, average retweets, average likes, and the percentage of verified users. Our analysis reveals a notable trend: as the representation of pronoun groups in the `Seed Set` decreases, the corresponding amount of attention also diminishes.

Figure 2 illustrates the distributions of attention across different gender pronoun groups. Visual inspection suggests that smaller pronoun groups tend to receive comparatively lower levels of attention. We observe a substantial negative correlation between the lack of representation and the median user retweets (Spearman's $\rho = -0.215$, $p < 0.01$), with the pronoun group *she* exhibiting the highest median retweets (0.192) and *he/they/xe* displaying the lowest (0.045). Similar trends are evident in the correlation between median likes and representation (Spearman's $\rho = -0.229$, $p < 0.01$), with the pronoun group *he* receiving the highest median likes (2.510) and *he/they/xe* receiving the lowest (1.111). The number of followers demonstrates an even more pronounced version of this trend (Spearman's $\rho = -0.363$, $p < 0.01$). Our findings strongly indicate that pronoun groups with less representation tend to receive diminished attention online.

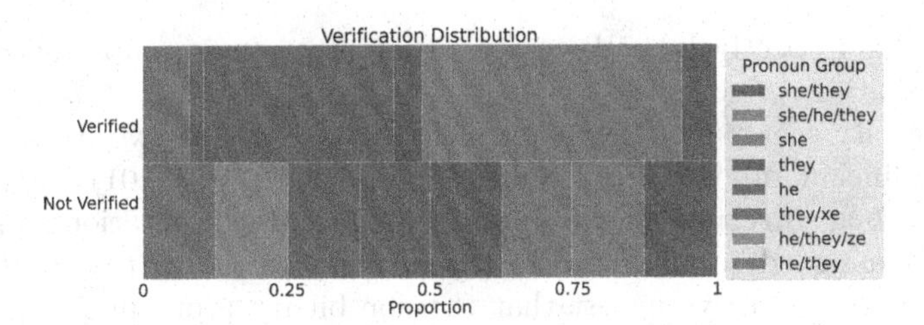

Fig. 3. Pronoun group decomposition or verified users in our sample. Groups *he/they/xe* and *she/he/they* have no verified users.

We analyze the pronoun group composition of verified users as displayed in Fig. 3. Verification status was obtained before X released an option for users to purchase verification status. Over 75% of verified users in our sample are in the pronoun groups *he* or *she*, the two most represented groups. The only groups with no verified users, *he/they/xe* and *she/he/they*, are non-binary. This strong disparity in verification indicates a gap in social validity and visibility of the users in our sample.

4.2 RQ2: Toxicity in Tweets and Replies

Next, we investigate the toxicity of tweets in both tweets posted (sent tweets) and a sample of replies received (received tweets). To generate received tweets, we randomly sample 100 users from each pronoun group in our dataset and collect replies to their original tweets using the tweet's `conversation_id`. This process results in 95,381 replies from 29,537 unique SPSVERBc2s. We look at incidence of highly toxic tweets, considering a tweet as highly toxic if its toxicity score exceeds the threshold of 0.9.

Figure 4 (a) displays the proportion of tweets deemed highly toxic among both sent and received tweets. Our analysis reveals that all groups send out tweets where a small fraction (less than 1%) are highly toxic. However, the share of tweets estimated to be highly toxic increases monotonically with lack of representation of the pronoun group in `Seed Set`. Notably, five out of eight pronoun groups receive more tweets estimated to be highly toxic than what they post, with only *they/xe*, *he/they/xe* and *they/them* posting more toxic tweets than they receive. We observe that these three groups are all non-binary.

In Fig. 4 (b) we present distributions of toxicity scores of tweets after removing outliers. Notably, no pronoun group exhibits a higher median toxicity score for sent tweets than for received tweets. The highest median toxicity scores for sent tweets are *he/they/xe* (.0032) and *they/xe* (.0028). These two particular groups also exhibit the highest median toxicity scores for received tweets as well (.0048, .0047, respectively). Remarkably, these are the two groups with the lowest initial representation. Simultaneously, the groups *he* and *she* boast the lowest median toxicity for both sent (.0009, .0010, respectively) and received tweets

(.0016, .0023, respectively). We observe that these two groups reflect binary conceptualizations of gender.

The toxicity scores for binary and non-binary users as two distinct groups exhibit significant differences (T-statistic = −125.72, $p < .01$). Additionally, there is a strong correlation between pronoun group representation and the toxicity scores assigned to their tweets (Spearman's $\rho = .16$, $p < 0.01$). These observations surprisingly suggest that the non-binary pronoun groups tend to post more highly toxic tweets than binary groups. Given that content moderation decisions often rely on automated toxicity classification, this finding implies that tweets by gender minorities are more likely to be flagged or removed by these algorithms.

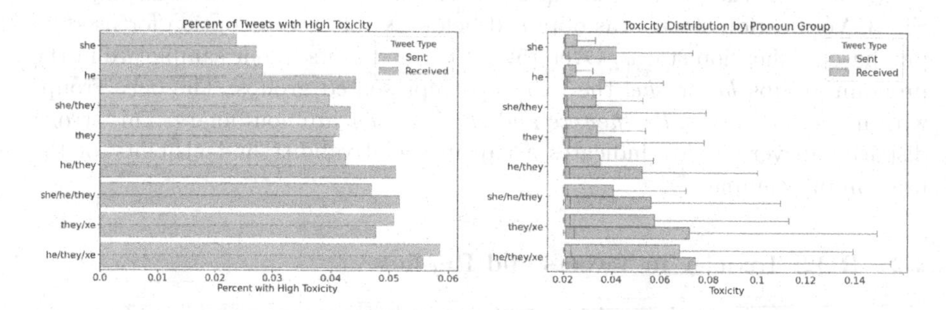

Fig. 4. (a) Percent of Tweets Posted and Percent of Replies Received Labeled as Toxic (toxicity > 0.9). (b) Toxicity distribution for tweets posted and received. Tweets from non-binary receive higher toxicity scores from Detoxify than those from binary users

5 Discussion

We compiled and analyzed the `NB-TwitCorpus3M` dataset, comprising of approximately 3 million tweets from users with pronouns in their biographies, primarily those using non-binary pronouns. Our exploratory analysis investigates online behavior for eight pronoun groups including both binary and non-binary pronoun series. Specifically, we examine outgoing activity, incoming attention, and toxicity levels in both posted tweets and received replies.

Our analysis reveals that non-binary pronoun groups exhibit lower activity levels on Twitter compared to binary pronoun groups. Given the sparse data focusing on gender-queer populations [22], this finding suggests a potential under-representation of users with non-binary pronouns in social media data. Addressing this disparity is crucial to ensure adequate representation of gender minority groups in online spaces.

Further, we observe that non-binary pronoun groups receive less attention through retweets, likes and followers when compared to binary pronoun groups. Attention plays a pivotal role in information spread in the digital age. The lower attention towards non-binary groups may indicate reduced social influence within Twitter's online ecosystem, potentially hindering the political power and visibility of non-binary communities. This underscores the importance of amplifying the voices of gender-queer users to bolster activist causes within the non-binary community.

Surprisingly, we find that non-binary users exhibit higher levels of toxicity detected in their posted tweets compared to users with binary pronouns. We posit that this discrepancy may stem from dialect bias within the toxicity classifier, where dialect commonly used in queer communities is erroneously flagged as expressing toxicity. This would align with prior evidence suggesting that social media content from gender-variant groups, such as drag queens, is disproportionately classified as hate speech [8], highlighting the need for further investigation into the impact of dialect bias on toxicity detection mechanisms.

Ethical Limitations. Inherently, this data set is sensitive due to its collection of individuals with historically marginalized gender identities. To safeguard privacy, we focus analyses on aggregated data and remove identifiable information from provided examples. Our analysis is constrained by the specificity of the population under study, which may inadvertently exclude certain binary and non-binary Twitter users who do not include pronouns in their biographies. Additionally, our study does not differentiate between pronoun order (e.g., *she/they* versus *they/she*).

This study could be enhanced by incorporating a group of users without pronouns in their biographies, or using multiple seed topics to ensure better representation of active X users. We acknowledge the limitations stemming from the relatively low number of replies collected and recognize the importance of further research into the dynamics of reply senders. Exploring how results evolve with an increased number of replies would contribute to a more nuanced understanding of user interactions. This study was reviewed by authors' IRB and designated exempt. Authors declare no competing interests.

Acknowledgments. This work was funded in part by Defense Advanced Research Projects Agency (DARPA) and Army Research Office (ARO) under Contract No. W911NF-21-C-0002 and Contract No. HR00112290021.

References

1. Allen, K., Cuthbert, K., Hall, J.J., Hines, S., Elley, S.: Trailblazing the gender revolution? Young people's understandings of gender diversity through generation and social change. J. Youth Stud. **25**(5), 650–666 (2022)
2. Aperocho, M.D., Aliñabon, G., Camia, K., Tenorio, A.J.: Exploring the hate language: an analysis of discourses against the LGBTQIA+ community. Psychol. Educ. Multidisc. J. **8**(6), 729–737 (2023)
3. Bamman, D., Eisenstein, J., Schnoebelen, T.: Gender identity and lexical variation in social media. J. Socioling. **18**(2), 135–160 (2014). https://doi.org/10.1111/josl.12080

4. Brody, E., Greenhalgh, S.P., Sajjad, M.: Free speech or free to hate?: Anti-LGBTQ+ discourses in LGBTQ+-affirming spaces on gab social. J. Homosexuality, 1–26 (2023)
5. Chen, E., Lerman, K., Ferrara, E.: Tracking social media discourse about the COVID-19 pandemic: development of a public coronavirus twitter data set. JMIR Publ. Health Surveill. **6**(2), e19273 (2020). https://doi.org/10.2196/19273
6. Conneau, A., et al.: Unsupervised cross-lingual representation learning at scale. arXiv preprint arXiv:1911.02116 (2019)
7. Deaux, K.: Reconstructing social identity. Pers. Soc. Psychol. Bull. **19**(1), 4–12 (1993)
8. Dias, O.T., Antonialli, D.M., Gomes, A.: Fighting hate speech, silencing drag queens? artificial intelligence in content moderation and risks to LGBTQ voices online. Sexuality Cult. **25**(2), 700–732 (2021). http://libproxy.usc.edu/login?url=https://www.proquest.com/scholarly-journals/fighting-hate-speech-silencing-drag-queens/docview/2495185828/se-2
9. Garcia, D., Rimé, B.: Collective emotions and social resilience in the digital traces after a terrorist attack. Psychol. Sci. **30**(4), 617–628 (2019). https://doi.org/10.1177/0956797619831964, pMID: 30865565
10. Hanu, L.: Unitary team. detoxify. github (2020). https://github.com/unitaryai/detoxify
11. Herrmann, L., Bindt, C., Hohmann, S., Becker-Hebly, I.: Social media use and experiences among transgender and gender diverse adolescents. Int. J. Transgender Health, 1–14 (2023)
12. Hicks, A., Rutherford, M., Fellbaum, C., Bian, J.: An analysis of WordNet's coverage of gender identity using Twitter and the national transgender discrimination survey. In: Proceedings of the 8th Global WordNet Conference (GWC), pp. 123–130. Global Wordnet Association, Bucharest, Romania, 27–30 January 2016. https://aclanthology.org/2016.gwc-1.19
13. Jiang, J., et al.: What are your pronouns? Examining gender pronoun usage on Twitter. arXiv preprint arXiv:2207.10894 (2022)
14. Locatelli, D., Damo, G., Nozza, D.: A cross-lingual study of homotransphobia on Twitter. In: Proceedings of the First Workshop on Cross-cultural Considerations in NLP (C3NLP), pp. 16–24 (2023)
15. Manago, A.M.: Identity development in the digital age: the case of social networking sites (2015)
16. McInroy, L.B., McCloskey, R.J., Craig, S.L., Eaton, A.D.: LGBTQ+ youths' community engagement and resource seeking online versus offline. J. Technol. Hum. Serv. **37**(4), 315–333 (2019)
17. Monro, S.: Non-binary and genderqueer: an overview of the field. Int. J. Transgender. **20**(2–3), 126–131 (2019)
18. Pedersen, M.A., Albris, K., Seaver, N.: The political economy of attention. Annu. Rev. Anthropol. **50**, 309–325 (2021)
19. Richards, C., Bouman, W.P., Barker, M.J.: Genderqueer and Non-binary Genders. Palgrave (2017). https://doi.org/10.1057/978-1-137-51053-2
20. Rogers, N., Jones, J.J.: Using Twitter bios to measure changes in self-identity: are Americans defining themselves more politically over time? J. Soc. Comput. **2**(1), 1–13 (2021). https://doi.org/10.23919/JSC.2021.0002
21. Rose, E., Winig, M., Nash, J., Roepke, K., Conrod, K.: Variation in acceptability of neologistic English pronouns. Proc. Linguist. Soc. Am. **8**(1), 5526–5526 (2023)

22. Ruberg, B., Ruelos, S.: Data for queer lives: how LGBTQ gender and sexuality identities challenge norms of demographics. Big Data Soc. **7**(1), 2053951720933286 (2020)
23. Sanz, V.: No way out of the binary: a critical history of the scientific production of sex. Signs: J. Women Cult. Soc. **43**, 1–27 (2017). https://doi.org/10.1086/692517
24. Selkie, E., Adkins, V., Masters, E., Bajpai, A., Shumer, D.: Transgender adolescents' uses of social media for social support. J. Adolesc. Health **66**(3), 275–280 (2020)
25. Tabaac, A., Perrin, P.B., Benotsch, E.G.: Discrimination, mental health, and body image among transgender and gender-non-binary individuals: constructing a multiple mediational path model. J. Gay Lesbian Soc. Serv. **30**(1), 1–16 (2018)
26. Wayne, L.D.: Neutral pronouns: a modest proposal whose time has come. Can. Woman Stud./les cahiers de la femme (2005)

Browsing Amazon's Book Bubbles

Paul Bouchaud[1,2(✉)]

[1] Center for Social Analysis and Mathematics (EHESS), Paris, France
paul.bouchaud@ehess.fr
[2] Complex Systems Institute of Paris Île-de-France (CNRS), Paris, France

Abstract. This study investigates Amazon's book recommendation system, uncovering cohesive communities of semantically similar books. The confinement within communities is extremely high, a user following Amazon's recommendations needs tens of successive clicks to navigate away. We identify a large community of recommended books endorsing climate denialism, COVID-19 conspiracy theories, and advocating conservative views on social and gender issues. Performing a collaborative filtering analysis, relying on Amazon users reviews, reveals that books reviewed by the same users tend to be co-recommended by Amazon. This study not only contributes to addressing a gap in the literature by examining Amazon's recommender systems, but also highlights that even non-personalized recommender systems may pose systemic risks by suggesting content with foreseeable negative effects on public health and civic discourse.

Keywords: Amazon · Recommender System · Filter Bubble · Collaborative Filtering

1 Introduction

"Recommendations are discovery, offering surprise and delight with what they help uncover for you. Every interaction should be a recommendation" in this study, we explore the claim made by Smith and Linden when discussing two decades of recommender systems at Amazon [21].

As early as the late 1990s, Amazon has embraced collaborative filtering [7,13]; the evaluation of items co-purchase likelihood [17], has been pivotal for the platform, driving up to 30% of Amazon.com's page views in 2015 [18]. However, concerns have been raised about algorithmic curation potentially reinforcing exposure to like-minded content and amplifying biases [14]. While collaborative filtering algorithms alone do not inherently narrow content diversity [26], their interaction with user preferences can create "echo chambers" [6,23]. Such effects have been observed on platforms like YouTube [8] and Twitter [2].

Despite Amazon's influence, serving over 181 million EU users [22], e-commerce platforms have received limited research attention from algorithmic auditors. Previous studies have highlighted partisan disparities in science book

L. M. Aiello et al. (Eds.): ASONAM 2024, LNCS 15212, pp. 96–106, 2025.
https://doi.org/10.1007/978-3-031-78538-2_8

consumption [19] and feedback loops reinforcing user interests [6]. During the COVID-19 pandemic, Amazon faced public criticism for promoting vaccine misinformation [4], and with studies revealing a prevalence of vaccine-hesitant books [20] and misleading search results [11].

This study employs a sock-puppet audit methodology [16] to characterize Amazon's non-personalized recommendations across a wide range of non-fiction books on the French Amazon website. We reveal tight communities of semantically similar books and identify a community co-recommending books supporting various contrarian viewpoints, from climate denialism to COVID-19 conspiracy theories. Through collaborative filtering analysis based on user reviews, we show that books reviewed by the same users tend to be recommended together. This research highlights potential systemic risks of non-personalized recommender systems on civic discourse and public health.

2 Methods

2.1 Data Collection

Amazon presents product recommendations in multi-page carousels under various labels such as 'Customers who viewed this item also viewed' and 'Frequently Bought Together'. This study focuses on non-personalized recommendations an unlogged user would encounter.

We employed a snowball sampling strategy, starting with the bestsellers from 18 non-fiction book categories on Amazon.fr, listed in [1]. Using an automated web browser that reset after each page visit, we collected recommendations and metadata for each book. We iterated the collection three times, considering the books recommended at least twice in the previous round.

In total, from the initial pool of 1 725 bestsellers, we gathered recommendations for 60 298 books between October 28 and November 4, 2023, capturing an average of 85.8% of Amazon's suggestions for each book. To assess temporal stability, we compared this dataset with a prior collection from August-September 2023, finding a 64.4% overlap in recommendations.

Our dataset comprises primarily French books (92.3%), with 81.4% as printed books, 12.3% as Kindle ebooks, and 3.6% as audiobooks. Amazon consistently recommends books in similar formats (97.9% on average). Our subsequent analysis focuses exclusively on printed books, removing redundant formats to avoid duplication.

2.2 Graph Construction

The recommendations gathered, we establish an unweighted directed graph, denoted as G, in which the vertices represent books (designated as v_i). A link between v_i and v_j is established if Amazon recommends book v_j on the page of book v_i. Filtering out non-fetched books, we end up with a graph $G = (V, E)$, with V the set of $|V| = 48\,636$ books, and E the set of $|E| = 391\,664$ edges.

2.3 Characterization

Community Structure. We employ the Leiden algorithm [25] with the Constant Potts Model [24] as the quality function to detect communities in G. This approach overcomes both the resolution limit inherent in modularity maximization [5] and Louvain's arbitrarily badly connected communities. Establishing a resolution profile we determine the resolution parameter, γ, imposing an upper limit on inter-community link density, which ensures the stability of our partitions.

Subsequently, we quantify book recommendation homophily within communities through both the fraction of a vertex's neighbors belonging to the same community [9] and, beyond their first-degree neighbors, through random walks. Specifically, we initiate 25 random walks starting from each book and compute the average length of the walks a random surfer need to transition out of the community they initiated the walk from.

Semantic Analysis. We employ TF-IDF and Non-Negative Matrix Factorization to analyze book titles and summaries. This classical approach was chosen over recent neural topic models due to its efficiency, interpretability, and simplicity, as discussed in [27]. Our analysis compares summary embeddings between books connected by recommendations in graph G and those that are not. Additionally, we measure semantic diversity within book communities relative to the entire corpus. This diversity is quantified using the geometric mean of the standard deviation of summary embeddings, a metric introduced in [12]. To ensure robustness, we conducted a sensitivity analysis, confirming that our results remain consistent across embedding dimensions ranging from 25 to 70.

2.4 Collaborative Filtering

Although the current implementation is unknown, Amazon historically used item-to-item collaborative filtering to recommend products [13]. This method suggest products by examining items frequently purchased together. To gain insight into Amazon recommender systems, and without access to purchase or rating data, we gathered the "verified purchase" reviews for 25 151 books from main communities in G. We then performed item-to-item collaborative filtering based on users reviews, while we acknowledge reviews offer stronger signals of (dis)agreement than simple ratings or purchases.

Overall, we collected 419 460 reviews by 245 734 unique reviewers, averaging 20.5 reviews per book (median 7.0). In contrast, Amazon's can leverage an average of 748.7 ratings per book (median 52.0), in addition to purchase and navigation data, to power its recommender systems. We then compare the overlap coefficient between user sets who reviewed each pair of books and Amazon's actual book recommendations.

2.5 Search Results

To investigate how users might enter a community through Amazon search results, we conducted searches on topics with clear scientific consensus: Climate Change and COVID-19. Specifically, we performed the following queries (in French): "Climate Change CO2," "Global Warming," "IPCC" for Climate Change, and "COVID-19," "COVID pandemics," "COVID vaccine" for COVID-19.

Starting on November 1, 2023, we performed these searches daily at noon for five consecutive days. For each query, we collected the first result page, sorted by Amazon's default algorithm or by decreasing average user ratings. Other rankers such as increasing/decreasing price or publication date yielded too few books in G for meaningful analysis.

3 Results

3.1 Characterization

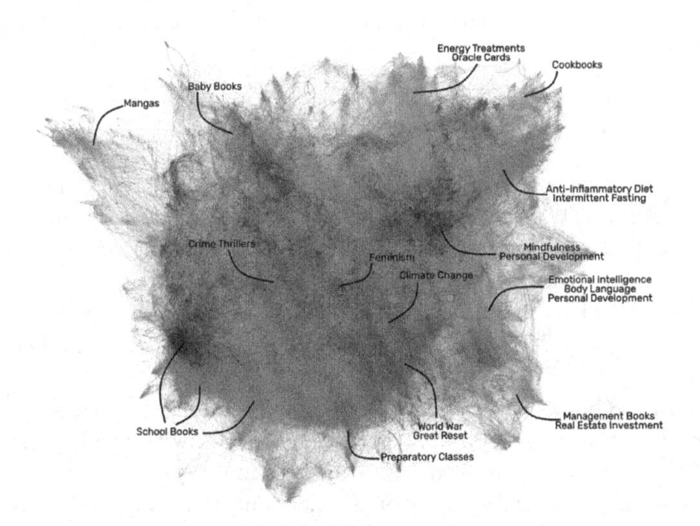

Fig. 1. Graph of Amazon book recommendations G [48 636 books, 391 664 edges], spatialized via ForceAtlas2 [10]. Vertices are color-coded by community, and their size is proportional to their in-degree.

The Leiden algorithm partitions G with high modularity ($Q = 0.86$), identifying 61 communities that encompass over 90% of the books. Figure 1 shows G with color-coded main communities. Homophily analysis reveals that 88.8% of recommended books are in the same community as the current book, rising to 94.9% for "Frequently Bought Together" items. Additionally, 53.2% of Amazon's suggested books share the same category as the current book. The likelihood that two randomly chosen books from a community belong to the same Amazon category is 5.1 times higher than for two random books. On average, 91.1% of books by authors with at least five books in G are in the same community.

Random walks based on Amazon's recommendations show strong confinement. From the 61 largest communities (covering 90% of books), 75.7% of surfers remain in their starting community after three clicks. On average, it takes 24.9 clicks (median 11) to leave a community. Confinement varies by community; for example, it takes an average of 6.9 clicks (median 4) to leave the social science community (356 books), while over 77.8 clicks (median 68) are needed to exit the coloring books community (380 books).

Figure 1 shows keywords extracted via TF-IDF for various communities in G, revealing diverse topics such as cartomancy, personal development, mangas and crime thrillers. Semantic diversity within a community is, on average, 57.9% lower than in the overall corpus. The average pairwise cosine similarity between books connected by an edge in G is 1.72 times higher than for books within the same community but not connected, and 5.41 times higher than for books from different communities without an edge. Similarly, along random walks, the similarity between a book and those recommended by Amazon after three clicks is 36.9% higher than the average similarity among books from the same community.

3.2 Collaborative Filtering

The user review data is sparse; among 25 100 books, users reviewed an average of 1.7 books (median 1.0). Books reviewed by the same user are 8.7 times more likely to belong to the same community in G and 6.1 times more likely to share the same Amazon book category compared to randomly selected books, considering users who reviewed at least 5 books. The cosine similarity between summary embeddings of books reviewed by the same user is 2.6 times higher than for random pairs.

For the 10 108 books with at least 10 "verified purchase" reviews, we analyzed pairwise reviewer overlap. In 58.1% of cases, the book with the highest reviewer overlap with a seed book is from the same community in G, and in 34.5% of cases, it is recommended by Amazon (i.e., connected by an edge in G). The average reviewer overlap is 15.6 times higher for books connected by an edge in G compared to random pairs from the same community, and 9.2 times higher for books within the same community than for books from different communities.

3.3 Case Studies

We manually curated lists of books in G discussing: Climate Change [146 books], Gender Issues (including gender identity, expression, and equality) [162 books] and COVID-19 [101 books]. These topics were chosen due to the relative abundance of available books and their social significance; aligning for instance, with the European Commission's topics of interest in their initiatives addressing misinformation [3], and the systemic risks defined in the Digital Services Act. We excluded books were these topics where not the main focus of the discussion, as well as fiction books.

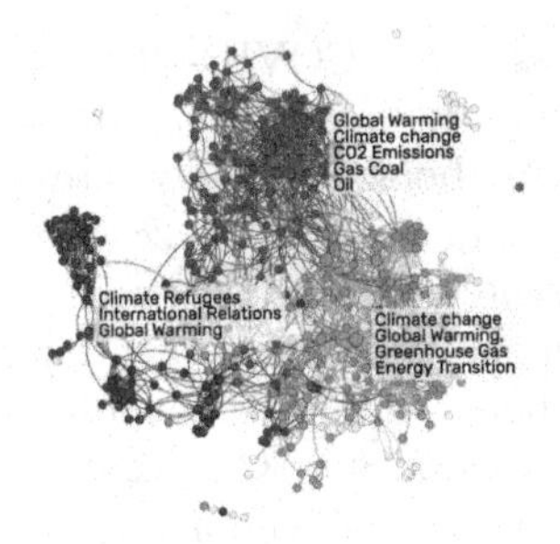

(a) Graph induced by 146 books discussing Climate Change [459 books, 1 991 edges]

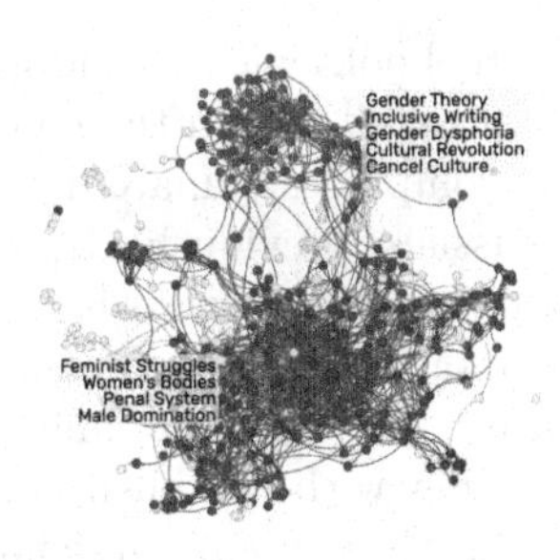

(b) Graph induced 117 books relating to Gender Issues [439 books, 1 862 edges]

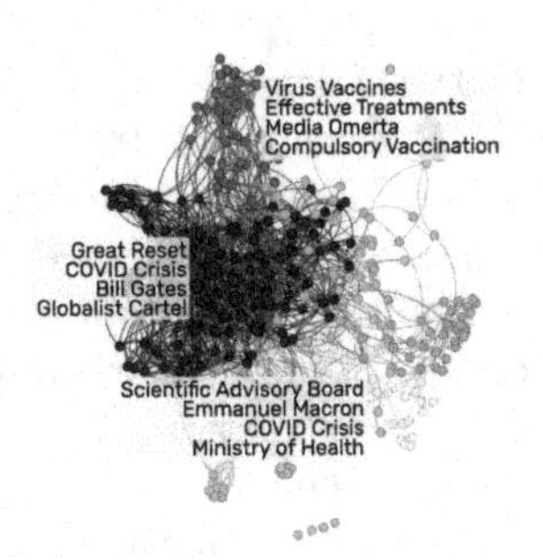

(c) Graph induced by 101 books discussing COVID-19 [304 books, 1 620 edges]

Fig. 2. Two-hop recommendation graphs. Vertices are color-coded by G community (smaller communities are in white), their size is proportional to their in-degree. (Color figure online)

Climate Change. The 146 books addressing Climate Change primarily fall into two G communities: 42.5% in the community depicted in green on Fig. 2a and 30.1% in the red community. A manual inspection reveals that books in the green community align with the scientific consensus on climate change, while 84.1% of those in the red community reject it. Apart from the two main communities, books discussing the geopolitical aspects of climate change (depicted in dark green) are accessible within two clicks from the Climate Change seed books. On average, when a user consults a climate-denialists book Amazon recommends 94.1% of books from the red community, and similarly, alongside pro-climate books, Amazon recommends 90.7% of books from the green community.

Gender Issues. The graph of Amazon's book recommendations induced by two-hops from 117 books related to gender issues is depicted in Fig. 2b. Again, two main communities emerge, encompassing 56.4% (in violet) and 20.5% (in red) of the books. While the books in the violet community address feminist struggles, male domination, women's rights, sexual violence, and engaged in discussions on gender identity and expression (hereafter designed as feminist/queer community), the books in the red community discuss cancel culture, inclusive writing, and "wokeism" (hereafter designed as conservative views). Interestingly, the community represented in red in Fig. 2b corresponds to the same community in G that embeds climate-denialist books shown in Fig. 2a. On average, when a user consults a feminist/queer book, Amazon recommends 95.7% of such books, and similarly, alongside conservative books, Amazon recommends 95.8% of conservative books.

COVID-19. The analysis of 101 books within G addressing the COVID-19 pandemic reveals that 92.1% are situated within the community, hereinafter called 'contrarian', that otherwise encompasses climate-denialist books and advocates conservative views on gender issues, previously depicted in red. In Fig. 2c, the recommendation graph derived from these books is presented with vertices color-coded based on their sub-communities, detected at a higher resolution than for G. The extraction of keywords from sub-community book summaries exposes distinct thematic focuses, aligning with established taxonomies of COVID-19-related disinformation [15]. The three main sub-communities: i) endorse New World Order and Great Reset conspiracy theories; ii) challenge the established scientific consensus on vaccinations and their side effects; and iii) discuss pandemic management.

We emphasize that the assignment of a book to a particular community is not of the author's will. For instance, the book "COVID-19: The Great Reset" by Klaus Schwab and Thierry Malleret lies in the same community as those relaying conspiracy theories, due to Amazon's algorithm, which suggests five such books alongside Schwab's works. This is likely because, among the fetched books, the top 10 books with the highest reviewer overlap with Schwab's book are conspiracy-related.

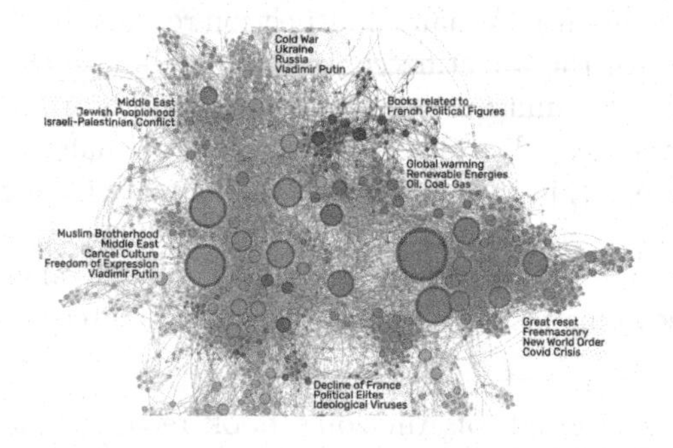

Fig. 3. Sub-graph of G induced by books of the contrarian community.

Contrarian Community. To gain further insight into this contrarian community, the third-largest community in G with 1 776 books, we isolated it, conducted a community detection analysis at a higher resolution than for G, and is displayed on Fig. 3. Employing TF-IDF to extract keywords from book summaries within sub-communities, reveals a broad range of topics, including Freemasonry, French Politics, Foreign Policy, Cancel Culture, and Great Reset conspiracy theories.

We can leverage the set of users book reviews to understand why various contrarian viewpoints coexist within the same recommendation community rather than being in distinct topic-specific communities. We observe that the average overlap among users who reviewed climate-denialist books and those reviewing books holding conservative views on gender is 6.8 times higher than the overlap between the sets of reviewers of pro-climate books and of feminist/queer books. Similarly, the reviewer overlap between COVID-19 related books and climate-denialist books (resp. conservative books) is 4.9 (resp. 4.2) times higher than the overlap between COVID-related books and pro-climate books (resp. feminist/queer books).

Beyond encompassing various disinformation narratives, this community stands out for its confinement. An average number of 15.5 (median 8) successive clicks are required for a random surfer following Amazon's recommendation to leave the contrarian communities, while 6.1 (median 4) and 6.3 (median 4) are required to leave, respectively, the feminism/queer and pro-climate communities. When the random surfers leave the contrarian community, 8.4% of them end up in the World War II and French history books community [625 books], 6.6% in community related to Personal Development and Communication [1 229 books]. Within the contrarian community, an average of 9.0 (median 5) clicks are required to leave the sub-communities, 76.6% of the random walks leaving the climate-denialist sub-community emerge in the COVID-19 sub-community, 11.2% emerge in the conservatism subcommunity.

3.4 Search Results

Performing search queries related to climate change with Amazon's default algorithmic ranking, it was observed that 51.1% of the first 10 results provided misleading information about the scientific consensus (52.5% were in the above identified contrarian community), this fraction increased to 64.1% when ranked by decreasing average user ratings. For COVID-19-related searches, when ranked according to Amazon's default algorithm, 71.7% of the top 10 results contain misinformation about COVID-19 pandemics, when ordered by decreasing average user ratings, the fraction increases to 91.1%.

4 Discussion

Our study of Amazon's book recommendation system revealed highly modular communities with strong homophily. The community of recommendation are made up of books that are semantically close, with a poorer semantic diversity than the overall book corpus; books by the same author tend to be embedded in the same community, and with users requiring many clicks to navigate away.

Examining recommendation graphs for Climate Change, Gender Issues, and COVID-19 books, we identified a community promoting climate denialism, COVID-19 conspiracy theories, and conservative views on social issues; readily accessible through search queries. Books arguing opposing viewpoints were in separate topic-specific communities. Collaborative filtering analysis showed that books reviewed by the same users tend to be recommended together, explaining why diverse contrarian viewpoints coexist within the same community.

While our data collection focused on the French Amazon market and excluded most fiction books, we believe this study provides valuable insights into a platform that significantly influences book distribution. We reveal that Amazon's non-personalized recommendation system tends to confine users within homogeneous communities, including one that spreads misinformation on climate change and COVID-19.

This study demonstrates that even non-personalized algorithms based on seemingly objective criteria can generate recommendations with potential negative effects on public health and civic discourse. Our findings contribute to discussions on algorithmic regulation, highlighting that explainability and transparency alone may not mitigate the systemic risks targeted by regulations like the Digital Services Act.

Acknowledgments.. Paul Bouchaud acknowledges the Jean-Pierre Aguilar fellowship from the CFM Foundation for Research and the resources provided by the Complex Systems Institute of Paris.

Disclosure of Interests. The author has no competing interests to declare that are relevant to the content of this article.

References

1. Bouchaud, P.: Browsing Amazon's Book Bubbles (2023). https://hal.science/hal-04308081, Working Paper or preprint
2. Bouchaud, P., Chavalarias, D., Panahi, M.: Crowdsourced audit of Twitter's recommender systems. Sci. Rep. **13**(1) (2023). https://doi.org/10.1038/s41598-023-43980-4
3. Comission, E.: Second call for the European narrative observatory to fight disinformation post-COVID 19
4. Dreisbach, T.: On Amazon, dubious "antiviral" supplements proliferate amid pandemic (2020)
5. Fortunato, S., Barth'elemy, M.: Resolution limit in community detection. Proc. Natl. Acad. Sci. U.S.A. **104**(1), 36–41 (2007). https://doi.org/10.1073/pnas.0605965104
6. Ge, Y., et al.: Understanding echo chambers in e-commerce recommender systems. In: Proceedings of the 43rd International ACM SIGIR Conference on Research and Development in Information Retrieval. ACM (2020). https://doi.org/10.1145/3397271.3401431
7. Hardesty, L.: The history of Amazon's recommendation algorithm (2022)
8. Haroon, M., Chhabra, A., Liu, X., Mohapatra, P., Shafiq, Z., Wojcieszak, M.: Youtube, the great radicalizer? Auditing and mitigating ideological biases in Youtube recommendations (2022)
9. Interian, R., Ribeiro, C.C.: An empirical investigation of network polarization. Appl. Math. Comput. **339**, 651–662 (2018). https://doi.org/10.1016/j.amc.2018.07.066
10. Jacomy, M., Venturini, T., Heymann, S., Bastian, M.: Forceatlas2, a continuous graph layout algorithm for handy network visualization designed for the Gephi software. PLoS ONE **9**(6), e98679 (2014). https://doi.org/10.1371/journal.pone.0098679

11. Juneja, P., Mitra, T.: Auditing e-commerce platforms for algorithmically curated vaccine misinformation. In: Proceedings of the 2021 CHI Conference on Human Factors in Computing Systems. ACM (2021). https://doi.org/10.1145/3411764.3445250
12. Lai, Y.A., Zhu, X., Zhang, Y., Diab, M.: Diversity, density, and homogeneity: quantitative characteristic metrics for text collections. In: Proceedings of the Twelfth Language Resources and Evaluation Conference, pp. 1739–1746. European Language Resources Association, Marseille, France (2020)
13. Linden, G., Smith, B., York, J.: Amazon.com recommendations: item-to-item collaborative filtering. IEEE Internet Comput. **7**(1), 76–80 (2003). https://doi.org/10.1109/MIC.2003.1167344
14. Pariser, E.: The Filter Bubble: What the Internet is Hiding from You. Penguin Books, London (2012)
15. Posetti, J., Kalina, B.: Disinfodemic: deciphering COVID-19 disinformation (2020). https://unesdoc.unesco.org/ark:/48223/pf0000374416
16. Sandvig, C., Hamilton, K., Karahalios, K., Langbort, C.: Auditing algorithms: research methods for detecting discrimination on internet platforms. Data Discrim. Convert. Crit. Concerns Product. Inquiry **22**(2014), 4349–4357 (2014)
17. Sarwar, B., Karypis, G., Konstan, J., Riedl, J.: Item-based collaborative filtering recommendation algorithms. In: Proceedings of the 10th International Conference on World Wide Web. ACM (2001). https://doi.org/10.1145/371920.372071
18. Sharma, A., Hofman, J.M., Watts, D.J.: Estimating the causal impact of recommendation systems from observational data. In: Proceedings of the Sixteenth ACM Conference on Economics and Computation. ACM (2015). https://doi.org/10.1145/2764468.2764488
19. Shi, F., Shi, Y., Dokshin, F.A., Evans, J.A., Macy, M.W.: Millions of online book co-purchases reveal partisan differences in the consumption of science. Nat. Hum. Behav. **1**(4) (2017). https://doi.org/10.1038/s41562-017-0079
20. Shin, J., Valente, T.: Algorithms and health misinformation: a case study of vaccine books on Amazon. J. Health Commun. **25**(5), 394–401 (2020). https://doi.org/10.1080/10810730.2020.1776423
21. Smith, B., Linden, G.: Two decades of recommender systems at amazon.com. IEEE Internet Comput. Internet Comput. **21**(3), 12–18 (2017). https://doi.org/10.1109/mic.2017.72
22. Team, A.A.: Amazon publishes first EU store transparency report, outlining our commitment to providing a trustworthy shopping experience (2023)
23. Thorburn, L., Stray, J., Bengani, P.: From "filter bubbles", "echo chambers", and "rabbit holes" to "feedback loops" (2023)
24. Traag, V.A., Dooren, P.V., Nesterov, Y.: Narrow scope for resolution-limit-free community detection. Phys. Rev. E **84**(1) (2011). https://doi.org/10.1103/physreve.84.016114
25. Traag, V.A., Waltman, L., van Eck, N.J.: From Louvain to Leiden: guaranteeing well-connected communities. Sci. Rep. **9**(1) (2019). https://doi.org/10.1038/s41598-019-41695-z

26. Vromman, F.V., Fouss, F.: Filter bubbles created by collaborative filtering algorithms themselves, fact or fiction? An experimental comparison. In: IEEE/WIC/ACM International Conference on Web Intelligence. ACM (2021). https://doi.org/10.1145/3498851.3498945
27. Zhang, Z., Fang, M., Chen, L., Rad, M.R.N.: Is neural topic modelling better than clustering? An empirical study on clustering with contextual embeddings for topics. In: Proceedings of the 2022 Conference of the North American Chapter of the Association for Computational Linguistics: Human Language Technologies. Association for Computational Linguistics (2022). https://doi.org/10.18653/v1/2022.naacl-main.285

Intertwined Biases Across Social Media Spheres: Unpacking Correlations in Media Bias Dimensions

Yifan Liu, Yike Li, and Dong Wang(✉)

School of Information Sciences, University of Illinois Urbana-Champaign, Champaign, IL 61820, USA
dwang24@illinois.edu

Abstract. Biased information on social media significantly influences public perception by reinforcing stereotypes and deepening societal divisions. Previous research has often isolated specific bias dimensions, such as *political* or *racial bias*, without considering their interrelationships across different domains. The dynamic nature of social media, with its shifting user behaviors and trends, further challenges the efficacy of existing benchmarks. Addressing these gaps, our research introduces a novel dataset derived from five years of YouTube comments, annotated for a wide range of biases including gender, race, politics, and hate speech. This dataset covers diverse areas such as politics, sports, healthcare, education, and entertainment, revealing complex bias interplays. Through detailed statistical analysis, we identify distinct bias expression patterns and intra-domain correlations, setting the stage for developing systems that detect multiple biases concurrently. Our work enhances media bias identification and contributes to the creation of tools for fairer social media consumption.

Keywords: Social Media · Bias Identification · Benchmark · Datasets

1 Introduction

In our digital era, the widespread distribution of information across social media often includes user-generated content that can perpetuate stereotypes, discrimination, and hatred. We categorize such content as a form of media bias [22], emphasizing its significant influence on these platforms. Beyond reinforcing existing social biases, online media bias interacts with cognitive biases [14], fostering information bubbles [16] that distort public perception and exacerbate social divisions. This highlights the urgency of developing robust systems for identifying online media bias to mitigate these effects. In our discussion, we specifically address online media bias arising from user-generated content, retaining the term 'media bias' for consistency. The concept of media bias, historically lacking a consensus definition, has recently been addressed through comprehensive literature reviews. Recent research proposes a unified definition that categorizes

L. M. Aiello et al. (Eds.): ASONAM 2024, LNCS 15212, pp. 107–116, 2025.
https://doi.org/10.1007/978-3-031-78538-2_9

skewed portrayals into distinct dimensions of media bias [22], reflecting its complex nature. To counteract media bias, various frameworks have been developed to support automated identification efforts.

The identification of media bias has primarily been addressed by the machine learning (ML) and natural language processing (NLP) communities [11,19]. In recent years, methods for identifying media bias have progressed from relying on hand-crafted features [10,18] to the employment of advanced transformer-based models [17,23,27]. However, many existing research efforts still focus predominantly on detecting a *single* type of media bias, typically evaluating models against benchmarks that assess only one dimension of bias [1,25]. Such a narrow focus presents significant challenges for developing comprehensive bias identification systems: 1) there is a disproportionate focus within the media bias detection community on different bias dimensions, resulting in a lack of high-quality benchmark datasets along some bias dimensions [26], and 2) without a thorough understanding of the various media bias dimensions, it is challenging to develop a bias identification system that is capable of jointly detecting and analyzing multiple bias dimensions.

To this end, we propose a new media bias identification benchmark that annotates and analyzes multiple dimensions of media bias across various topic domains. Drawing from prior work [26], we select specific bias dimensions tailored to our social media dataset. Our dataset comprises YouTube user comments collected over the last five years. Through rigorous statistical analysis, we find that the politics domain exhibits significantly higher proportions of biased content compared to other topics. Additionally, our analysis identifies domain-specific patterns in the expression of bias. For example, biased posts in politics often manipulate narratives to support specific agendas, while in sports, bias may manifest through specific word choices or jargon that convey prejudices. Additionally, temporal analysis shows that the correlations between different bias dimensions are dynamic, fluctuating in response to spikes in discussion volume.

2 Related Works

2.1 Media Bias Dimensions

Compared to previous studies on media bias, our work emphasizes the multi-dimensional aspects of media bias and the domain-specific occurrences of bias dimensions on social media platforms. Recent research has utilized the inter-relationships between different types of biases as a foundation for developing more robust bias identification systems [26]. Specifically, this approach is applied within the multi-task learning (MTL) framework [4,9], which provides a joint optimization framework for various media bias dimensions. It is important to note that both task selection and data selection play critical roles in the success of MTL [3,20]. To effectively address such challenges, automated task-selection algorithms are considered to be a promising enhancement [12].

However, gradient-based automated task selection schemes, which rely on monitoring the training dynamics of different sub-tasks, are susceptible to discrepancies between data sources and could require a large number auxiliary tasks to perform effectively [9].

2.2 Comparison with Other Media Bias Datasets

Within the field of media bias research, various datasets are specifically designed to support the analysis of a single media bias type, employing multi-level labeling to capture its nuances. For example, RTGender dataset adopt a multi-categorical labeling to characterize different aspects of gender bias [24]. Similarly, prior research on political bias categorizes texts into five distinct political tenancies, ranging from *left* to *right* [1]. For more nuanced labeling, CMSB employs a continuous sexism scale to measure subtleties in Twitter data [21]. While such fine-grained categorization in each bias dimension provides a detailed view of media bias, our work primarily focuses on exploring inter-correlations among different bias dimensions.

Similar to our work, multidimensional bias dataset collects dataset with a bias dimension specification based on hidden assumptions, subjectivity and representation tendencies [7]. However, multidimensional bias dataset has a focus on news articles with a special emphasis on political tendencies, while our work focuses on the social media space with a wider range of topics. Recent media bias identification benchmark (MBIB) summarizes a rich list of publicly available datasets following a set of media bias dimension specification [22,26]. Despite the well rounded bias dimensions discussed, MBIB provides a single label for each post, limiting the analysis of across-dimension analysis of media bias. Compared to other existing bias identification datasets, our dataset is the first to account for the joint occurrence of multiple bias dimensions with a joint labeling scheme. Additionally, we have intentionally segregated data collections across different domains using general keyword choices.

3 Dataset Creation

In this section, we elaborate on our dataset creation process, which includes the following steps: 1) Retrieval of domain-specific data; 2) Examination of bias dimensions leading to the development of our bias specification framework; and 3) Generation of multi-dimensional bias labels. Furthermore, we compare our dataset with existing datasets to underscore the research gap in the study of intertwined media biases. Our data collection process is designed to investigate dimensions of media bias within and across various domains, namely politics, healthcare, sports, entertainment and job & education.

To analyze media bias across social media, we collected comments from specific YouTube domains, using Google Trends to select representative keywords for each domain. For example, "COVID" was chosen for the healthcare sector due to its high search frequency over the past five years. The keywords we used are listed in Table 1. We employed YouTube's API to gather relevant posts from the last five years, collecting approximately 2,000 comments per domain. Our dataset includes only comments directly related to video topics, excluding replies to other comments to maintain focus on domain-specific content and avoid off-topic discussions. For preparing our data for media bias annotation, we filtered out non-English and overly long posts (over 200 words). Unlike previous studies, we converted emojis into text tokens to preserve the semantic content in our bias annotations, detailed further in Sect. 3.2.

Table 1. Data Collection by Domain

Domain	Keywords	Comment Count
Politics	election contest, election result, voting	1993
Healthcare	COVID, pain injury, symptom	2580
Sports	NBA, NFL, MLB	2398
Entertainment	film, lyrics, episodes	998
Job & Education	career, college, job	1455

3.1 Bias Dimension Specifications

In our analysis, we ground our investigation under the umbrella of media bias introduced in recent works [22,26]. Our focus is on dimensions of media bias that are defined solely by post-level social media contents. In our exploration, we investigate a subset of media bias types summarized in prior work [26] including: linguistic bias [2], political bias [8], gender bias [26], hate speech [5,13], racial bias [6] and text-level context bias [26]. We defer to the related works for the specific definitions of these bias dimensions.

3.2 Bias Annotation

To annotate the social media posts we collected, we employed a mixed approach combining manual and automated annotations. For each domain dataset, 100 samples were annotated by two annotators. Each annotator was responsible for annotating 60 samples, with 20 samples overlapping between the two annotators for consistency checks. In Table 2, we report the performance of automated annotation, together with the inter-rater agreement scores (Cohen's κ) for each domain. Across all bias dimensions, we observe substantial inter-rater agreement (Cohen's $\kappa > 0.8$), with the conflicting annotations being reviewed and inspected. Building on the manual annotations, we evaluated two groups of automated annotations: 1) a shallow pretrained models trained on existing benchmark bias evaluation datasets; and 2) zero/few-shot annotations from pretrained large language model [15]. Our results indicate that while smaller-scale transformer baselines achieve good robustness and generalizability with cross-validated train-test splits, they exhibit poor generalization when applied to the noisy social media posts we collected along some bias dimensions. Overall, we utilize the predictions generated by the best performing considering both shallow models and LLMs evaluated on our manually annotated test set in our further analysis of bias dimensions.

Table 2. Weighted F1-Scores of Automated Annotation & Inter-Rater Agreement for Bias Dimensions (HS: Hate Speech, PB: Political Bias, GB: Gender Bias, RB: Racial Bias, LB: Linguistic Bias, TLCB: Text-level Context Bias)

Bias Dimension	Politics	Sports	Healthcare	Job & Education	Entertainment	Model	Cohen's κ Score
GB	0.90	0.99	0.99	0.97	0.88	GPT-Turbo-3.5	1.00
RB	0.97	0.98	1.00	0.93	0.96	GPT-Turbo-3.5	1.00
HS	0.77	0.85	0.92	0.95	0.91	Roberta-Twitter	0.82
LB	0.63	0.75	0.81	0.84	0.63	GPT-Turbo-3.5	0.93
TLCB	0.56	0.77	0.86	0.84	0.87	ConvBert	0.89
PB	0.64	1.00	0.94	0.98	1.00	GPT-Turbo-3.5	0.81

Table 3. Number of Biased Content Along Different Dimensions from Automated Annotation with Percentage of Total Posts Shown in Brackets (%)

Bias Dimension	Politics	Sports	Healthcare	Job & Education	Entertainment
Gender Bias	91 (4.61)	76 (3.21)	77 (3.09)	72 (5.15)	44 (4.53)
Racial Bias	99 (5.02)	39 (1.64)	30 (1.21)	35 (2.50)	14 (1.44)
Hate Speech	464 (23.54)	330 (13.91)	244 (9.82)	172 (12.29)	101 (10.40)
Linguistic Bias	499 (25.32)	501 (21.13)	382 (15.37)	239 (17.08)	192 (19.77)
Text-level Context Bias	480 (24.35)	204 (8.60)	247 (9.94)	123 (8.79)	56 (5.77)
Political Bias	578 (29.32)	35 (1.48)	60 (2.42)	18 (1.29)	3 (0.31)
Total Posts	1971	2371	2484	1399	971

4 Experiments

4.1 Distribution Shift

For all annotated data, we report the number of samples and their percentage occurrences across various bias dimensions in Table 3. Notably, the politics domain exhibits a significantly higher proportion of biased content, especially in *text-level context bias* and *political bias*. Additionally, biased content in most domains primarily manifests as *linguistic biases*, characterized by discriminatory word usage. In the politics domain, biased content appears through both specific word choices (*linguistic bias*) and skewed descriptions (*text-level context bias*), occurring in similar proportions as detailed in Table 3.

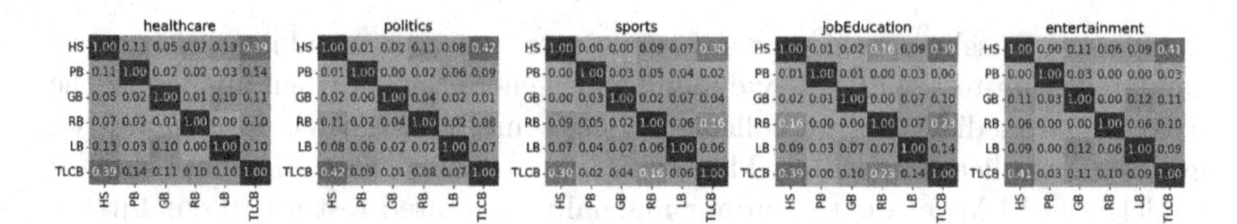

Fig. 1. Correlation heatmap for each bias dimension of different domains, calculated using Cramér's $\mathcal{V}$. Higher values indicate stronger correlations. (HS: Hate Speech, PB: Political Bias, GB: Gender Bias, RB: Racial Bias, LB: Linguistic Bias, TLCB: Text-level Context Bias)

4.2 Correlated Bias Dimensions

Based on the definitions of our media bias dimensions, we divide media bias dimensions into two categories: i) style-based bias (*linguistic bias, text-level context bias*), which focuses on the phrasing of biased texts, and ii) content-based bias (*hate speech, gender bias, political bias, racial bias*), concerning the topics of bias within the content. Noting that these groups contribute differently to the media bias spectrum and may require distinct identification frameworks, we analyze correlations within and between these categories. For each bias dimension, we examine their co-occurrence within topic domains through pairwise chi-square tests using binary labels. We report the results as Cramer's $\mathcal{V}$ values in Fig. 1. Our analysis confirms strong statistical significance across all pairwise correlations, with the highest p-value at 1.87×10^{-3}.

In order to understand how media bias are expressed in different domains, we first investigate the correlations we observe associated with style-based bias dimensions. Specifically, we observe that for all topic domains, *text-level context bias* is mostly associated with *hate speech.* In sports and job & education domain, we observe *hate speech* also has a moderate positive correlation with racial bias with Cramer's $\mathcal{V} > 0.15$. Unlike *text-level context bias*, *linguistic bias*, which primarily focuses on specific word choices, does not exhibit a clear positive correlation with some specific types of biases, but more evenly correlated with all content-based bias dimensions. This observation suggests that content-based bias dimensions, particularly hate speech, are more often expressed through biased descriptions rather than specific biased terms.

For content-based biases, we focus exclusively on correlations that are relatively strong, specifically where Cramer's $\mathcal{V} > 0.1$. Among the content-based bias dimensions, unlike other content-based biases, we observe *hate speech* often coexists with other types of content-based biases. Across all topic domains, we observe that *hate speech* is most significantly correlated with *political bias* in healthcare, *racial bias* in sports, *racial bias* in politics, *racial bias* in job & education, and *gender bias* in entertainment. The aforementioned correlations may reflect the nuanced ways in which content creators and social media users engage with topics sensitive to identity and political context. For instance, in the healthcare domain, political discussions often intersect with deeply polarized issues such as healthcare policy and reproductive rights, which often incite hate speech. In politics and job & education, racial discussions usually evoke strong biases, potentially escalating into hate speech.

4.3 Time Series Analysis

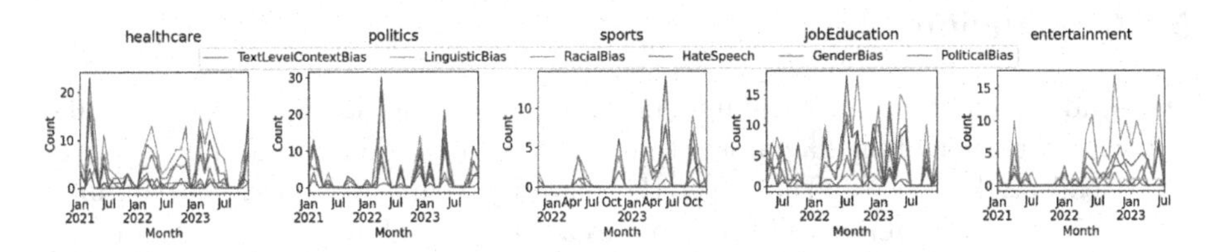

Fig. 2. Line plot visualizations of monthly aggregated counts of bias dimensions. Key observations include: 1) Hate speech manifests in varying proportions of the two types of style-based bias dimensions across different domains. 2) Notable surges in aggressive biases are observed in specific months within the politics domain, supporting our hypothesis that biases in these domains are more event-driven compared to others.

Social media is a rapidly evolving field with frequent shifts in content and interaction patterns. Recognizing the importance of temporal analysis, we monitor the dynamics of biased content by aggregating data within each bias dimension monthly. This data is analyzed across different topic domains and presented in Fig. 2. We standardize the time series for each domain to facilitate pairwise comparisons and perform t-tests to investigate differences across domains. Additionally, we conduct intra-domain analysis, reporting on the top-2 correlated bias dimensions using Pearson's correlation.

We first investigate the strong (style, content)-based correlation pairs. Most notably, diverging from the typically dominant correlation of (*hate speech, text-level content bias*), *linguistic bias* emerges as the most closely correlated style-based bias with *hate speech* with a Pearson's coefficient of 0.86 in the aggregated time series for the entertainment domain. Furthermore, the month-interval aggregation significantly strengthens the (*hate speech, linguistic bias*) correlation across all domains, with the lowest Pearson's coefficient observed being 0.77 in the politics domain. The observation that temporal aggregation enhances the visibility of (*hate speech, linguistic bias*) correlation is likely due to the smoothing of outliers and noise in the data. The observed correlation in the entertainment domain, in particular, might reflect a trend where hate speech is more frequently expressed through nuanced language choices rather than content cues.

For content-based bias dimensions, we observe that aggregation introduces distinct correlations between bias dimensions. For clarity, we refer to correlations observed in monthly aggregated bias counts as 'short-term correlations,' and those across the entire dataset as 'long-term correlations.' We summarize the differences in the highest correlated pairs of biases as follows: 1) In the politics domain, the most closely correlated pair is (*hate speech, gender bias*) with a

Pearson's coefficient of 0.92, which is significantly higher than the correlation between (*hate speech, political bias*) with a coefficient of 0.69. 2) In the sports domain, *gender bias* and *racial bias* show the closest correlation. 3) In job & education domain, *hate speech* is closely correlated with both *gender bias* and *racial bias* with Pearson's coefficients of 0.54 and 0.53 respectively.

5 Conclusion

Our study aims to advance the understanding of media bias by introducing and investigating a social media dataset that spans multiple domains and bias dimensions, collected from YouTube over the past five years. Moreover, our findings reveal significant differences in how biases are expressed across various domains such as politics, sports, and healthcare. We also discover fluctuations in the correlations between bias dimensions in response to surges in social media posts. These findings underscore the complex and evolving nature of media bias and lay the foundation for the future development of multi-dimensional bias identification systems. By advancing investigations into media bias, we hope to equip both researchers and practitioners with the tools necessary to address and mitigate the impacts of media bias, ultimately fostering a fairer media environment.

Acknowledgement. This research is supported in part by the National Science Foundation under Grant No. IIS-2202481, CHE-2105032, IIS-2130263, CNS-2131622, CNS-2140999. The views and conclusions contained in this document are those of the authors and should not be interpreted as representing the official policies, either expressed or implied, of the U.S. Government. The U.S. Government is authorized to reproduce and distribute reprints for Government purposes notwithstanding any copyright notation here on.

References

1. Aksenov, D., Bourgonje, P., Zaczynska, K., Ostendorff, M., Moreno-Schneider, J., Rehm, G.: Fine-grained classification of political bias in German news: a data set and initial experiments. In: WOAH (2021). https://api.semanticscholar.org/CorpusID:236486143
2. Beukeboom, C.J., Burgers, C.: Linguistic bias (2017). https://doi.org/10.1093/acrefore/9780190228613.013.439
3. Bingel, J., Søgaard, A.: Identifying beneficial task relations for multi-task learning in deep neural networks. In: Lapata, M., Blunsom, P., Koller, A. (eds.) Proceedings of the 15th Conference of the European Chapter of the Association for Computational Linguistics: Volume 2, Short Papers, pp. 164–169. Association for Computational Linguistics, Valencia, Spain (2017). https://aclanthology.org/E17-2026
4. Chen, S., Zhang, Y., Yang, Q.: Multi-task learning in natural language processing: an overview (2021)
5. Davidson, T., Warmsley, D., Macy, M.W., Weber, I.: Automated hate speech detection and the problem of offensive language. CoRR abs/1703.04009 (2017). http://arxiv.org/abs/1703.04009

6. Dixon, T.L., Azocar, C.L.: Priming crime and activating blackness: understanding the psychological impact of the overrepresentation of Blacks as lawbreakers on television news. J. Commun. **57**(2), 229–253 (2007). https://doi.org/10.1111/j.1460-2466.2007.00341.x. Blackwell Publishing, UK
7. Färber, M., Burkard, V., Jatowt, A., Lim, S.: A multidimensional dataset based on crowdsourcing for analyzing and detecting news bias. In: CIKM '20, pp. 3007–3014. Association for Computing Machinery, New York, NY, USA (2020). https://doi.org/10.1145/3340531.3412876
8. Feldman, S.: Political ideology. In: The Oxford Handbook of Political Psychology, 2nd edn, pp. 591–626. Oxford University Press, New York (2013)
9. Horych, T., et al.: Magpie: multi-task media-bias analysis generalization for pre-trained identification of expressions (2024)
10. Hube, C., Fetahu, B.: Detecting biased statements in Wikipedia. In: Companion Proceedings of the The Web Conference 2018. WWW '18, International World Wide Web Conferences Steering Committee, Republic and Canton of Geneva, CHE, pp. 1779–1786 (2018). https://doi.org/10.1145/3184558.3191640
11. Kou, Z., Shang, L., Zeng, H., Zhang, Y., Wang, D.: Exgfair: a crowdsourcing data exchange approach to fair human face datasets augmentation. In: 2021 IEEE International Conference on Big Data (Big Data), pp. 1285–1290. IEEE (2021)
12. Ma, W., Lou, R., Zhang, K., Wang, L., Vosoughi, S.: GradTS: a gradient-based automatic auxiliary task selection method based on transformer networks. In: Moens, M.F., Huang, X., Specia, L., Yih, S.W.t. (eds.) Proceedings of the 2021 Conference on Empirical Methods in Natural Language Processing, pp. 5621–5632. Association for Computational Linguistics, Online and Punta Cana, Dominican Republic (2021). https://aclanthology.org/2021.emnlp-main.455
13. Mathew, B., Saha, P., Yimam, S.M., Biemann, C., Goyal, P., Mukherjee, A.: Hatexplain: a benchmark dataset for explainable hate speech detection. In: Proceedings of the AAAI Conference on Artificial Intelligence, vol. 35, pp. 14867–14875 (2021)
14. Nickerson, R.S.: Confirmation bias: a ubiquitous phenomenon in many guises. Rev. Gen. Psychol. **2**(2), 175–220 (1998). https://doi.org/10.1037/1089-2680.2.2.175, https://psycnet.apa.org/record/2018-70006-003, cited by: 4362
15. Ouyang, L., et al.: Training language models to follow instructions with human feedback (2022)
16. Pariser, E.: The Filter Bubble: What the Internet is Hiding from You. Penguin UK, London (2011)
17. Raza, S., Reji, D.J., Ding, C.: Dbias: detecting biases and ensuring fairness in news articles. Int. J. Data Sci. Anal. **17**(1), 39–59 (2024). https://doi.org/10.1007/s41060-022-00359-4
18. Recasens, M., Danescu-Niculescu-Mizil, C., Jurafsky, D.: Linguistic models for analyzing and detecting biased language. In: Schuetze, H., Fung, P., Poesio, M. (eds.) Proceedings of the 51st Annual Meeting of the Association for Computational Linguistics (Volume 1: Long Papers), pp. 1650–1659. Association for Computational Linguistics, Sofia, Bulgaria (2013). https://aclanthology.org/P13-1162
19. Rodrigo-Ginés, F.J., de Albornoz, J.C., Plaza, L.: A systematic review on media bias detection: what is media bias, how it is expressed, and how to detect it. Expert Syst. Appl. **237**, 121641 (2024). https://doi.org/10.1016/j.eswa.2023.121641, https://www.sciencedirect.com/science/article/pii/S0957417423021437

20. Ruder, S., Plank, B.: Learning to select data for transfer learning with Bayesian optimization. In: Palmer, M., Hwa, R., Riedel, S. (eds.) Proceedings of the 2017 Conference on Empirical Methods in Natural Language Processing, pp. 372–382. Association for Computational Linguistics, Copenhagen, Denmark (2017). https://doi.org/10.18653/v1/D17-1038, https://aclanthology.org/D17-1038
21. Samory, M., Sen, I., Kohne, J., Flöck, F., Wagner, C.: "unsex me here": Revisiting sexism detection using psychological scales and adversarial samples. CoRR abs/2004.12764 (2020). https://arxiv.org/abs/2004.12764
22. Spinde, T., et al.: The media bias taxonomy: a systematic literature review on the forms and automated detection of media bias. ACM Comput. Surv. (2023)
23. Spinde, T., et al.: Exploiting transformer-based multitask learning for the detection of media bias in news articles. In: Smits, M. (ed.) iConference 2022. LNCS, vol. 13192, pp. 225–235. Springer, Cham (2022). https://doi.org/10.1007/978-3-030-96957-8_20
24. Voigt, R., Jurgens, D., Prabhakaran, V., Jurafsky, D., Tsvetkov, Y.: RtGender: a corpus for studying differential responses to gender. In: Calzolari, N., et al. (eds.) Proceedings of the Eleventh International Conference on Language Resources and Evaluation (LREC 2018). European Language Resources Association (ELRA), Miyazaki, Japan (2018). https://aclanthology.org/L18-1445
25. Wang, W.Y.: "liar, liar pants on fire": A new benchmark dataset for fake news detection. In: Barzilay, R., Kan, M.Y. (eds.) Proceedings of the 55th Annual Meeting of the Association for Computational Linguistics (Volume 2: Short Papers), pp. 422–426. Association for Computational Linguistics, Vancouver, Canada (2017). https://doi.org/10.18653/v1/P17-2067, https://aclanthology.org/P17-2067
26. Wessel, M., Horych, T., Ruas, T., Aizawa, A., Gipp, B., Spinde, T.: Introducing mbib - the first media bias identification benchmark task and dataset collection. In: Proceedings of the 46th International ACM SIGIR Conference on Research and Development in Information Retrieval. SIGIR '23, pp. 2765–2774. Association for Computing Machinery, New York, NY, USA (2023). https://doi.org/10.1145/3539618.3591882
27. Zhang, D.Y., Kou, Z., Wang, D.: FairFL: a fair federated learning approach to reducing demographic bias in privacy-sensitive classification models. In: 2020 IEEE International Conference on Big Data (Big Data), pp. 1051–1060. IEEE (2020)

Improving the Accuracy of Community Detection in Social Network Through a Hybrid Method

Mahsa Nooribakhsh[1](✉), Marta Fernández-Diego[1], Fernando González-Ladrón-De-Guevara[1], and Mahdi Mollamotalebi[2]

[1] Instituto Universitario Mixto de Tecnología Informática, Universitat Politècnica de València, Camino de Vera, s/n, 46022 Valencia, Spain
mnoorib@doctor.upv.es

[2] Department of Computer and Information Technology Engineering, Qazvin Branch, Islamic Azad University, Qazvin, Iran

Abstract. The inherent complexity of social networks in terms of topological properties requires sophisticated methodologies to detect communities or clusters. Community detection in social networks is essential for understanding organizational structures and patterns in complex interconnected systems. Traditional methods face challenges in handling the scale and complexity of modern social networks, such as local optima trapping and slow convergence. This paper proposes a hybrid method to improve the accuracy of community detection, leveraging stacked auto-encoder (SAE) for dimensionality reduction and the Shuffled Frog Leaping (SFLA) as memetic algorithm for enhanced optimization alongside k-means clustering. The proposed method constructs a hybrid similarity matrix combining structural information and community-related features, followed by SAE to reduce dimensionality and facilitate efficient processing of high-dimensional data. SFLA optimizes the k-means clustering process, introducing adaptability and diversity to exploration of the solution space. Experimental results indicated its superior performance in terms of normalized mutual information (NMI) and modularity compared to existing approaches.

Keywords: community detection · social networks · stacked auto-encoder · shuffled frog leaping algorithm

1 Introduction

Community detection in social networks represents an area of research within the broader domain of network science that aims to reveal the organizational structures and patterns that characterize complex systems of interconnected nodes [1]. A social network can be conceptualized as a graph, where nodes correspond to individual entities (e.g. people, organizations, or web pages) and edges represent relations or interactions between these entities [2]. Traditional methods for community detection have relied on various algorithms, such as graph theory, modularity optimization, spectral clustering, and label

L. M. Aiello et al. (Eds.): ASONAM 2024, LNCS 15212, pp. 117–126, 2025.
https://doi.org/10.1007/978-3-031-78538-2_10

propagation. However, the ever-increasing scale and complexity of social networks have prompted the exploration of innovative approaches, including the integration of deep learning techniques like auto-encoders. A classification offers an organized framework for understanding and exploring various approaches to community detection in social networks including graph partitioning algorithms, density-based algorithms, hierarchical clustering algorithms, spectral clustering, optimization algorithms, game-theoretic algorithms, and deep learning algorithms [3, 4].

By using iterative improvement processes and balancing exploration and exploitation, memetic algorithms can identify exact groups of individuals within social networks, considering connectivity patterns and community densities. These algorithms provide robustness to noise, and flexibility in customization which makes them suitable to discover meaningful community structures in social network. Examples of memetic algorithms include the Shuffled Frog Leaping Algorithm (SFLA), genetic algorithms, and, etc. [5].

Deep learning-based methods such as auto-encoders, are increasingly adopted for community detection due to their ability to discern intricate patterns and representations from network data [3]. A stack auto-encoder is a neural network architecture employed for unsupervised learning tasks such as community detection. It acts by compressing input data into a lower-dimensional representation through an encoder and then reconstructing the original input using a decoder [6, 7]. Recently, the hybrid methods have combined the capabilities of auto-encoders with meta heuristic algorithms such as memetic algorithms to detect communities on social networks. This paper provides a hybrid method using a stacked auto-encoder and SLFA as a memetic algorithm for detecting the communities accurately in social networks. The remainder of this paper is organized as follows: Sect. 2 includes the related works. The proposed method is described in Sect. 3. We present the simulation results and discussion in Sect. 4. Finally, Sect. 5 concludes the research and presents future works.

2 Related Work

Community detection in social networks handles distinguishing the structures inside an organization that is more thickly associated inside than with the rest of the arrangement [4]. In the following, the recent works related to community detection on social networks is reviewed concisely. Xu et al. [8] have proposed community detection method using ensemble clustering with a stack auto-encoder for low-dimensional feature representation. It integrates multiple clustering results via nonnegative matrix factorization for reliability. However, its performance depends on similarity representation and initial clustering. Pierezan et al. [9] have proposed the Coyote Optimization Algorithm (COA) as a metaheuristic inspired by the social behavior of coyotes. It features a simple parameter set involving the number of packs and coyotes per pack. Despite its potential, COA is still in its early stages and has limitations such as the need for further refinement, potential scalability issues, and the absence of adaptive mechanisms.

Wang et al. [10] have introduced a proximity-based group formation game model for detecting communities. It formulates community formation as a two-step non-cooperative game and introduces a community interaction probability matrix to improve detection performance. However, the model's current application is limited to specific social networks, and its effectiveness in attributed social networks. Zhang et al. [11] have proposed a community detection algorithm based on core nodes and layer-by-layer label propagation, extended to detect overlapping communities. Layered label propagation starting from core nodes improves detection accuracy, and node labels are calibrated to reduce early misclassification. However, the algorithm's evaluation relies heavily on modularity.

Aslan et al. [12] presented a modified Coot bird model named MCOOT to detect the community in social networks which is inspired by the collective manners of coots on water surfaces. The update process improves the ratio between search and exploitation capabilities, leading to better detection. On the other hand, it suffers from the weakness of sensitiveness such that the performance of the MCOOT depends on parameter settings. Recent approaches to community detection highlight the use of learning methods to enhance clustering accuracy. While these methods show promise, challenges remain such as noise sensitivity, interpretability, etc. Research gaps enclose the need for better optimization techniques and using hybrid methods like stack auto-encoder and memetic algorithm is a promising way to improve these gaps and attain higher accuracy.

3 Proposed Method

In this section, our proposed hybrid method is presented in detail, and it is evaluated in the next section. Our method is designed using a stacked auto-encoder along with SFLA as a memetic algorithm to detect the communities more accurately. The SFLA as a memetic algorithm can be used for community detection and leverages its population-based cooperative search. In this algorithm, a virtual population of frogs is separated into memeplexes, each defining a cultural unit of evolution known as a meme. Also, the frogs are occasionally shuffled among memeplexes [5]. The components of the proposed method are shown in Fig. 1:

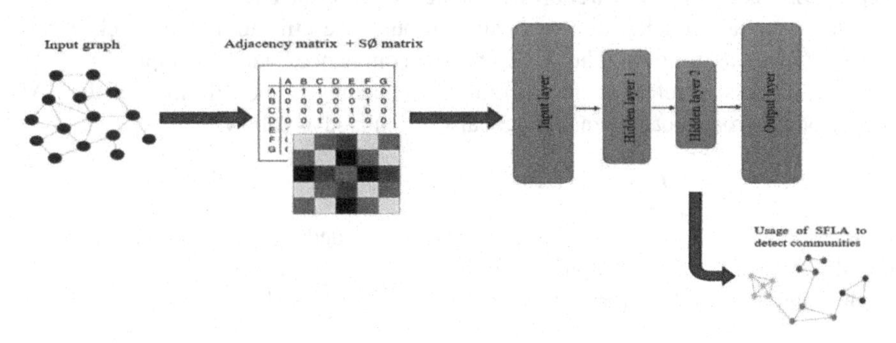

Fig. 1. The components of proposed method

According to Fig. 1 the proposed method comprises three steps. The steps are presented in the following:

Step1: Matrix Construction

Graphs can be effectively represented through the utilization of an adjacency matrix, The adjacency matrix comprehensively captures direct connections between nodes, which can be expressed mathematically as Eq. (1):

$$A_{ij} = \begin{cases} 1 \text{ if } v_{ij} \in E \\ 0 \quad \text{otherwise} \end{cases} \tag{1}$$

On the other hand, the use of a similarity matrix, known as SØ (SØ-similarity), offers a unique approach to comprehending the relationships between nodes [13]. Equation (2) presents the SØ matrix:

$$S\emptyset = \frac{2\text{comNeig}(v_i, v_j)}{d(v_i) + d(v_j)} S\emptyset \in R^{n*n} \tag{2}$$

where $comNeig(v_i, v_j)$ is the number of common neighbors between two vertices v_i and v_j, and $d(v_i) + d(v_j)$ denotes the degrees of vertices vi and vj respectively.

In order to overcome the limitations of individual matrices, the enriched hybrid matrix *H* represents a combined relationship measure between nodes V_i (*i*-th node in the graph) and V_j, (j-th node in the graph) aiming to balance the direct connectivity captured by adjacency and the community-based relationships highlighted by SØ. This combination is achieved through a weighted sum of the two matrices, resulting in a synthesized representation that enhances the accuracy of community detection in social networks. Equation (3) shows the structure of H:

$$H = \alpha.A + (1 - \alpha).S\emptyset \tag{3}$$

where α is a weighting parameter (between 0 and 1) that balances the influence of A and SØ.

Step 2: Dimensionality Reduction with Stacked Auto-encoder

This step aims at reducing dimensionality through the utilization of a stacked auto-encoder (SAE) as follows: The SAE process commences with the input layer. The encoding process E_i refers to the operation of encoding in the i-th layer of the SAE. The encoding process can be mathematically articulated as follows:

$$Zi = Ei(Zi - 1), \text{for } i = 1, 2, \ldots, NumberOfLayers \tag{4}$$

where $Z_0 = X$, and $Z_{NumberOfLayers}$ represents the encoded low-dimensional representation, Z_0 is considered as the input matrix X.

In a simple linear transformation, Eq. (5) represents a SAE:

$$Z_{ij} = \sigma(W_{ij}.(Z_{i-1}) + b_{ij}) \tag{5}$$

where Z_{ij} is the encoded representation of the j-th node in layer i. W_{ij} is weight matrix associated with the connection between the j-th node in layer i-1 and the j-th node in layer i and b_{ij} is the bias term associated with the j-th node in layer i. Z_{i-1} is output from the previous layer (i-1). Moreover, σ is the activation function (commonly a nonlinear function like the sigmoid or ReLU). We employed the He initialization function as described in Eq. (6):

$$W \sim Normal\left(0, \sqrt{\frac{2}{numbr\ of\ input\ units}}\right) \tag{6}$$

Simultaneously, for the sigmoid activation function in the output layer, we chose the Xavier initialization as shown in Eq. (7):

$$W \sim Normal\left(0, \sqrt{\frac{2}{numbr\ of\ input\ units + number\ of\ output\ units}}\right) \tag{7}$$

where the *number of input units* refers to the total number of input nodes or neurons in the layer for which you are initializing the weights. It represents the dimensionality of the input data that is fed into that layer.

After encoding, the SAE follows with decoding D_i. The decoding is presented as:

$$X_{reconstructed} = D_i(Z_i),\ \text{for}\ i = num_{layers,\ldots,1} \tag{8}$$

where $X_{Reconstructed}$ is the reconstructed input.

In the training stage of our stacked auto-encoder, we used Adam optimization to update the weights, biases, or descent loss functions, depth, and number of layers (as described in Sect. 4). The output of one of the encoding layers (Z_{final}) acts as a low-dimensional representation of the input hybrid matrix (H). The low-dimensional representation (Z_{final}) integrates in to with a community detection algorithm to provide an improved feature space for higher accuracy and efficiency.

Step 3: Using a Memetic Metaheuristic Algorithm for Community Detection
In order to select the most efficient centroid point and increase the accuracy of detection, we employed the SFLA [5]. Unlike the pure SFLA, where frogs typically represent solutions to optimization problems, in our proposed adaptation, frogs represent potential centroids for the K-means algorithm. The presented process incorporates evolutionary operators along with a local search function to refine the positions of frogs. The fitness function evaluates the quality of centroids based on the alignment with the underlying community structure. Such adaptation enhances the efficiency of centroid initialization for community detection on social networks distinguishing it from the pure SFLA. Algorithm1 presents the customized version of SFLA used in the proposed method.

Algorithm 1. Customized version of SFLA used in proposed method

Input
P: population size (number of frogs/centroids), N: number of dimensions, Max Iterations, Crossover Rate: Rate of crossover for evolutionary operators, Mutation Rate: Rate of mutation for evolutionary operators
Output: Objective function (centroid fitness)
1. **For** i = 1 to P
2. Frogs[i] = Initialize random centroid(N)
3. End
4. **For** iteration = 1 to Max Iterations
5. Evaluate fitness of each frog (centroid)
6. **For** i = 1 to P
7. Frogs[i]. Fitness = Fitness function (Frogs[i])
8. **End**
9. Sort Frogs by Fitness frog
10. Memeplexes = Divide frog to memeplexes
11. Shuffle memeplexes for diversity
12. **For** each memeplex in Shuffled memeplexes
13. **For** i = 1 to size(memeplex)
14. memeplex[i] = Local search(memeplex[i])
15. **End**
16. **End**
17. Frogs = Combine Memeplexes (Shuffled memeplexes)
18. **For** i = 1 to P
19. **if** Random () < Crossover Rate
20. j = Random Integer (1, P)
21. Offspring = Crossover (Frogs[i], Frogs[j])
22. Frogs[i] = Offspring
23. **End**
24. **if** Random () < Mutation Rate
25. Frogs[i] = Mutation (Frogs[i])
26. **End**
27. **End**
28. Frogs = Shuffle (Frogs)
29. **Return** centroid fitness

Algorithm 2. Community detection via a stack auto-encoder & SFLA

Input: social network as graph representation, adjacency matrix A=[a_ij] R^(N*N) , Parameters of SAE, Parameters of SFLA
Output: NMI, Modularity, Clustering result
1. Create hybrid matrix H: Construction of similarity matrix by Equation (3) based on adjacency A
2. Initialize the weight matrix and bias by Equation (6), (7)
3. Calculate the activation of each layer by Equation (5), z_ (final)
4. **Repeat**
5. Update iteratively the weight, bias
6. **Until** maximum number of iterations
7. Initialize the Shuffled Frog Leaping (SFLA)
8. Apply evolutionary operators (crossover and mutation) and local search to refine the positions of the centroids
9. **Repeat**
10. Update iteratively the variables
11. **Until** convergence
12. Aquire the result of clustering, NMI, Modularity
13. **Return** clustering result, NMI, Modularity

The process is terminated after a specified number of iterations, ensuring adaptation to the evolving characteristics of the network while preventing excessive computational costs. An objective function connected to k-means performance is used in fitness evaluation. SFLA dynamically evolves candidate centroids through iterative loops and finishes according to predetermined criteria. The optimal set f centroids for the k-means algorithm are determined by the final solution, which uses the frog with the highest fitness. Algorithm 2 presents the steps of community detection via combination of stack auto-encoder and SFLA.

4 Experimental Results

To evaluate the performance of the proposed method, we used three real-world datasets including Football, Dolphin, and Political books (Polbooks). The experiments were carried out in the Google Collaboratory platform (GoogleColab). The details of the datasets and layer setting are shown in Table 1:

Table 1. The features of the dataset and layer setting of autoencoder on the selected dataset

Datasets	Network Format	Number of Nodes/Edges	Number of Communities	Layer Setting
Football[1]	undirected	115/613	12	N, 64,32
Dolphin[2]	undirected	62/159	2	N,32,16
Polbooks[3]	undirected	105/441	3	N 64,32

The learning rate for each auto-encoder was 0.001 trained up to 1000 epochs.

Each dataset has a known number of communities, as listed in Table 1. When we run our proposed algorithm on a given dataset, our proposed algorithm can effectively identify the same number of communities as specified in Table 1 for each dataset. The quality of the detected communities is then assessed using Normalized Mutual Information (NMI) and Modularity, demonstrating the accuracy and reliability of our method in community detection. Two sets of partitions are compared for similarity using the NMI which is determined using the Eq. (9), and its value can range from 0 to 1 [5]:

$$NMI(Y, C) = \frac{2 \times I(Y; C)}{[H(Y) + H(C)]} \tag{9}$$

where Y refers to class labels, C refers to cluster labels, H is entropy, and I (Y, C) denote the mutual information between Y and C.

The analysis for Figs. 2, 3 and 4 emphasizes the importance of both the configuration and the number of basic communities to have optimal performance based on NMI in community detection schemes. Figures 2, 3 and 4 illustrate the results of NMI using selected layer settings of SAEs.

Figures 2, 3 and 4 present the effect of the designed stacked auto-encoder on 50 communities using three different datasets to measure average values of NMI. The results indicate that different configurations of stacked auto-encoders can affect the performance with regard to NMI. Further, it highlights the number of basic communities required to achieve the best NMI in each configuration. In order to evaluate the performance of the proposed method, its results were compared with recent methods SAECF [8], AECD-COA [9], PFGM [10], and CNLLP [11] in terms of the NMI measures. In this paper, the experimental results are reported by the average NMI for datasets of real communities. Table 2 shows the NMI for all algorithms evaluated:

[1] http://konect.cc/networks/dimacs10-football/
[2] https://networkrepository.com/
[3] https://networkrepository.com/

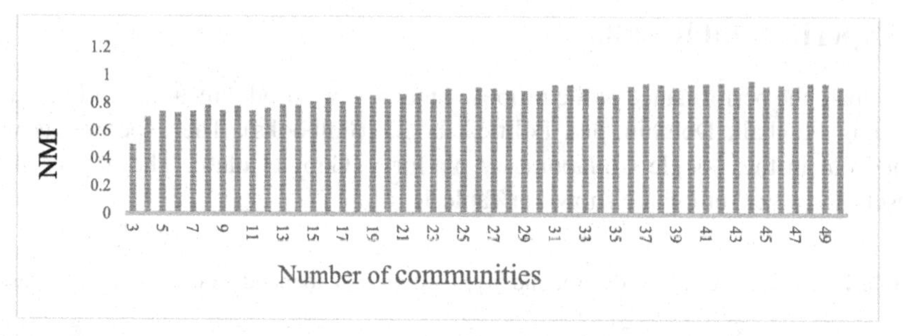

Fig. 2. The results of NMI using 62,32,16 stacked auto-encoders with different numbers of communities on the Dolphin dataset

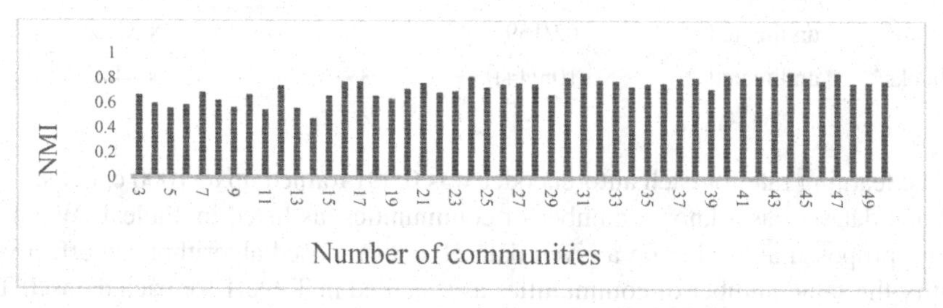

Fig. 3. The results of NMI using 105,64,32 stacked auto-encoders with different numbers of communities on the pollbook dataset

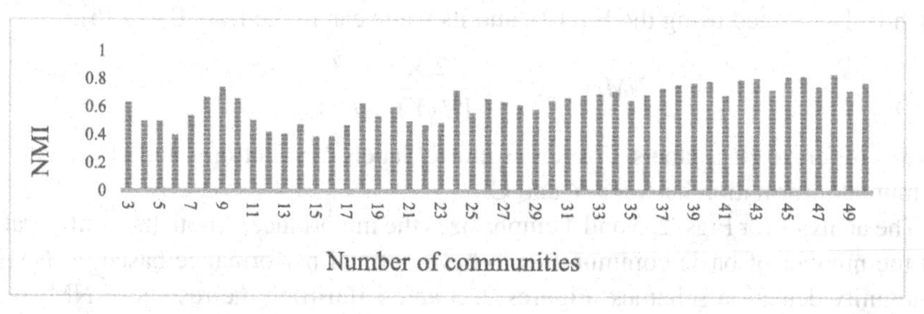

Fig. 4. The results of NMI using 115,64,32 stacked auto-encoders with different numbers of communities on the football dataset

Using SFLA to improve the k-means function in our method resulted in more diversity in population selection and avoided trapping in a local optimum. The algorithms like SFLA are able to adapt to environment changes; therefore, they can act appropriately in complex conditions with high dynamicity over time.

Another advantage of the proposed method is that it acts efficiently to search/explore within local and global scopes of the network structure. This is because the proposed method converges to optimal solutions quickly. Notably, our findings indicate that while the results of our proposed method using Football dataset for 50 clusters is comparatively lower than other methods, scaling up to 100 clusters results better outcomes for our method. It is noteworthy that if we consider very large number of clusters and use small

Table 2. NMI values for all the algorithms evaluated according to the selected datasets

Method	Football	Dolphin	Pollbooks
SAECF	0.67	0.49	0.39
AECD-COA	0.89	0.63	0.41
PFGM	0.72	0.7	0.34
CNLLP	0.27	0.60	0.48
Proposed method	0.84	0.96	0.85

datasets, the probability of trapping in local optimum increases. Modularity is known as a measure for quantifying the strength of the division of a network into communities. The modularity Q is computed as:

$$Q = \frac{1}{2m} \sum_{ij} \left(Aij - \frac{kikj}{2m} \right) \delta(ci, cj) \tag{10}$$

where A_{ij} is the adjacency matrix denoting the presence or absence of an edge between nodes i and j, k_i and k_j are the degrees of nodes i and j, respectively located in the same community c_i and c_j are the community assignments of nodes i and j will be equal to 1. Table 3 show the modularity value:

Table 3. Modularity values for all the algorithms evaluated according to the selected datasets

Method	Football	Dolphin	Pollbooks
SAECF	0.57	0.48	0.59
AECD-COA	0.59	0.30	0.43
PFGM	0.52	0.40	0.54
CNLLP	0.56	0.52	0.45
Proposed method	0.58	0.57	0.61

Comparing our proposed algorithm to other community detection algorithms on all of the selected datasets, the above results demonstrate that it performs more optimally in terms of modularity score. As opposed to other algorithms, the shuffled frog leaping algorithm clusters the nodes in the networks much more effectively, and this is responsible for the optimized performance. It is also paired with stacked auto-encoders. The complexity analysis of the proposed method involves considerations of both computational complexity and iterative processes. Constructing the hybrid has a time complexity of $O(N^2)$. If the SAE has L layers and each layer has d_i nodes, the time complexity for forward propagation is $O(L.N.d^2)$, , where d is the maximum dimensionality across all layers. Initializing the population and memeplexes has a time complexity of O (P.N), where P is the population size.

5 Conclusion

The proposed method addressed the challenges posed by the complex topological properties, heterogeneous node properties, and dynamic evolution mechanisms of social networks. Using stacked auto-encoders for dimensionality reduction can reduce the complexity of high-dimensional data while preserving important features. Moreover, using SFLA provides adaptability and diversity to the optimization process, thus enabling effective exploration of the solution space. The proposed method exhibited robustness in dynamic environments, making it suitable for real-world applications where community detection is pivotal for understanding network structures and patterns. The evaluation of the proposed method showed its superiority in terms of the average NMI factor and modularity compared to recent works. In future works, we intend to study the scalability and consider large networks.

References

1. Newman, M.E.: Modularity and community structure in networks. Proc. Natl. Acad. Sci. **103**(23), 8577–8582 (2006)
2. Fortunato, S.: Community detection in graphs. Phys. Rep. **486**(3–5), 75–174 (2010)
3. Souravlas, S., et al.: A classification of community detection methods in social networks: a survey. Int. J. Gen. Syst. **50**(1), 63–91 (2021)
4. Su, X., et al.: A comprehensive survey on community detection with deep learning. IEEE Trans. Neural Netw. Learn. Syst. **35**(4), 4682–4702 (2022)
5. Maaroof, B.B., et al.: Current studies and applications of shuffled frog leaping algorithm: a review. Arch. Comput. Methods Eng. **29**(5), 3459–3474 (2022)
6. Souravlas, S., Anastasiadou, S., Katsavounis, S.: A survey on the recent advances of deep community detection. Appl. Sci. **11**(16), 7179 (2021)
7. Wu, L., et al.: Deep learning techniques for community detection in social networks. IEEE Access **8**, 96016–96026 (2020)
8. Xu, R., et al.: Stacked autoencoder-based community detection method via an ensemble clustering framework. Inf. Sci. **526**, 151–165 (2020)
9. Pierezan, J., Coelho, L.D.S.: Coyote optimization algorithm: a new metaheuristic for global optimization problems. In: 2018 IEEE Congress on Evolutionary Computation (CEC). IEEE (2018)
10. Wang, Y., et al.: Proximity-based group formation game model for community detection in social network. Knowl.-Based Syst..-Based Syst. **214**, 106670 (2021)
11. Zhang, W., Shang, R., Jiao, L.: Large-scale community detection based on core node and layer-by-layer label propagation. Inf. Sci. **632**, 1–18 (2023)
12. Aslan, M., Koç, İ: Modified Coot bird optimization algorithm for solving community detection problem in social networks. Neural Comput. Appl.Comput. Appl. **36**(10), 5595–5619 (2024)
13. Jin, D., et al.: A survey of community detection approaches: from statistical modeling to deep learning. IEEE Trans. Knowl. Data Eng.Knowl. Data Eng. **35**(2), 1149–1170 (2021)

A Model of Net Flaming Caused by News Propagation in Online Social Networks

Harumasa Tada[1(✉)], Masayuki Murata[2], and Masaki Aida[3]

[1] Kyoto University of Education, Kyoto, Kyoto 612-8522, Japan
htada@kyokyo-u.ac.jp
[2] Osaka University, Suita, Osaka 565-0871, Japan
murata@ist.osaka-u.ac.jp
[3] Tokyo Metropolitan University, Hino, Tokyo 191-0065, Japan
aida@tmu.ac.jp

Abstract. Net flaming that occurs on Online Social Networks (OSNs) has become a serious problem. Net flaming occurs when some news is spread on OSNs and users react strongly to it. In OSNs, however, net flaming can occur even though there is nothing particularly wrong with the news itself that is spread. In this paper, we use a network oscillation model to analyze user dynamics caused by news propagation in OSNs, and show that net flaming may be caused by user comments added to the news during the propagation process. Based on these observations, we discuss countermeasures to suppress net flaming.

Keywords: Online flaming · Oscillation model · Spectral graph theory · Laplacian matrix

1 Introduction

The spread of social networking services (SNS) has been remarkable in recent years and has now become a major communication tool. As the influence of existing mass media such as TV and newspapers is declining, more and more people are getting most of their information from SNS, and the influence of SNS on people's thinking and behavior is becoming increasingly significant.

On the other hand, net flaming is a major problem in OSNs. Many cases of net flaming that seriously affected the social activities of individuals and companies in the real world have been reported [6].

Net flaming in OSNs occur when users react to certain news spread on OSNs. However, whether or not a given news item causes net flaming in OSNs is not necessarily determined by the news itself. For example, in the case of the advertisement posted by Marks & Spencer on Instagram [1], the intent of the ad was not controversial, but some users added critical comments based on their own interpretation, which many users agreed with, resulting in net flaming. In this case, whether or not net flaming occurs depends largely on the propagation path of the news, i.e., through which users the news is transmitted to other users. This

L. M. Aiello et al. (Eds.): ASONAM 2024, LNCS 15212, pp. 127–136, 2025.
https://doi.org/10.1007/978-3-031-78538-2_11

paper focuses on net flaming caused by the propagation of news through users, which is unique to OSNs.

In order to analyze OSNs, we have proposed the oscillation model with a graphical representation of the relationship between users [2,3]. To apply the oscillation model to OSN analysis, each OSN user is mapped to a node in the network, so that the user dynamics are represented by the oscillation of the node. By considering the net flaming as a divergence of user dynamics and relating the explosive dynamics that occur in the oscillation model to it, we aim to clarify the factors that cause net flaming and the mechanisms behind it.

Our previous study [5] has shown that imposing a periodic external force as an external stimulus to a node in the oscillation model can cause a phenomenon called resonance, which leads to explosive dynamics. Corresponding the propagation of an external stimulus to the propagation of news in OSNs, this phenomenon represents how the news spread causes net flaming. In this case, whether or not net flaming occurs depends on the angular frequency of the external stimulus, which indicates that the cause of the net flaming is the news itself.

On the other hand, in this paper, we model net flaming in which the propagation path of news affects its occurrence more than the news itself. The propagation path of news in OSNs is represented by a directed acyclic graph (DAG). Therefore, we consider the oscillation model represented by DAG and examine the behavior of nodes when an external stimulus is imposed to one node. We show that net flaming may be caused by user comments added to the news during the propagation process. Based on these observations, we discuss measures to suppress net flaming.

The rest of this paper is organized as follows. Section 2 describes related works. Section 3 provides an overview of the network oscillation model. Section 4 describes the proposed model and shows that this model represents the net flaming targeted in this paper, and Sect. 5 discusses countermeasures against such net flaming. Finally, we state conclusions and future work in Sect. 6.

2 Related Work

The phenomenon of information spreading in a short period of time in OSNs is known as the information cascade. There are many studies dealing with information cascades. For example, Zhou et al. [8] analyzed the dynamics of the information cascade on Twitter during the 2009 Iranian election. Also, Li et al. [7] analyze the information diffusion that occurred on Twitter during the Fukushima nuclear accident. Goel et al. [4] proposed a model of information cascades based on the SIR model. On the other hand, in net flaming, information changes the behavior of users, which further influences other users. This paper focuses on changes in user behavior caused by information rather than information spreading itself. While the SIR model represents information propagation by node state transitions, the oscillation model representing node states as oscillation, thereby represents not only information propagation but also the changes in node dynamics that it causes. This allows us to represent explosive dynamics such as net flaming.

3 Overview of Network Oscillation Model

3.1 Laplacian Matrix of the Network with Directed Links

Let $\mathcal{G}(V, E)$ be a directed graph representing the structure of a network with n nodes, where $V = \{1, ..., n\}$ is the set of nodes and E is the set of directed links. Also, let $w_{ij} > 0$ be the weight of the directed link $(i \to j) \in E$ from node i to node j. The (weighted) adjacency matrix $\mathcal{A}:=[\mathcal{A}_{ij}]_{1 \le i,j \le n}$ is an $n \times n$ matrix defined as

$$\mathcal{A}_{ij} := \begin{cases} w_{ij}, & (i \to j) \in E, \\ 0, & (i \to j) \notin E. \end{cases} \tag{1}$$

For the weighted out-degree $d_i := \sum_{j=1}^{n} \mathcal{A}_{ij}$ of node i, the degree matrix $\mathcal{D}$ is an $n \times n$ matrix defined as $\mathcal{D}:=\text{diag}(d_1, \ldots, d_n)$. The Laplacian matrix $\mathcal{L}$ of the (weighted) directed graph is defined as $\mathcal{L}:=\mathcal{D} - \mathcal{A}$.

3.2 Oscillation Model on Directed Networks

Let $x_i(t)$ be the state of node i at time t, and each node is subjected to a force from each adjacent node. That is, node i is subjected to the restoring force that is represented as $f_{i \to j} = -w_{ij}(x_i(t) - x_j(t))$, where $f_{i \to j}$ is the force acting on node i from adjacent node j; w_{ij} is a positive constant. Note that the direction of the directed links and the direction of force propagation are opposite.

We consider the situation that we impose a periodic external stimulus with angular frequency ω and amplitude F on a certain node, s. Also, each node is subjected to a damping force that is proportional to its own velocity. The equation of motion of the node state vector $\boldsymbol{x}(\omega, t) := {}^t(x_1(\omega, t), x_2(\omega, t), ..., x_n(\omega, t))$ for the forced oscillation of a directed graph can be written by using its Laplacian matrix $\mathcal{L}$ as follows:

$$\frac{\partial^2 \boldsymbol{x}(\omega, t)}{\partial t^2} + \gamma \frac{\partial \boldsymbol{x}(\omega, t)}{\partial t} + \mathcal{L}\boldsymbol{x}(\omega, t) = F \cos(\omega t) \mathbf{1}_{\{s\}}, \tag{2}$$

where $\gamma \ge 0$ is the damping coefficient and $\mathbf{1}_{\{s\}}$ is an n-dimensional vector whose s-th component is 1 and all others are 0.

Let $\lambda_\mu (\mu = 0, 1, \ldots, n-1)$ be the eigenvalue of $\mathcal{L}$ and $\boldsymbol{v}_\mu$ be the eigenvector of $\mathcal{L}$ associated with λ_μ. We assume λ_μs are different from each other. We expand $\boldsymbol{x}(\omega, t)$ and $\mathbf{1}_{\{s\}}$ by using $\boldsymbol{v}_\mu$ as follows:

$$\boldsymbol{x}(\omega, t) = \sum_{\mu=1}^{n} a_\mu(\omega, t) \boldsymbol{v}_\mu, \tag{3}$$

$$\mathbf{1}_{\{s\}} = \sum_{\mu=1}^{n} b_\mu \boldsymbol{v}_\mu. \tag{4}$$

Substituting these into Eq. (2), we obtain the equation of motion for the oscillation mode $a_\mu(\omega, t)$ as follows:

$$\frac{\partial^2 a_\mu(\omega,t)}{\partial t^2} + \gamma \frac{\partial a_\mu(\omega,t)}{\partial t} + \lambda_\mu a_\mu(\omega,t) = F\cos(\omega t) b_\mu. \tag{5}$$

The equation of motion (5) means that the oscillation dynamics on directed networks can be expressed by superposing the oscillations of each oscillation mode.

The solution of Eq. (5) is given by

$$a_\mu(\omega,t) = c_\mu e^{-\frac{\gamma}{2}t} \cos\left(\sqrt{\lambda_\mu - \left(\frac{\gamma}{2}\right)^2} t + \phi_\mu\right) + A_\mu(\omega)\cos(\omega t + \theta_\mu(\omega)), \tag{6}$$

where c_μ and ϕ_μ are constants, and $A_\mu(\omega)$ and $\theta_\mu(\omega)$ are the amplitude and the initial phase, respectively. They are expressed as

$$A_\mu(\omega) = \frac{F b_\mu}{\sqrt{(\lambda_\mu - \omega^2)^2 + (\gamma\omega)^2}}, \tag{7}$$

$$\theta_\mu(\omega) = \arctan\left(-\frac{\gamma\omega}{\lambda_\mu - \omega^2}\right). \tag{8}$$

4 Model of Net Flaming Caused by News Propagation in OSNs

In this section, we model the unidirectional propagation of news in OSNs using the oscillation model and show that the path of news propagation affects the occurrence of net flaming. Since the path of news propagation in OSNs is represented by DAG that maps users to nodes, we focus the oscillation model represented by DAG. The external stimulus imposed to one node propagate to other nodes through links, which corresponds to the news propagation in an OSN. When a user receives news directly from another user, a directed link exists between the corresponding nodes. The larger the weight of the link, the more strongly the user is influenced by the other user.

4.1 Assumptions

We assume that nodes are topologically sorted and node IDs are assigned in descending order. That is, if $i < j$, then $w_{ij} = 0$ because there is no link from node i to node j. Since node 1 has no outgoing link, $d_1 = 0$. Also, assume that $d_1, \ldots, d_n$ are different from each other. Thus, since $d_i \geq 0$ for any $i \geq 2$, all nodes except node 1 have outgoing links. This also means that node 1 is reachable from all nodes. An external stimulus is imposed to node 1, which is transmitted to all other nodes.

4.2 Mapping of Oscillation Modes to Nodes

Since the Laplacian matrix of DAG is a triangular matrix, each eigenvalue of $\boldsymbol{\mathcal{L}}$ is equal to one of its diagonal components. From the definition of the Laplacian matrix described in Sect. 3.1, each diagonal component of $\boldsymbol{\mathcal{L}}$ is d_i of some node i. Therefore, for each node i, d_i is equal to one of the eigenvalues of $\boldsymbol{\mathcal{L}}$, which means that the oscillation mode μ such that $\lambda_\mu = d_i$ can be mapped to node i. Hereafter, the oscillation mode corresponding to node i will be referred to as oscillation mode i.

From Eq. (6), the solution of oscillation mode i is

$$a_i(\omega, t) = c_i e^{-\frac{\gamma}{2}t} \cos(\omega_i t + \phi_i) + A_i(\omega) \cos(\omega t + \theta_i(\omega)), \tag{9}$$

where $\omega_i{:=}\sqrt{d_i - \left(\frac{\gamma}{2}\right)^2}$.

4.3 Addition of Oscillation at Nodes

Let $\boldsymbol{v}_i = {}^t(v_{i1}, v_{i2}, \ldots, v_{in})$ be the eigenvector associated with the eigenvalue d_i of the oscillation mode i.

From the definitions of eigenvalues and eigenvectors, $\boldsymbol{\mathcal{L}}\boldsymbol{v}_i = d_i \boldsymbol{v}_i$, which becomes

$$\begin{pmatrix} d_1 v_{i1} \\ w_{21} v_{i1} + d_2 v_{i2} \\ w_{31} v_{i1} + w_{32} v_{i2} + d_3 v_{i3} \\ \vdots \\ w_{n1} v_{i1} + \cdots + w_{n(n-1)} v_{i(n-1)} + d_n v_{in} \end{pmatrix} = \begin{pmatrix} d_i v_{i1} \\ d_i v_{i2} \\ d_i v_{i3} \\ \vdots \\ d_i v_{in} \end{pmatrix}, \tag{10}$$

since $\boldsymbol{\mathcal{L}}$ is a triangular matrix.

Lemma 1. *For any $j < i$, $v_{ij} = 0$.*

Proof. We prove by mathematical induction that $v_{ij} = 0$ for any $j < i$.
When $j = 1$, the j-th component of Eq. (10) is $d_j v_{ij} = d_i v_{ij}$. Since $d_j = 0 \neq d_i$, $v_{ij} = 0$ holds.
Assume that $v_{ij} = 0$ holds when $j \leq k$, where $k < i - 1$.
When $j = k + 1$, using the inductive hypothesis, the j-th component of Eq. (10) is $d_j v_{ij} = d_i v_{ij}$. Since $d_j \neq d_i$ from $j = k + 1 \neq i$, $v_{ij} = 0$ holds.

By induction, we have shown $v_{ij} = 0$ for any $j < i$. □

Lemma 2. *For any i, $v_{ii} \neq 0$.*

Proof. Applying Lemma 1 to Eq. (10), $v_{ii} = 0$ implies $\boldsymbol{v}_i = \boldsymbol{0}$, which contradicts the fact that $\boldsymbol{v}_i$ is an eigenvector of $\boldsymbol{\mathcal{L}}$.
Therefore, $v_{ii} \neq 0$. □

Lemma 3. *If $v_{ij} \neq 0$, node i is reachable from node j.*

Proof. Let i be an arbitrary node. We show by mathematical induction that for any j, if $v_{ij} \neq 0$, node i is reachable from node j.
When $j = 1$, if $i = 1$, node i is reachable from node j, otherwise, $v_{ij} = 0$ from Lemma 1 since $j < i$.
Assume that for $j \leq k$, if $v_{ij} \neq 0$, node i is reachable from node j.
When $j = k + 1$, from the j-th component of Eq. (10),

$$v_{ij} = \frac{w_{j1}v_{i1} + \cdots + w_{j(j-1)}v_{i(j-1)}}{d_i - d_j}. \tag{11}$$

From Eq. (11), if $v_{ij} \neq 0$, there exists $l \leq j - 1$ such that $w_{jl} \neq 0 \wedge v_{il} \neq 0$. Using the inductive hypothesis, node i is reachable from node l since $v_{il} \neq 0$. Also, there is a directed link $(j \to l)$ since $w_{jl} \neq 0$. Therefore, node i is reachable from node j.

By induction, we have shown that for any node j, if $v_{ij} \neq 0$, node i is reachable from node j. □

Lemma 4. *At any node j, the amplitude of oscillation mode i is proportional to v_{ij} and b_i.*

Proof. From Eq. (3), the oscillation of node j can be expressed as a superposition of each oscillation mode as follows:

$$x_j(\omega, t) = \sum_{i=1}^{n} a_i(\omega, t)v_{ij}. \tag{12}$$

From Eq. (12), the amplitude of oscillation mode i is proportional to v_{ij}.

From Eq. (7), $A_i(\omega)$ in Eq. (9) is proportional to b_i. Under $a_i(\omega, 0) = 0$ as an initial condition, c_i in Eq. (9) is

$$c_i = -\frac{A_i(\omega)\cos(\theta_i(\omega))}{\cos(\phi_i)}, \tag{13}$$

which implies that c_i is also proportional to b_i. Therefore, from Eq. (9), $a_i(\omega, t)$ is proportional to b_i, that is, the amplitude of oscillation mode i is proportional to b_i. □

From Lemma 4, oscillation mode i does not affect node j where $v_{ij} = 0$. Therefore, from Lemma 3, oscillation mode i affects only nodes those are reachable to node i. Equation (9) indicates that oscillation mode i consists of two oscillations, one with angular frequency ω_i and the other with angular frequency ω. Of these, oscillation with angular frequency ω_i is only included in oscillation mode i and is therefore observed only at nodes that are reachable to node i. This indicates that as the external stimulus propagates over the network, a new oscillation is added at each node, which then propagates to other nodes.

This new oscillation represents information that is different from the original news and corresponds to a comment added by the user to the news.

We show by simulation how new oscillations are added at each node during the propagation of an external stimulus on the network. As the response of node i to the external stimulus, we observe the oscillation energy $E_i(\omega) = \frac{1}{2}\omega^2 \sum_{k=1}^{n}(a_k(\omega))^2(v_{ki})^2$. The network used in the simulation is shown in Fig. 1. In the figure, the numbers next to the links indicate the link weights.

Figure 2 shows the spectrum of oscillation energy at nodes 2,3, and 4 when the damping factor $\gamma = 0.1$ and an external stimulus with an angular frequency of 100 is imposed to node 1. The peak of angular frequency 100 seen at each node corresponds to the oscillation of the external stimulus propagated from node 1. A peak at an angular frequency of 20 is seen at node 2, which represents the addition of oscillation with angular frequency of $\omega_2 = \sqrt{d_2 - (\frac{\gamma}{2})^2}$, since the eigenvalue of node 2 is $d_2 = 400$. In addition, the peak around an angular frequency of 60 seen at node 4 is not seen at nodes 2 and 3. This indicates that this oscillation is added at node 4. The eigenvalue of node 4 is $d_4 = 3600$, which means that the angular frequency $\omega_4 \simeq 60$.

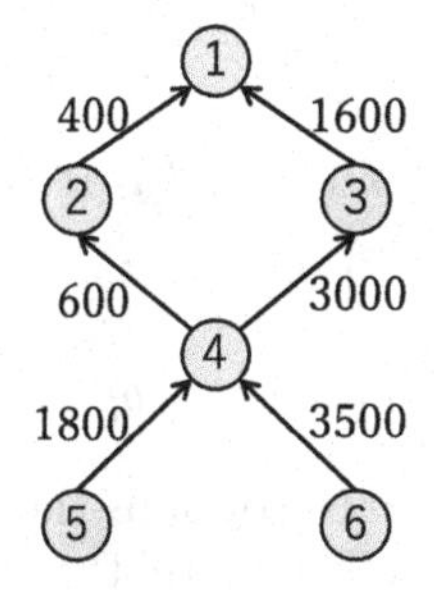

Fig. 1. A Simple Network

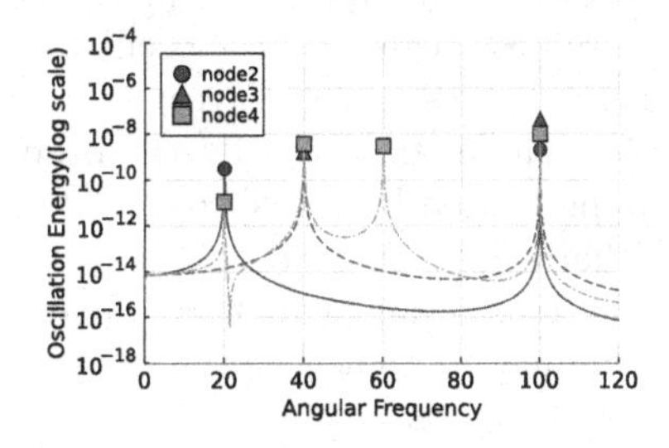

Fig. 2. Adding Oscillations at Nodes

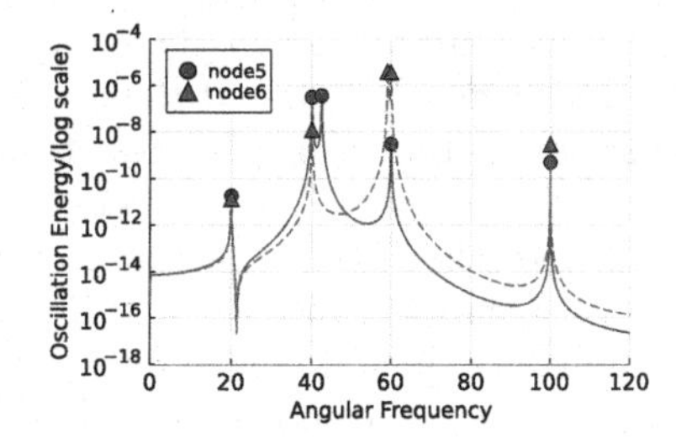

Fig. 3. Occurrence of Resonance at Nodes

4.4 Resonance Caused by Added Oscillation

We show that the phenomenon of nodes resonating with the added oscillation occurs, and that this model represents net flaming that occur when a user's comment prompts other users.

Suppose that $d_i \simeq d_j$ for two nodes i, j where $i < j$.

Since $|d_i - d_j| \simeq 0$, Eq. (11) implies that $|v_{ij}|$ is very large unless $v_{ij} = 0$. Therefore, the amplitude of oscillation mode i at node j is very large from Lemma 4. Equation (9) indicates that oscillation mode i consists of two oscillations, one with angular frequency ω_i and the other with angular frequency ω. The oscillation with angular frequency ω is included in all oscillation modes and may cancel each other out due to superposition. On the other hand, the oscillation with angular frequency ω_i is included only in oscillation mode i. Therefore, at node j, the amplitude of oscillation with angular frequency ω_i increases.

From Lemma 2, for any i, we can choose an eigenvector $\boldsymbol{v}_i$ so that $v_{ii} = 1$. Since the external stimulus is imposed to node 1, Eq. (4) becomes

$$\begin{pmatrix} 1 \\ 0 \\ \vdots \\ 0 \end{pmatrix} = \begin{pmatrix} b_1 \\ b_1 v_{12} + b_2 \\ \vdots \\ b_1 v_{1n} + \cdots + b_{n-1} v_{(n-1)n} + b_n \end{pmatrix}. \tag{14}$$

From the j-th component of Eq. (14),

$$-b_j = b_1 v_{1j} + \cdots + b_{(j-1)} v_{(j-1)j}. \tag{15}$$

Since $i < j$, the right side of Eq. (15) contains the term $b_i v_{ij}$, which implies that $|b_j|$ becomes very large with $|v_{ij}|$. Therefore, the amplitude of oscillation mode j at node j is very large from Lemma 4. From Eq. (9), oscillation mode j includes the oscillation with angular frequency ω_j, which is included only in oscillation mode j and do not cancel each other out by superposition. Therefore, at node j, the amplitude of oscillation with angular frequency ω_j increases.

Thus, when $d_i \simeq d_j$, the amplitudes of oscillations with angular frequency ω_i and ω_j increase at node j. These amplitude increases mean that the oscillation added at node i has caused resonance at node j. This resonance corresponds to the situation that a user reacts strongly to the comments added by the other user, and represents the occurrence of net flaming caused by users' comments.

On the other hand, when node i is not reachable from node j, no amplitude increase occurs because $v_{ij} = 0$ from Lemma 3. This means that user j reacts strongly only when the news arrives via user i, indicating that the occurrence of net flaming depends on the propagation path of the news.

Figure 3 shows the oscillation energy spectrum at nodes 5 and 6 in the same simulation as Fig. 2. At node 5, two high peaks are seen around the angular frequency of 40. This indicates that node 5 resonates with the oscillation added at node 3 because the eigenvalue $d_5 = 1800$ of node 5 is close to the eigenvalue $d_3 = 1600$ of node 3. Similarly, at node 6, there are two high peaks around the angular frequency of 60, which indicates that node 6 resonates with the oscillation added at node 4.

5 Measures to Suppress Network Resonance

5.1 Manipulating Link Weights

As mentioned in Sect. 4.4, whether resonance occurs at node i depends on its eigenvalue, which is equal to its weighted out-degree d_i, i.e. the sum of weights of all links from node i. Therefore, if a resonating node can be identified, a possible countermeasure is to change the eigenvalue of the node by manipulating the weights of any of the links from the node. The link weights in the oscillation model reflect the strength of the association between users in the OSN, such as the speed of information transfer and trust in other users. Manipulation of these may be able to keep resonance to a small scale and prevent net flaming. However, since it is difficult to predict the occurrence of resonance in advance, this measure can be taken only after catching the omen of resonance. Figure 4 shows the results of the simulation under the same conditions as Fig. 3, except that the weight of the link $(6 \rightarrow 4)$ changed from 3500 to 3000. The peak around the angular frequency of 60 is lower than that of Fig. 3. This means that the resonance was suppressed because the eigenvalue of node 6 changed.

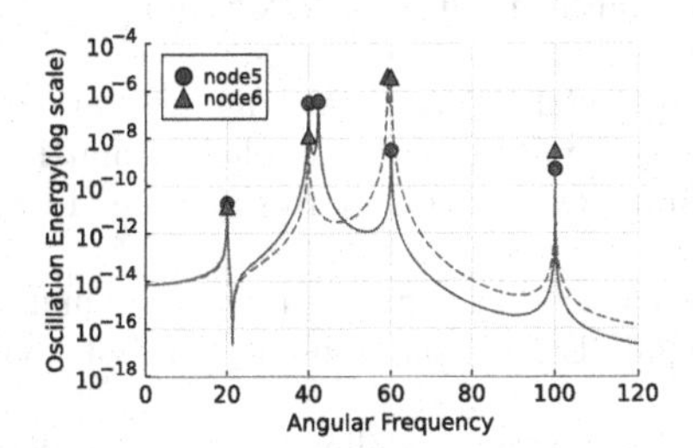

Fig. 4. Manipulating a Link Weight

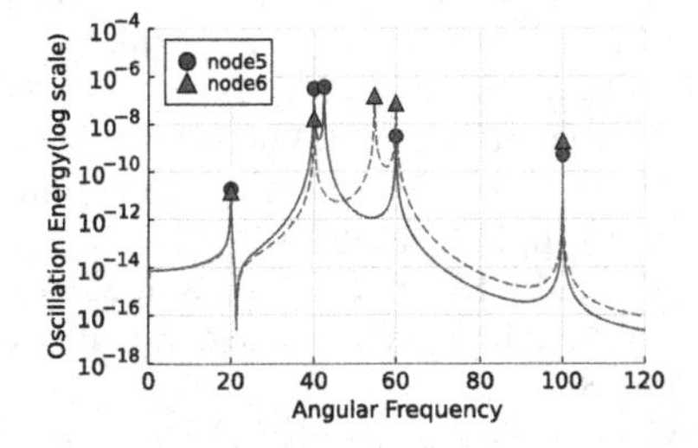

Fig. 5. Increase in Damping Coefficient

5.2 Increase in Dumping Coefficient

Equation (9) shows that the oscillation with angular frequency ω_i added at node i is damped with time by the damping coefficient γ. This implies that an increase in γ is effective in reducing the occurrence of node resonance. γ reflects the degree to which OSN user dynamics decrease over time, i.e. the fickleness of users. Although γ is difficult to artificially manipulate, it is likely to increase in the future, since users are becoming increasingly fickle with the explosive increase in information contents, and this trend is expected to further accelerate. As a result, the number of net flaming caused by node resonance may become smaller and shorter in the future. Figure 5 shows the results of a simulation of the network in Fig. 1 with a damping factor of $\gamma = 0.5$ and other conditions identical to those in Fig. 3. Due to the effect of damping, all peaks are lower than in Fig. 3.

6 Conclusion

In this paper, we proposed a theoretical model of net flaming that occurs in the process of unidirectional propagation of news in OSNs. In the oscillation model represented by DAG, another oscillation is added at a node during the propagation of an external stimulus, which may cause other nodes to resonate with it. This phenomenon represents the mechanism by which net flaming is triggered by user comments during news propagation. Based on these observations, we discussed countermeasures to suppress net flaming.

Acknowledgments. This work was supported by JSPS KAKENHI (grant number 20H04179).

Disclosure of Interests. The authors have no competing interests to declare that are relevant to the content of this article.

References

1. M&S pulls Christmas advert post after Palestinian flag criticism, 2 November 2023. https://www.bbc.com/news/uk-67294809
2. Aida, M., Takano, C., Murata, M.: Oscillation model for network dynamics caused by asymmetric node interaction based on the symmetric scaled Laplacian matrix. In: The 12th International Conference on Foundations of Computer Science (FCS 2016), pp. 38–44 (2016)
3. Aida, M., Takano, C., Murata, M.: Oscillation model for describing network dynamics caused by asymmetric node interaction. IEICE Trans. Commun. **101**(1), 123–136 (2018)
4. Goel, A., Munagala, K., Sharma, A., Zhang, H.: A note on modeling retweet cascades on Twitter. In: Gleich, D., Komjathy, J., Litvak, N. (eds.) Algorithms and Models for the Web Graph: 12th International Workshop, WAW 2015, Eindhoven, The Netherlands, 10–11 December 2015, Proceedings 12, pp. 119–131. Springer, Cham (2015). https://doi.org/10.1007/978-3-319-26784-5_10
5. Kinoshita, T., Aida, M.: A new model of flaming phenomena in online social networks that considers resonance driven by external stimuli. In: 2020 Eighth International Symposium on Computing and Networking Workshops (CANDARW), pp. 28–34. IEEE (2020)
6. Lamba, H., Malik, M.M., Pfeffer, J.: A tempest in a teacup? Analyzing firestorms on Twitter. In: Proceedings of the 2015 IEEE/ACM International Conference on Advances in Social Networks Analysis and Mining 2015, pp. 17–24 (2015)
7. Li, J., Vishwanath, A., Rao, H.R.: Retweeting the Fukushima nuclear radiation disaster. Commun. ACM **57**(1), 78–85 (2014)
8. Zhou, Z., Bandari, R., Kong, J., Qian, H., Roychowdhury, V.: Information resonance on Twitter: watching Iran. In: Proceedings of the First Workshop on Social Media Analytics, pp. 123–131 (2010)

Gradient Descent Clustering with Regularization to Recover Communities in Transformed Attributed Networks

Soroosh Shalileh[1,2](✉)

[1] Laboratory of Artificial Intelligence for Cognitive Sciences (AICS), HSE University, Moscow, Russia
sr.shalileh@gmail.com

[2] Center for Language and Brain (CLB), HSE University, Moscow, Russia

Abstract. Community detection in attributed networks aims to recover clusters in which the within-community nodes are as interconnected and as homogeneous as possible, while the between-communities nodes are as disconnected and as heterogeneous as possible. The current research proposes a straightforward data-driven model with an integrated regularization term to recover communities. For further improvement of the quality of detected communities we also propose a softmax-scaled-dot-product to transform the data spaces into more cluster-friendly data spaces. We adopt the gradient descent optimization strategy to optimize our proposed clustering objective function. We compare the performance of the proposed method using both real-world and synthetic data sets with three state-of-art algorithms. Our results showed that the proposed method obtains promising result.

Keywords: Attributed Network · Feature-Rich Network · Community Detection · Gradient Descent · Clustering

1 Introduction

Community Detection (CD) is a popular topic in context of social and complex networks analysis. The current research focuses on detecting communities in attributed networks. An attributed network is data structure in which, in addition to the conventional between-nodes links, each node is associated with a set of attributes. The objective of CD methods, in such a data structure, is to recover communities (clusters) using both sources of data, namely, the links and attributes information, such that the within-community nodes are as interconnected and as homogeneous as possible, while the between-communities nodes are as disconnected and as heterogeneous as possible.

Various CD methods have been proposed in the literature. For instance, in the period from 2020-2023, three comprehensive review papers were published

L. M. Aiello et al. (Eds.): ASONAM 2024, LNCS 15212, pp. 137–148, 2025.
https://doi.org/10.1007/978-3-031-78538-2_12

on this subject [1–3]. The authors of [3], the most recent review, focused on the healthcare applications of CD. Prior to that, in the middle of 2021, Al-Andoli et al. [2] provided a more comprehensive review of CD methods, focusing on deep learning and meta-heuristic approaches. Before these two reviews, in the middle of 2020, another comprehensive review was published [1]. In our opinion, this review introduced a more comprehensive taxonomy of the previous methods, which is still valid and applicable to even very recent developments, such as [4] and [5]. Thus we adopted its taxonomy.

In this taxonomy, concerning how the two sources of data are fused, the methods are categorized into: (i) early fusion, where the link and attribute data are fused before the cluster recovery process starts, (ii) simultaneous fusion, where the two sources of data are fused during the cluster recovery process, and (iii) late fusion, where the cluster recovery results of the two sources are fused after the process. We think the simultaneous fusion methods can be further classified into (ii-a) theory-based and (ii-b) data-driven methods. While (ii-a) derives a probability distribution of the process that generated the data, (ii-b) deals with the data as is. This research belongs to (ii-b), i.e., the data-driven simultaneous fusion category.

To be more precise, the current research continues the research line started in [6]. In that work, inspired by the triumph and recent advancements of deep learning (DL) methods, the authors adopted gradient descent (GD) optimization, the working horse of DL, for the task of clustering tabular data. Their results showed that, in the majority of the experiments, their proposed method, i.e., gradient descent clustering (GDC), had an edge over other benchmark clustering methods. Later, in [7], they utilized a non-summable data-driven model [8], and extended GDC to the attributed networks. However, that extension, in its simplest version did not lead to promising results, and thus they proposed a filter mechanism to cull out the misleading objects during GDC and to improve the quality of the results.

This research, unlike the methods reviewed in [2], pursues another direction to improve the performance of the GDC method extended to attributed networks. More precisely, we proposed two modifications to GDC: (a) we integrate a regularization term into our clustering objective function, to impose Occam's razor; and (b) we used a softmax-scaled-dot-product to transform the data spaces into more cluster-friendly spaces. We empirically evaluated and compared the performance of our proposed methods with three state-of-the-art (SOTA) algorithms using five real-world and 320 synthetic data sets.

2 Methodology

2.1 Softmax-Scaled-Dot-Product to Transform the Data Space

Given (an) $n \times m$ data matrix $Y = (y_{ij})$, let us denote its i-th row with $\mathbf{y}_i$, we transform Y to an $n \times n$ matrix $Y' = (y_{ij})$ such that:

$$y'_{ij} = \psi(\frac{< \mathbf{y}_i, \mathbf{y}_j >}{|m|}) \tag{1}$$

where $\psi(y_{ij}) = \frac{e^{(y_{ij})}}{\sum_{j=1}^{m} e^{(y_{ij})}}$ is the softmax function and the division by $|m|$ scales down the dot product to avoid numerical instabilities (and gradient explosions).

The rows of the transformed matrix, $Y^{'}$, now express the probability of relation of the rows with one another. We expect that this transformed space manifests the relationship between the data points more clearly, and the data points will be more clustering-friendly.

2.2 Notation and Problem Formulation

Let I represents the set of N nodes, we define an attributed network (AN) over this set as $D = \{X, A\}$, where $X \in \mathbb{R}^{N \times V}$ represents the features matrix, V is the number of features, and $A \in \mathbb{R}^{N \times N}$ represents the adjacency matrix. We aim to partition D into K crisp clusters using both sources of information, such that the within-group nodes are as interconnected and homogeneous as possible. To this end, each cluster, $\mathbf{s}_k$, is associated with its center in the feature space $\mathbf{c}_k$, and in the network space $\boldsymbol{\lambda}_k$, to form the set of centers in the feature space, $C = \{\mathbf{c}_k\}_{k=1}^{K}$, and in the network space, $\Lambda = \{\boldsymbol{\lambda}_k\}_{k=1}^{K}$, and the set of clusters, $S = \{\mathbf{s}_k\}_{k=1}^{K}$. Thus, we formulate the clustering objective function as follows:

$$J(D, C, \Lambda) = \sum_{i=1}^{N} \sum_{k=1}^{K} \rho f(\mathbf{c}_k, \mathbf{x}_i) + \xi h(\boldsymbol{\lambda}_k, \mathbf{a}_i) + \tau ||\mathbf{c}_k||_2^2 + \tau ||\boldsymbol{\lambda}_k||_2^2 \tag{2}$$

where $f(.) : \mathbb{R}^V \longrightarrow \mathbb{R}$ ($h(.) : \mathbb{R}^N \longrightarrow \mathbb{R}$) represents a generic (continuous) distance function that will be used to measure the distance between the i-th data point in the feature (network) space and the corresponding centroid. The coefficients $\rho, \xi \in [0, 1]$ are meant to adjust the trade between the two data sources during the clustering procedure. In this research, we merely fix them to unity. The terms $||\mathbf{c}_k||_2^2$ and $||\boldsymbol{\lambda}_k||_2^2$ are $L2$ norm regularization terms; they are added to impose Occam's razor on the centroids' values, that is, to encourage smaller values of the centers (and decrease their variance). The coefficient $\tau \in [0, 1]$ controls the strength of the regularization terms. The larger its value, the higher the impact of the regularization term. We postponed applying other regularization terms such as $L1$ norm, Elastic Net and separating this term per each data source to our future study.

By applying the softmax-scaled-dot-product transformation, Eq. (1), a transformed attributed network (TAN) will be obtained, i.e., $D^{'} = \{X^{'}, A^{'}\}$, where $X^{'} \in \mathbb{R}^{N \times N}$ represents the transformed features matrix, and $A^{'} \in \mathbb{R}^{N \times N}$ represents the transformed adjacency matrix. A similar clustering objective function (as per Eq. (2)) can be defined. In the remainder of this section for the sake of notational brevity we avoid define this objective function and will be using Eq. (2), unless stated.

2.3 Optimization of the Clustering Objective Function

To optimize the clustering objective function, Eq. (2), or its counterpart in TAN, we assume that the distance functions, $f(.), h(.)$, are continuous and differentiable, and therefore we concentrate on the first-order derivative-based optimization algorithm, namely, gradient descent (GD) algorithm. GD is an iterative approach, therefore, we need to update our notation to demonstrate the concept of iterations. We did that by adding subscript (t). More precisely, we denote the set of centers in the feature space, in the network space, and the set of detected clusters, at iteration (t), with $C^{(t)} = \{\mathbf{c}_k^{(t)}\}_{k=1}^K$, $\Lambda^{(t)} = \{\boldsymbol{\lambda}_k^{(t)}\}_{k=1}^K$ and $S^{(t)} = \{\mathbf{s}_k^{(t)}\}_{k=1}^K$, respectively.

Our adoption of GD to detect clusters in AN or TAN, consists of three components: (i) a criterion to assign the objects to the communities (community assignment), (ii) a rule to update cluster centers, and (iii) a stop condition.

The first component (i), the community assignment criterion, is explained in Eq. (3):

$$\underset{k}{argmin} f(\mathbf{x}_i, \mathbf{c}_k^{(t)}) + h(\mathbf{a}_i, \boldsymbol{\lambda}_k^{(t)}) < f(\mathbf{x}_i, \mathbf{c}_j^{(t)}) + h(\mathbf{a}_i, \boldsymbol{\lambda}_j^{(t)}), \forall j \neq k. \quad (3)$$

According to this criterion, at each iteration, the i-th vertex will be assigned to the k-th cluster for which the total sum of the distances between the i-th data point in the feature space, $\mathbf{x}_i$ and the corresponding center, $\mathbf{c}_k$, and the i-th data point vector, $\mathbf{a}_i$ and the corresponding center, $\boldsymbol{\lambda}_k$, is minimum.

The rules to update the centers (ii), in the feature and network spaces, are explained in Eqs. (4) and (5), respectively:

$$\mathbf{c}_k^{(t+1)} = \mathbf{c}_k^{(t)} - \alpha[\nabla_{\mathbf{c}_k^{(t)}} f(\mathbf{x}_i, \mathbf{c}_k^{(t)}) + 2\tau \mathbf{c}_k] \quad (4)$$

and

$$\boldsymbol{\lambda}_k^{(t+1)} = \boldsymbol{\lambda}_k^{(t)} - \alpha[\nabla_{\boldsymbol{\lambda}_k^{(t)}} h(\mathbf{x}_i, \mathbf{c}_k^{(t)}) + 2\tau \boldsymbol{\lambda}_k] \quad (5)$$

where α denotes the step size, and $\nabla_{\mathbf{c}_k^{(t)}}$ is the gradient vector of the distance metric, $f(.)$, w.r.t the k-th feature center at iteration (t) which is evaluated with the data point $\mathbf{x}_i$. And $2\tau\mathbf{c}_k$ represents the gradient of the $L2$ norm regularization term. Similar description is valid for Eq. (5). By equating τ to zero, the vanilla gradient descent update rule will be obtained. Using other values of τ in the range $(0, 1]$ adjusts the impact of the regularization terms, i.e., the larger the values of τ, the larger its impact. We treated τ as a hyper-parameter and empirically studied its impact in Sect. 4 to propose a default value for it.

Our preliminary experimental results were aligned with the results of [9], and showed that the on-line update rule is more efficient than the batch version of the GD algorithm; therefore, in the rest of this paper, we update the centroids in an on-line fashion.

In principles, there are various options for the last component of our proposed method (iii), i.e., the stop condition, for instance, applying first- and second-order necessary optimality conditions, etc., in this work, we relied on reaching the predefined maximum number of iterations. We treated it as a hyperparameter and empirically scrutinized its impact in Sect. 4 to propose a default value.

In the current research, to prove the efficiency of the introduced regularization and the data transformation, we proceed with the two versions of our proposed method. In the first version, we simply apply GD to optimize Eq. (2) with $\tau = 0$. We called this version Gradient Descent Clustering in Attributed Network (GDCinAN). Clearly, GDCinAN represents the vanilla adoption of gradient descent clustering without any regularization or data space transformation, and forms a baseline for our empirical studies. In the second version we adopt GD to TAN with regularization. We called this version Gradient Descent Clustering with Regularization in Transformed Attributed Network (GDCRinTAN). We outline our proposed clustering methods in Algorithm (1).

Algorithm 1: GDCinAN and GDCRinTAN

Input: $D = \{X, A\}$: AN or $D' = \{X', A'\}$: TAN and K: number of clusters.
Hyperparameters: α: step size; T: maximum number of iterations; τ: regularization strength coefficients.
Result: $S = \{\mathbf{s}_k^{(t)}\}_{k=1}^K$ % set of K binary cluster membership vectors;
$C = \{\mathbf{c}_k^{(t)}\}_{k=1}^K$ % set of K centroids in feature space;
$\Lambda = \{\boldsymbol{\lambda}_k^{(t)}\}_{k=1}^K$ % set of K centroids in network space.
Initialize: Randomly initialize C, Λ and S.
for $t \in Range(T)$ **do**
  for $(\mathbf{x}_i, \mathbf{a}_i) \in D$ **do**
    find k using Eqn. (3) and set i-th entry of the $\mathbf{s}_k^{(t)}$ to one;
    update the centroids using the equations (4) and (5);
  end
end

We implemented our proposed methods using JAX, an automatic differentiation library [10]. The source code of the proposed methods are publicly available at https://github.com/Sorooshi/GDCRinTAN/tree/main.

3 Experimental Setting

3.1 Competitors

We compared the performance of our proposed methods with DMoN [11], EVA [12], and KEFRIN [13]. DMoN is a clustering method based on graph convolutional networks and modularity criterion. EVA also utilizes modularity to model network data and the purity equation to model categorical features. KEFRIN is a meaningful extension of K-Means clustering to feature-rich networks.

We use Adjusted Rand Index [14], ARI as the evaluation metric. The closer the ARI to one, the better is the match between the recovered clusters and the ground truth.

3.2 Data Sets

We compared the performance of the methods using synthetic and real-world data sets. We provide the summary of the five real-world data sets in Table 1.

Table 1. Real world data sets: the symbols N, E, and F represent the number of nodes, the number of links, and the number of attributes, respectively.

Name	Nodes	Edges	Features	Number of Communities	Ground Truth	Ref.
Malaria HVR6	307	6526	6	2	Cys Labels	[15]
Lawyers	71	339	18	6	Derived out of office and status features	[16,17]
Parliament	451	11646	108	7	Political parties	[18]
COSN	46	552	16	2	Region	[19]
SinaNet	3490	30282	10	10	Users of same forum	[20]

To generate the synthetic attributed networks we applied the mechanism introduced in [8], which was later was applied in several works such as [13,21]. It ought to be added that we generated two types of features, namely, categorical, and mixed-scale features. Refer to any of the earlier-mentioned papers for more details on the mechanism applied to generate synthetic data. Each attributed network with categorical or mixed scale features had eight settings and each setting was repeated ten times, leading to 320 such synthetic data sets.

4 Study the Impact of the Hyperparameters of the Proposed Methods

4.1 Step Size

We studied the impact of step size on the performance of our proposed methods and reported the results in Table 2.

Considering the number of wins, one can conclude that step size equal to 0.01 is the most suitable value for GDCinAN. Regarding GDCRinTAN, although this value also won four settings, however, a closer look at the results reveals that a step size equal to 0.1 is a better default value for this version of our proposed clustering method.

4.2 Regularization Strength Coefficient

We reported the impact various regularization strength coefficients, τ, on the performance of GDCRinTAN in Table 3. These results imply that the best performance GDCRinTAN is obtained when $\tau = 0.001$.

Table 2. Impact of step size (α) with fixed $n\_init = max\_iter = 10$. The best results are bold-faced.

Method	Configuration	$\alpha = 0.0001$	$\alpha = 0.001$	$\alpha = 0.01$	$\alpha = 0.1$
	p, q, ζ	ARI	ARI	ARI	ARI
GDCinAN	(0.6, 0.2, 0.6)	0.854 ± 0.104	0.893 ± 0.015	**0.899 ± 0.002**	0.834 ± 0.131
	(0.6, 0.2, 0.8)	0.895 ± 0.068	0.913 ± 0.014	**0.925 ± 0.003**	0.781 ± 0.145
	(0.6, 0.4, 0.6)	**0.175 ± 0.035**	0.003 ± 0.001	0.003 ± 0.001	0.003 ± 0.001
	(0.6, 0.4, 0.8)	**0.174 ± 0.091**	0.005 ± 0.001	0.004 ± 0.002	0.004 ± 0.001
	(0.8, 0.2, 0.6)	0.959 ± 0.032	0.966 ± 0.028	**0.968 ± 0.026**	0.951 ± 0.016
	(0.8, 0.2, 0.8)	0.942 ± 0.065	**0.967 ± 0.016**	0.949 ± 0.057	0.954 ± 0.059
	(0.8, 0.4, 0.6)	**0.774 ± 0.085**	0.762 ± 0.095	0.727 ± 0.153	0.551 ± 0.221
	(0.8, 0.4, 0.8)	0.817 ± 0.061	**0.838 ± 0.007**	0.767 ± 0.163	0.645 ± 0.204
GDCinTAN	(0.6, 0.2, 0.6)	0.969 ± 0.056	0.916 ± 0.033	**1.000 ± 0.000**	**1.000 ± 0.000**
	(0.6, 0.2, 0.8)	0.893 ± 0.065	0.894 ± 0.035	**0.999 ± 0.003**	**0.999 ± 0.002**
	(0.6, 0.4, 0.6)	0.768 ± 0.112	0.595 ± 0.082	0.878 ± 0.086	**0.903 ± 0.075**
	(0.6, 0.4, 0.8)	**0.707 ± 0.184**	0.318 ± 0.028	0.518 ± 0.019	0.669 ± 0.213
	(0.8, 0.2, 0.6)	0.953 ± 0.045	0.960 ± 0.064	**1.000 ± 0.000**	0.940 ± 0.092
	(0.8, 0.2, 0.8)	0.915 ± 0.064	**0.998 ± 0.003**	0.971 ± 0.059	0.961 ± 0.061
	(0.8, 0.4, 0.6)	0.860 ± 0.076	**0.927 ± 0.039**	0.905 ± 0.150	0.889 ± 0.169
	(0.8, 0.4, 0.8)	0.933 ± 0.087	0.935 ± 0.063	**1.000 ± 0.000**	**1.000 ± 0.000**

Table 3. GDCRinTAN: impact of τ using ARI values with $n\_init = T = 10$ and the step size $\alpha = 0.1$. The best results are bold-faced.

Configuration p, q, ζ	$\tau = 0.0001$	$\tau = 0.001$	$\tau = 0.01$	$\tau = 0.1$	$\tau = 0.3$	$\tau = 0.9$
(0.6, 0.2, 0.6)	**1.000 ± 0.000**	**1.000 ± 0.000**	**1.000 ± 0.000**	**1.000 ± 0.000**	0.978 ± 0.047	0.739 ± 0.242
(0.6, 0.2, 0.8)	**0.999 ± 0.002**	**0.999 ± 0.002**	**0.999 ± 0.002**	0.998 ± 0.005	0.955 ± 0.059	0.593 ± 0.371
(0.6, 0.4, 0.6)	0.909 ± 0.067	**0.918 ± 0.066**	0.911 ± 0.072	0.792 ± 0.154	0.581 ± 0.283	0.427 ± 0.334
(0.6, 0.4, 0.8)	0.966 ± 0.026	**0.974 ± 0.024**	0.669 ± 0.214	0.819 ± 0.149	0.715 ± 0.292	0.485 ± 0.294
(0.8, 0.2, 0.6)	**1.000 ± 0.000**	**1.000 ± 0.000**	**1.000 ± 0.000**	**1.000 ± 0.000**	0.991 ± 0.021	0.975 ± 0.033
(0.8, 0.2, 0.8)	0.992 ± 0.023	**1.000 ± 0.000**	0.957 ± 0.066	0.865 ± 0.106	0.848 ± 0.074	0.823 ± 0.068
(0.8, 0.4, 0.6)	**0.995 ± 0.012**	0.977 ± 0.021	0.979 ± 0.039	0.739 ± 0.104	0.792 ± 0.115	0.741 ± 0.225
(0.8, 0.4, 0.8)	0.987 ± 0.040	0.987 ± 0.040	**1.000 ± 0.000**	0.999 ± 0.004	0.986 ± 0.026	0.767 ± 0.159

4.3 Impact of the Number of Iterations

We studied the impact of the number of iterations on the performance of our proposed methods and reported the results in Table 4. Considering the number of wins, one can conclude that one iteration is sufficient for GDCinAN. The best performance of GDCRinTAN was obtained when we used 100 iterations. However, a closer look at the results implies that even ten iterations usually led to acceptable performance; and selecting the final best value should be decided by the user according to their preferences considering the speed and accuracy trade-off.

Table 4. Studying the impact of the number of iterations using ARI values with tuned hyperparameters. The best results are bold-faced.

Configuration	GDCinAN			GDCRinTAN		
p, q, ζ	$T = 1$	$T = 10$	$T = 100$	$T = 1$	$T = 10$	$T = 100$
(0.6, 0.2, 0.6)	0.840 ± 0.098	0.854 ± 0.104	**0.856 ± 0.030**	0.982 ± 0.037	**1.000 ± 0.000**	**1.000 ± 0.000**
(0.6, 0.2, 0.8)	0.838 ± 0.045	0.895 ± 0.068	**0.903 ± 0.004**	0.966 ± 0.041	0.992 ± 0.002	**0.999 ± 0.002**
(0.6, 0.4, 0.6)	**0.227 ± 0.042**	0.175 ± 0.035	0.005 ± 0.002	0.852 ± 0.070	**0.918 ± 0.002**	0.909 ± 0.062
(0.6, 0.4, 0.8)	**0.304 ± 0.054**	0.174 ± 0.091	0.017 ± 0.004	0.943 ± 0.022	0.918 ± 0.066	**0.966 ± 0.028**
(0.8, 0.2, 0.6)	0.932 ± 0.060	**0.959 ± 0.032**	0.937 ± 0.041	0.999 ± 0.001	0.974 ± 0.024	**1.000 ± 0.000**
(0.8, 0.2, 0.8)	**0.953 ± 0.050**	0.942 ± 0.065	0.940 ± 0.030	0.986 ± 0.024	**1.000 ± 0.000**	0.992 ± 0.023
(0.8, 0.4, 0.6)	**0.883 ± 0.064**	0.774 ± 0.085	0.757 ± 0.003	0.980 ± 0.021	0.977 ± 0.021	**0.986 ± 0.020**
(0.8, 0.4, 0.8)	**0.833 ± 0.080**	0.817 ± 0.061	0.789 ± 0.106	0.995 ± 0.014	0.987 ± 0.040	**1.000 ± 0.000**

4.4 Tuned Hyperparameters

We propose using a step size of 0.001 for GDCinAN and 0.1 for GDCRinTAN. To obtain the best performance for GDCRinTAN, the regularization strength coefficient, τ, is 0.001. For the number of iterations, the situation is less straightforward. More precisely, our experiments showed that any number between one to ten can be adopted for GDCinAN. And any number between ten and 100 can be adopted for GDCRinTAN. We took a closer look to the performance of GDCRinTAN during the optimization procedure, and we realized that setting the number iterations to 20 usually leads to acceptable results. Therefore, in the rest of our experiments we set this number to 20.

5 Experimental Results

5.1 Comparison over Real-Word Data Sets

We compared the performance of the proposed methods using five real-world data sets with three SOTA algorithms and reported the results in Table 5.

KEFRIN, by wining three out of five cases, is the winner of the real-world competition. DMoN is the winner of the Lawyers data set. And GDCRinTAN wins the competition for the parliament data set. Although GDCRinTAN does not win all cases; however, the improvements obtained by this method, compared to GDCinAN, can be considered as evidence for the effectiveness of the data space transformation and the adopted regularization.

Table 5. Real-world data sets comparison: average values of ARI are presented over 10 random initialization. The best results are boldfaced.

data set	DMoN	EVA	KEFRiN	GDCinAN	GDCRinTAN
HRV6	0.64 ± 0.00	0.04 ± 0.00	**0.69 ± 0.38**	0.51 ± 0.18	0.60 ± 0.00
Lawyers	**0.60 ± 0.04**	0.16 ± 0.03	0.44 ± 0.14	0.36 ± 0.11	0.45 ± 0.03
Parliament	0.48 ± 0.02	0.01 ± 0.0	0.41 ± 0.05	0.40 ± 0.07	**0.51 ± 0.04**
COSN	0.91 ± 0.00	−0.0 ± 0.00	**1.00 ± 0.00**	0.73 ± 0.09	0.74 ± 0.12
SinaNet	0.28 ± 0.01	0.00 ± 0.00	**0.34 ± 0.02**	0.08 ± 0.05	0.25 ± 0.00

Table 6. Comparison using synthetic data sets with categorical features: the average and standard deviation of ARI using ten random seed initializations are reported. The best results are boldfaced.

Network size	p, q, ζ	EVA	DMoN	KEFRiN	GDCinAN	GDCRinTAN
Small	0.9, 0.3, 0.9	0.185 ± 0.046	0.709± 0.101	**0.922 ± 0.119**	0.836 ± 0.160	0.849 ± 0.067
	0.9, 0.3, 0.7	0.211 ± 0.053	0.380 ± 0.107	0.819 ± 0.142	**0.885 ± 0.118**	0.541 ± 0.185
	0.9, 0.6, 0.9	0.266 ± 0.080	0.412 ± 0.109	0.726 ± 0.097	0.232 ± 0.044	**0.754 ± 0.057**
	0.9, 0.6, 0.7	0.321 ± 0.060	0.213 ± 0.051	**0.711 ± 0.145**	0.192 ± 0.048	0.314 ± 0.060
	0.7, 0.3, 0.9	0.126 ± 0.039	0.566 ± 0.105	**0.877 ± 0.130**	0.461 ± 0.100	0.735 ± 0.066
	0.7, 0.3, 0.7	0.126 ± 0.025	0.292 ± 0.077	**0.795 ± 0.117**	0.464 ± 0.096	0.308 ± 0.069
	0.7, 0.6, 0.9	0.015 ± 0.015	0.345 ± 0.064	**0.834 ± 0.132**	0.031 ± 0.010	0.740 ± 0.056
	0.7, 0.6, 0.7	0.008 ± 0.007	0.115 ± 0.058	**0.540 ± 0.107**	0.024 ± 0.01	0.306 ± 0.071
Medium	0.9, 0.3, 0.9	0.121 ± 0.031	0.512 ± 0.137	0.724 ± 0.097	0.388 ± 0.210	**0.842 ± 0.022**
	0.9, 0.3, 0.7	0.076 ± 0.038	0.272 ± 0.073	**0.742 ± 0.182**	0.475 ± 0.112	0.633 ± 0.110
	0.9, 0.6, 0.9	0.159 ± 0.046	0.370 ± 0.063	0.652 ± 0.110	0.001 ± 0.001	**0.837 ± 0.033**
	0.9, 0.6, 0.7	0.109 ± 0.046	0.168 ± 0.030	**0.733 ± 0.083**	0.002 ± 0.001	0.430 ± 0.020
	0.7, 0.3, 0.9	0.078 ± 0.036	0.446 ± 0.099	0.641 ± 0.111	0.007 ± 0.014	**0.814 ± 0.015**
	0.7, 0.3, 0.7	0.059 ± 0.010	0.228 ± 0.077	**0.797± 0.088**	0.002 ± 0.004	0.443 ± 0.041
	0.7, 0.6, 0.9	0.002 ± 0.002	0.332 ± 0.051	0.591 ± 0.094	0.001 ± 0.001	**0.800 ± 0.031**
	0.7, 0.6, 0.7	0.002 ± 0.002	0.133 ± 0.016	**0.773 ± 0.070**	0.001 ± 0.000	0.395 ± 0.025

5.2 Comparison Using Synthetic Data with Categorical Features

We reported the results of synthetic attributed networks with categorical features in Tables 6.

In the small-size synthetic data competition KEFRIN is the winner and the proposed methods of this work and specially GDCRinTAN are its close followers. In the medium-size data sets, KEFRIN and GDCRinTAN both win four cases and can be considered as the winner of this competition, while the rest of the methods under consideration performed rather poorly. The improvement obtained by GDCRinTAN in row 3 and row 4 compared to GDCinAN at those settings, can be considered as evidence for the effectiveness of the proposed transformation and regularization.

5.3 Comparison over Synthetic Data with Mixed Scale Features

We reported the results of synthetic data with mixed scale features in Tables 7.

Table 7. Comparison using synthetic data with mixed scale features: the average and standard deviation of ARI using ten random seed initializations are reported. The best results are boldfaced.

	small-size			medium-size		
p, q, ζ	KEFRiN	GDCinAN	GDCRinTAN	KEFRiN	GDCinAN	GDCRinTAN
0.9, 0.3, 0.9	0.752 ± 0.096	0.781 ± 0.118	**0.808 ± 0.054**	**0.834 ± 0.044**	0.415 ± 0.210	0.786 ± 0.013
0.9, 0.3, 0.7	0.769 ± 0.101	**0.803 ± 0.110**	0.636 ± 0.161	**0.801 ± 0.051**	0.385 ± 0.126	0.524 ± 0.028
0.9, 0.6, 0.9	**0.809 ± 0.138**	0.225 ± 0.048	0.768 ± 0.057	0.747 ± 0.071	-0.000 ± 0.004	**0.770 ± 0.024**
0.9, 0.6, 0.7	**0.716 ± 0.122**	0.175 ± 0.046	0.508 ± 0.071	**0.722 ± 0.059**	0.001 ± 0.001	0.492 ± 0.039
0.7, 0.3, 0.9	**0.750 ± 0.078**	0.542 ± 0.105	0.746 ± 0.058	**0.853 ± 0.048**	0.012 ± 0.029	0.763 ± 0.011
0.7, 0.3, 0.7	**0.681 ± 0.078**	0.458 ± 0.072	0.474 ± 0.040	**0.773 ± 0.060**	0.050 ± 0.050	0.518 ± 0.024
0.7, 0.6, 0.9	0.704 ± 0.139	0.028 ± 0.013	**0.736 ± 0.067**	0.726 ± 0.058	0.005 ± 0.004	**0.755 ± 0.032**
0.7, 0.6, 0.7	**0.540 ± 0.135**	0.028 ± 0.012	0.475 ± 0.087	**0.608 ± 0.037**	0.003 ± 0.002	0.499 ± 0.038

We observe similar patterns to the synthetic data sets with categorical attributes, that is, KEFRIN is the winner of small-size network and GDCRinTAN is its close follower. Similarly, in the medium-size networks we observe better performance of GDCRinTAN and the competitiveness of its obtained results.

6 Conclusion and Future Work

The current research reports our intermediate results towards developing an effective clustering method to recover communities in attributed networks. To achieve this objective, we relied on the data-driven model proposed in [8], and adopted the gradient descent optimization strategy, similar to what was done in [6], to optimized our clustering objective function. We called this version of our proposed method Gradient Descent Clustering in Attributed Network (GDCinAN). However, this straightforward method did not lead to promising results. To improve the quality of our results, we made two modifications: (a) we transformed the data spaces, using a softmax-scaled-dot-product, to a more cluster friendly data spaces; (b) we integrated the regularization term into our clustering objective function. We called this version as Gradient Descent Clustering with Regularization in Transformed Attributed Network (GDCRinTAN). With these two modifications we significantly improved the quality of the recovered clusters, compared to GDCinAN.

We evaluated and compared the performance of our proposed methods with three SOTA algorithms using five real-world and 320 synthetic networks. Our results showed that although GDCRinTAN was not the sole-winner, still it obtained comparable results and more so, in various experimental settings it outperformed the competitors. The current research is not without limits and those limits shape the directions of our future research.

1. adopting the L1 norm or elastic-net regularization,
2. adopting other distance functions, such as cosine distance function,
3. separating the regularization and the step size per each data source,
4. adopting different update rules,
5. proposing more sophisticated data-driven models to incorporated the topology of data, etc.

Acknowledgments. Support from the Basic Research Program of HSE University is gratefully acknowledged. This research was supported in part through computational resources of HPC facilities at HSE University.

References

1. Chunaev, P.: Community detection in node-attributed social networks: a survey. Comput. Sci. Rev. **37**, 100286 (2020)
2. Al-Andoli, M.N., Tan, S.C., Cheah, W.P., Tan, S.Y.: A review on community detection in large complex networks from conventional to deep learning methods: a call for the use of parallel meta-heuristic algorithms. IEEE Access **9**, 96 501–96 527 (2021)
3. Rostami, M., Oussalah, M., Berahmand, K., Farrahi, V.: Community detection algorithms in healthcare applications: a systematic review. IEEE Access (2023)
4. Cai, J., et al.: A new community detection method for simplified networks by combining structure and attribute information. Expert Syst. Appl. **246**, 123103 (2024)
5. Morand, J., Yip, S., Velegrakis, Y., Lattanzi, G., Potestio, R., Tubiana, L.: Quality assessment and community detection methods for anonymized mobility data in the Italian Covid context. Sci. Rep. **14**(1), 4636 (2024)
6. Shalileh, S.: An effective partitional crisp clustering method using gradient descent approach. Mathematics **11**(12), 2617 (2023)
7. Shalileh, S., Mirkin, B.: Community detection in feature-rich networks using gradient descent approach. In: Cherifi, H., Rocha, L.M., Cherifi, C., Donduran, M. (eds.) Complex Networks & Their Applications XII. COMPLEX NETWORKS 2023. Studies in Computational Intelligence, vol. 1142, pp. 185–196. Springer, Cham (2023). https://doi.org/10.1007/978-3-031-53499-7_15
8. Shalileh, S., Mirkin, B.: Summable and nonsummable data-driven models for community detection in feature-rich networks. Soc. Netw. Anal. Mining **11**, 1–23 (2021)
9. Wilson, D.R., Martinez, T.R.: The general inefficiency of batch training for gradient descent learning. Neural Netw. **16**(10), 1429–1451 (2003)
10. Bradbury, J., et al.: JAX: composable transformations of Python+NumPy programs (2018). http://github.com/google/jax
11. Müller, E.: Graph clustering with graph neural networks. J. Mach. Learn. Res. **24**, 1–21 (2023)
12. Citraro, S., Rossetti, G.: Identifying and exploiting homogeneous communities in labeled networks. Appl. Netw. Sci. **5**(1), 1–20 (2020)
13. Shalileh, S., Mirkin, B.: Community partitioning over feature-rich networks using an extended k-means method. Entropy **24**(5), 626 (2022)

14. Hubert, L., Arabie, P.: Comparing partitions. J. Classif. **2**, 193–218 (1985)
15. Larremore, D., Clauset, A., Buckee, C.O.: A network approach to analyzing highly recombinant malaria parasite genes. PLoS Comput. Biol. **9**(10), e1003268 (2013)
16. Lazega, E.: The Collegial Phenomenon: The Social Mechanisms of Cooperation Among Peers in a Corporate Law Partnership, 1st edn. GB: Oxford University Press, Oxford (2001)
17. Snijders, T.: Lawyers data set. Siena (2001). https://www.stats.ox.ac.uk/~snijders/siena/
18. Bojchevski, A., Günnemanz, S.: Bayesian robust attributed graph clustering: joint learning of partial anomalies and group structure. In: Thirty-Second AAAI Conference on Artificial Intelligence, pp. 1–10, California, USA. AAAI Press (2018)
19. Cross, R., Parker, A.: The Hidden Power of Social Networks: Understanding How Work Really Gets Done in Organizations, 1st edn. Harvard Business Press, USA (2004)
20. Jia, C., Li, Y., Carson, M., Wang, X., Yu, J.: Node attribute-enhanced community detection in complex networks. Sci. Rep. **7**(1), 1–15 (2017)
21. Shalileh, S., Mirkin, B.: Least-squares community extraction in feature-rich networks using similarity data. PLoS ONE **16**(7), e0254377 (2021)

OutlineGen: Multi-lingual Outline Generation for Encyclopedic Text in Low Resource Languages

Shivansh Subramanian[1(✉)], Dhaval Taunk[1], Manish Gupta[1,2], and Vasudeva Varma[1]

[1] IIIT Hyderabad, Hyderabad, India
{shivansh.s,dhaval.taunk}@research.iiit.ac.in,
{manish.gupta,vv}@iiit.ac.in
[2] Microsoft, New Delhi, India

Abstract. Lack of encyclopedic text contributors, especially on Wikipedia, makes automated text generation for low resource (LR) languages a critical problem. A step towards enabling this is to generate the structural outline of Wikipedia pages for LR languages. Hence, in this work, we propose and study OutlineGen, the problem of generating the outline of Wikipedia pages for LR languages using minimal information in the form of entity name, language and domain. The OutlineGen task is challenging because even within a (language, domain) pair, the outlines vary a lot across entities. Further, given the diversity of Wikipedia editors and audience, the outlines are not consistent across languages. First, we create a dataset, WikiOutlines, which contains Wikipedia section outlines from ∼166K Wikipedia pages across 8 domains and 10 languages. Then, we investigate the effectiveness of non-neural weighted finite state automata as well as Transformer-based methods for this task. We make the code and data publicly available.

Keywords: OutlineGen · WikiOutlines · deep learning · multi-lingual generation, Wikipedia text generation · low resource natural language generation

1 Introduction

Wikipedia has been the most popular source of factual and neutral encyclopedic information for millions of users. Although English Wikipedia is rich with ∼7M articles, number of Wikipedia pages in nine low resource (LR) languages which we consider in this work add to ∼100K. Unfortunately, recent efforts towards enriching LR Wikipedia over the years have also not been as encouraging as for English [25]. Hence, automated text generation for low-resource Wikipedia is critical.

Recently, some studies [8,25,26] have focused on generating text corresponding to a specific section for LR Wikipedia. However, to generate an entire

L. M. Aiello et al. (Eds.): ASONAM 2024, LNCS 15212, pp. 149–159, 2025.
https://doi.org/10.1007/978-3-031-78538-2_13

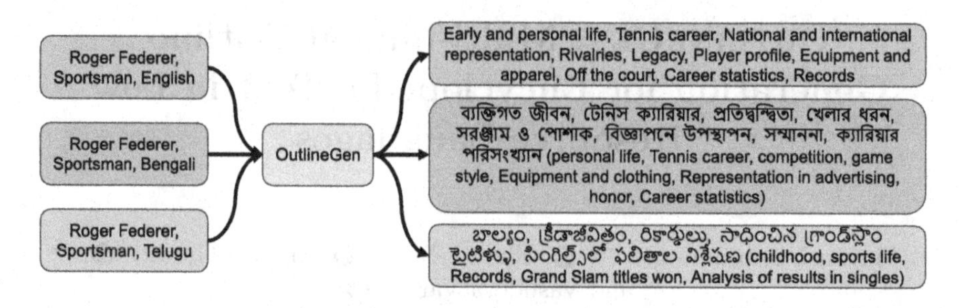

Fig. 1. OUTLINEGEN examples: Generating outlines for the "Roger Federer" entity (which belongs to the sportsmen domain) for English, Bengali and Telugu Wikipedia pages.

Wikipedia page, it is important to first generate the structural outline and then fill the sections with LR language text using these existing methods. In Table 1, we show percentage of articles with outline same as most frequently occurring article outline per (language, domain) for 8 domains and 10 languages. We observe that, on average, only 26.6% of all articles follow the templatized outline. Thus, we need to design a method which generates the Wikipedia section outline by conditioning on (entity, language, domain) triple. Translating the outline from corresponding English Wikipedia page is not effective because (1) several LR pages on Wikipedia do not have equivalent pages on English Wikipedia, and (2) LR Wikipedia pages are written by LR language editors for LR language users and hence their outlines differ significantly from outlines for corresponding Wikipedia pages (if they exist).

Hence, in this paper, we propose the task of Outline Generation, OUTLINEGEN, for Wikipedia articles, which is a novel task to generate Wikipedia-styled outlines given an article's (entity, language, domain) triple. Since our goal is to

Table 1. Percentage of articles with outline same as the most-frequent outline of (language, domain) pair

	hi	mr	bn	or	ta	en	ml	pa	kn	te	Avg
politicians	53.44	63.70	36.55	35.25	32.40	9.36	20.50	27.94	17.74	13.00	30.99
cities	17.46	49.82	15.46	22.76	19.98	18.00	37.00	56.08	24.22	10.69	27.15
books	74.09	40.54	17.76	31.03	12.58	10.63	29.73	32.69	30.48	7.91	28.74
writers	33.91	36.77	15.67	15.25	11.91	7.26	14.96	16.66	11.70	8.29	17.24
companies	27.96	50.82	26.51	36.84	25.00	28.75	26.22	45.65	15.57	20.56	30.39
sportsmen	44.23	69.18	22.63	22.79	39.33	14.69	22.73	39.15	16.21	10.47	30.14
films	17.38	20.81	19.95	38.37	17.13	21.80	17.11	27.99	40.90	30.27	25.17
animals	34.96	22.53	18.87	18.31	28.47	12.41	38.41	29.00	16.83	12.68	23.25
Avg	37.9	44.3	21.7	27.6	23.3	15.4	25.8	34.4	21.7	14.2	26.6

Table 2. #samples per domain per language in WIKIOUTLINES

	hi	mr	bn	or	ta	en	ml	pa	kn	te	Total
politicians	6,617	3,815	8,071	1,336	5,885	566	3,405	1,589	699	1,808	33,791
cities	1,048	827	854	268	851	3,550	554	526	256	290	9,024
books	1,428	148	805	87	1,988	762	740	468	105	215	6,746
writers	3,474	1,882	3,605	564	3,005	758	3,475	3,320	1,128	1,339	22,550
companies	683	366	679	38	644	4,546	431	138	212	180	7,917
sportsmen	9,476	11,556	13,154	408	9,808	177	2,583	2,327	660	640	50,789
films	4,959	1,033	3,655	920	6,504	1,165	3,934	618	1,704	3,621	28,113
animals	472	395	1,261	142	1,556	427	2,317	200	315	205	7,290
Total	28,157	20,022	32,084	3,763	30,241	11,951	17,439	9,186	5,079	8,298	166,220

generate Wikipedia outlines for entities where no Wikipedia page already exists, we take minimal inputs (entity, language, domain) for the task. Figure 1 shows examples of OUTLINEGEN task for the "Roger Federer" entity (which belongs to the sportsmen domain) for English, Bengali and Telugu Wikipedia pages. These outlines could help human editors to plan the article content better. These outlines could also help improve the quality of the automatically generated text (using methods like [2,14,23,25]) and hence reduce human post-editing efforts.

Lack of availability of a relevant dataset makes the OUTLINEGEN task further challenging. Hence, as part of this work, we contribute a novel dataset, WIKIOUTLINES, which contains Wikipedia section outlines from ~166K Wikipedia pages across 8 domains and 10 languages. The domains include politicians, cities, books, writers, companies, sportsmen, films and animals. Languages include Hindi (hi), Marathi (mr), Bengali (bn), Oriya (or), Tamil (ta), English (en), Malayalam (ml), Punjabi (pa), Kannada (kn) and Telugu (te). The dataset can serve as a great benchmark for the OUTLINEGEN task.

Overall, we make the following contributions in this paper: (1) We define and motivate the need for the novel OUTLINEGEN task, where the input is minimal (entity, target language, and domain). The output is the wikipedia-style outline. (2) We contribute a novel dataset, WIKIOUTLINES, with outlines for ~166K articles across 8 domains and 10 languages. (3) We model OUTLINEGEN using Weighted Finite State Automata (WFSA) and multi-lingual Transformer encoder-decoder models. (4) Our experiments show that mT5 [28] outperforms mBART [15] as well as a WFSA-based solution for the outline generation task.

2 Related Work

Wikipedia Text Generation: Initial efforts in the fact-to-text (F2T) line of work focused on generating short text, typically the first sentence of Wikipedia pages using structured fact tuples. Seq-2-seq neural methods [12,17], such as LSTM-based [19,24,27] and pretrained Transformer-based [21] like BART [13] and T5 [20], have been widely adopted for F2T tasks. Most of the previous efforts on F2T focused on English F2T only. Recently, the Cross-lingual F2T (XF2T) problem was proposed in [1] and [22]. Besides generating short Wikipedia text, there have also been efforts to generate Wikipedia articles by summarizing long sequences [7–9,14,26]. For all of these datasets, the generated text is either the full Wikipedia article or text for a specific section [9]. Most of these studies [7, 9,14] have been done on English only. Compared to all of these pieces of work which have focused on Wikipedia article text generation, the focus of the current paper is on generating outlines.

Wikipedia Outline Generation: Outline Generation for documents was first proposed in [29], where the goal was to identify the boundaries for sections and to generate the required section heading for the specified section. They proposed a model based on LSTMs [10] that performs both tasks simultaneously. Recently, Maheshwari et al. [16] proposed the related task of creating a Table of Contents (or Outline) for long documents like contracts, financial documents etc. given the section distinctions. Both of these efforts focused on English only. Outline generation for multiple low resource languages brings a different set of challenges which we tackle in this paper.

Multi-lingual Generation: Popular multi-lingual and cross-lingual natural language generation (NLG) tasks include machine translation [5], question generation [6,18], blog title generation [3], summarization [11,30], and style transfer [4]. In this work, we propose the novel multilingual Wikipedia outline generation task and propose strong baseline solutions for the task.

3 WikiOutlines: Data Collection, Pre-processing and Analysis

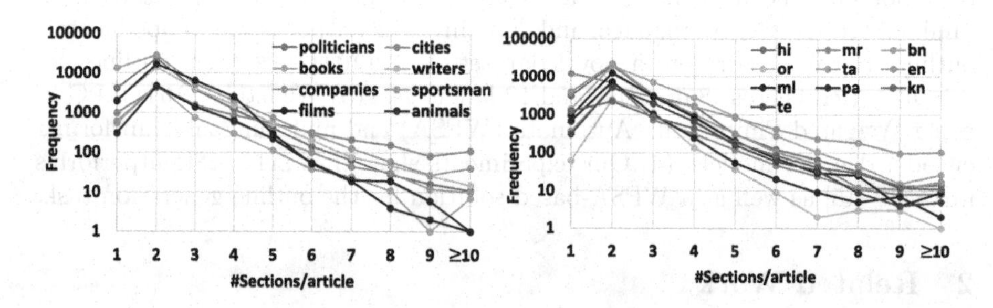

Fig. 2. Distribution of number of sections across various domains and languages in the WikiOutlines dataset

Data Collection and Pre-processing: WIKIOUTLINES contains Wikipedia sections related to eight distinct domains spanning across ten languages. To begin, we utilize the Wikidata API[1] to initially narrow down the domains of interest. Then, we retrieve entities that have corresponding Wikipedia pages in our chosen languages. Afterward, we use Wikipedia's language-specific 20221201 XML dumps, to acquire the Wikipedia pages of the filtered entities. The text in Wikipedia pages follows a standardized structure, with sections and subsections. We extract these sections and subsections from the text. Overall, each sample in the dataset consists of the domain, language, and section title. This dataset is then split into train, validation, and test in the 80:10:10 ratio, stratified by domain and language. We make these standard splits publicly available as part of the dataset[2].

Fig. 3. Word clouds of most frequent Wikipedia section titles per domain. Each word cloud contains titles across all languages. Section titles for one language are shown using a single color. Font size indicates relative frequency.

Data Analysis: Table 2 shows the overall distribution of the ~166K samples in the dataset across all domains and languages. The number of samples differ significantly across various (domain, language) pairs. Sportsmen and politicians have largest number of samples while books and animals have lowest number of samples. Oriya and Kannada have lowest number of samples from language perspective while Bengali and Tamil are the richest languages. Figure 2 shows the distribution of number of sections across various domains and languages in the WIKIOUTLINES dataset. Notice that the y-axis is drawn in log scale. The

[1] https://query.wikidata.org/.
[2] https://github.com/AurumnPegasus/OutlineGen.

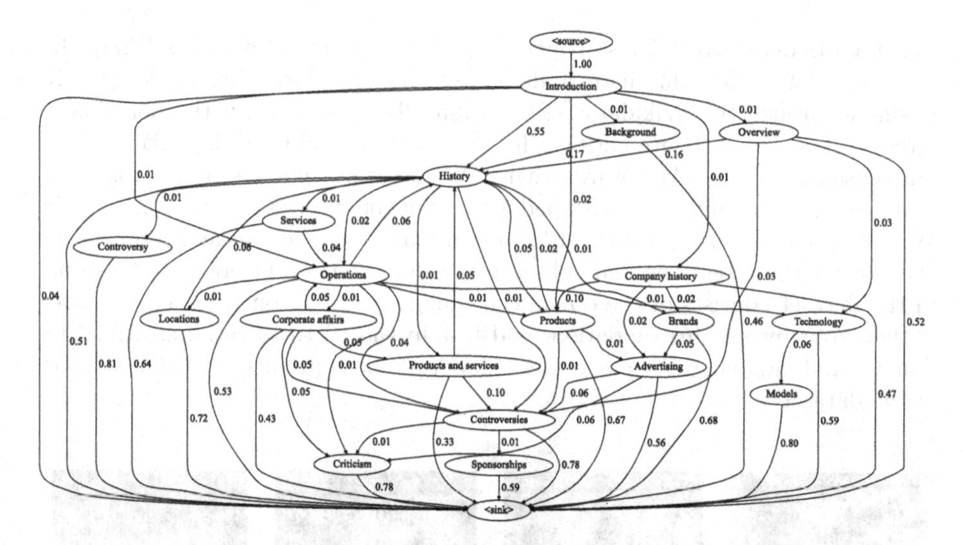

Fig. 4. Examples of generated weighted finite state automata for (en, companies) where section-titles are the nodes, and transition probability is written on the edges.

figures show that for every language and every domain, most samples have 2 sections, except for Marathi where most samples have just 1 section. Amongst the languages, the distribution is flatest for English, where the number of samples with ≥10 sections is the highest. Amongst the domains, cities has a similar behavior.

Finally, we show word clouds of the most frequent Wikipedia section titles for each of the eight domains in Fig. 3. Each word cloud contains the five most frequent titles per language. Section titles for one language are shown using a single color. Font size indicates relative frequency. The word clouds show the variety of section titles per (language, domain) pair.

4 Approaches for OutlineGen Task

Weighted Finite State Automata (WFSA): Table 1 shows that many articles share the same outline. These article outlines are often specific for a language over a particular domain, and the section transition patterns can potentially be found via simple statistical models instead of large generative ones. Hence, instead of defining static outlines based simply on frequency, we learn a weighted finite state automata for all articles belonging to a (language, domain) pair. The source node for the WFSA is ⟨source⟩, and the sink node is represented by ⟨sink⟩. The nodes between the source and the sink contain the section titles, and the transition probability from node A to node B is the conditional probability of section title B following section title A in an article outline. Figure 4 shows an example of WFSA learned for (en, companies) pair. These are drawn using top 20 most frequent section titles as nodes. Also, only the edges with weight more

Table 3. Comparison of WFSA and Transformer-based methods for multi-lingual outline generation.

	XLM-Score	BLEU	METEOR	ROUGE-L
WFSA (section-level)	70.0	45.0	37.1	56.8
WFSA (word-level)	69.1	43.4	36.1	55.9
mBART	70.2	39.1	31.9	52.2
mT5	**76.2**	**48.5**	**40.3**	**59.4**

than 0.005 are shown. WFSA involves two hyper-parameters: (i) a *beam-size* (samples from top-k instead of choosing the most probable next state), and (ii) *token-level* (word or section-title level WFSA).

The WFSA is used for inference as follows. We start from ⟨source⟩ state and select *beam-size* number of next possible states. We base our selection either greedily (selecting the most probable next states) or by sampling them from a probability distribution over the next states. We repeat these instructions recursively (in a breadth-first manner), maintaining the visited part of the outline in a queue and the total accumulated transition probability. Once we reach the ⟨sink⟩ state, we terminate the recursion and store the generated outline with the geometric mean of transition probabilities signifying the probability of that outline occurring. We terminate when the breadth-first search queue is empty. Outline with the highest probability is selected as the output. Of course, we ensure that the generated outline does not have repeated section titles.

Multi-lingual Transformer Encoder-Decoder Generative Models: WFSA based methods are not entity-specific. This restricts them to predict the same outline for all entities belonging to the same (language, domain) pair. Hence, we also experiment with popular multi-lingual Transformer encoder-decoder generative models like mT5 [28] and mBART [15]. The language and domain are passed as input with a separator token. The models are fine-tuned to generate outlines. The model is now supposed to automatically decide the number of sections in the outline and the actual section titles in the outline as well.

5 Experiments and Results

We trained both WFSA as well as Transformer-based models using training data and tuned hyper-parameters on the validation set. For WFSA, we found beam size = 4 to provide best result on validation set. For mT5 and mBART, we trained using AdamW optimizer for 10 epochs on a machine with 4 NVIDIA V100 GPUs. We used a batch size of 8, a learning rate of 2e−5, and a beam size of 3.

Table 4. XLM-Score, BLEU, ROUGE-L and METEOR scores for mT5 across various (language, domain) pairs.

	en	mr	hi	kn	ta	bn	pa	te	ml	or	Avg
companies	80.4	80.7	64.8	69.1	82.2	73.4	92.2	72.6	69.8	80.4	76.6
writers	73.4	70.7	71.3	63.3	75.2	66.1	86.9	68.4	72.2	84.0	73.2
cities	77.0	72.3	62.9	66.0	78.6	66.9	92.0	63.5	76.1	92.0	74.7
politicians	71.6	81.8	79.2	65.4	81.0	75.2	87.5	69.2	70.5	88.4	77.0
books	71.6	73.3	87.9	70.1	78.9	69.4	89.9	63.4	72.8	83.3	76.1
films	80.7	76.1	72.7	81.1	80.3	71.3	91.1	76.5	72.5	91.8	79.4
animals	71.9	62.0	67.0	69.0	80.9	68.9	88.9	68.2	78.0	81.4	73.6
sportsmen	81.1	88.2	83.0	66.7	86.1	74.2	90.8	64.2	75.0	83.5	79.3
Avg	75.9	75.6	73.6	68.9	80.4	70.7	89.9	68.2	73.4	85.6	76.2
	en	mr	hi	kn	ta	bn	pa	te	ml	or	Avg
companies	52.8	66.7	43.8	44.1	57.9	57.0	67.2	45.6	47.5	37.6	52.0
writers	31.5	46.2	54.3	29.6	37.1	40.6	45.5	33.0	50.2	41.1	40.9
cities	39.4	48.0	28.9	33.3	48.0	35.0	67.7	26.8	60.5	62.4	45.0
politicians	35.3	68.3	67.5	31.9	53.0	58.2	50.1	37.6	46.6	51.9	50.0
books	38.3	54.2	81.5	46.4	48.3	44.4	58.0	38.2	53.9	49.2	51.2
films	52.6	57.8	40.2	64.9	52.0	49.2	62.0	56.8	52.6	55.5	54.4
animals	32.0	37.4	37.3	35.6	54.9	38.6	50.9	44.7	59.3	31.9	42.3
sportsmen	41.9	79.2	72.0	31.7	62.4	45.5	62.9	32.8	53.2	44.4	52.6
Avg	40.5	57.2	53.2	39.7	51.7	46.0	58.1	39.4	53.0	46.7	48.5
	en	mr	hi	kn	ta	bn	pa	te	ml	or	Avg
companies	64.9	80.5	54.3	58.1	64.4	62.6	69.3	57.0	54.6	45.2	61.1
writers	44.1	62.5	64.0	44.8	46.8	51.0	52.6	44.9	56.8	50.0	51.7
cities	59.9	68.5	43.7	48.6	55.9	46.7	73.2	44.9	66.7	77.3	58.5
politicians	44.3	80.9	74.3	52.1	59.8	66.8	57.6	52.0	54.4	64.5	60.7
books	51.1	72.8	84.9	64.3	55.7	51.2	64.4	46.8	58.5	56.1	60.6
films	67.1	66.7	59.0	75.8	57.7	59.1	65.4	63.9	59.6	70.7	64.5
animals	49.8	55.8	47.5	46.7	59.8	50.0	58.1	51.6	65.2	44.0	52.9
sportsmen	73.5	86.7	77.5	54.4	71.9	59.2	68.9	45.7	60.1	53.5	65.1
Avg	56.8	71.8	63.2	55.6	59.0	55.8	63.7	50.9	59.5	57.7	59.4
	en	mr	hi	kn	ta	bn	pa	te	ml	or	Avg
companies	49.3	38.6	35.0	39.9	47.9	45.2	52.8	39.5	32.9	24.9	40.6
writers	28.5	33.5	45.1	26.0	29.5	31.7	32.5	30.6	37.6	29.9	32.5
cities	46.4	30.6	34.5	23.4	37.2	31.1	58.1	25.7	49.8	70.9	40.8
politicians	26.6	39.3	59.0	26.7	42.5	52.7	39.6	36.7	34.3	53.4	41.1
books	40.1	31.7	74.1	28.2	37.0	31.4	46.7	24.1	39.8	37.9	39.1
films	52.3	53.0	54.1	42.8	36.5	36.2	47.2	45.9	37.2	68.4	47.4
animals	43.9	27.1	39.5	25.7	41.7	29.6	41.3	32.0	48.1	22.6	35.1
sportsmen	63.8	46.6	62.9	31.5	58.6	42.8	54.0	25.7	39.6	34.3	46.0
Avg	43.9	37.6	50.5	30.5	41.4	37.6	46.5	32.5	39.9	42.8	40.3

Table 5. Generated outlines using our best method

Entity Name	Language	Domain	Ground-truth outline	Generated outline
Takin	en	animals	Introduction, Appearance, Distribution and habitat, Behaviour and ecology	Introduction, Taxonomy, Distribution and habitat, Behaviour and ecology
Dresden	en	cities	Introduction, History, Geography, Governance, Culture, Economy, Education and science, Notable people	Introduction, History, Geography, Demographics, Culture, Notable people
इन्साफ का मन्दिर (1969 फ़िल्म)	hi	films	परिचय, मुख्य कलाकार, परिणाम, नामांकन और पुरस्कार	परिचय, मुख्य कलाकार, परिणाम, नामांकन और पुरस्कार
ତତେ ଭଲ ପାଉଛି ବୋଲି	or	films	ପରିଚୟ, ଅଭିନୟ, ଗୀତ ଏବଂ ସଙ୍ଗୀତ	ପରିଚୟ, ଅଭିନୟ, ଗୀତ ଏବଂ ସଙ୍ଗୀତ
জনার্দন নবলে	bn	sportsman	ভূমিকা, আন্তর্জাতিক ক্রিকেট, মূল্যায়ন	ভূমিকা, প্রথম-শ্রেণীর ক্রিকেট, আন্তর্জাতিক ক্রিকেট

Table 3 shows the comparison of WFSA and Transformer-based methods for multi-lingual outline generation using popular natural language generation metrics like XLM-Score, BLEU, METEOR and ROUGE-L. We observe that (1) mT5 outperforms other methods by large margins across all metrics. (2) WFSA at section-title level leads to better results compared to WFSA at word level.

We show detailed results for our best model (mT5) at a (language, domain) level using the four metrics (XLM-Score, BLEU, METEOR, and ROUGE-L) in Table 4. From these tables, we observe that (1) The model performs best for films and sportsmen domains, and worst for writers and animals domains. This is largely justified because of the large number of training samples in films and sportsmen domains and low number of samples in animals domain. However, it is surprising that the model does not perform well on writers domain inspite of the large number of training samples. (2) The model performs best for Punjabi, and worst for Telugu and Kannada. The worse performance for Telugu and Kannada can perhaps be because of low number of training samples for those languages. Examples of generated outlines using our best method in Table 5 show that our method can generate reasonably usable outlines.

6 Conclusion

In this paper, we proposed OUTLINEGEN, which is the problem of multi-lingual outline generation. We contributed a novel dataset, WIKIOUTLINES, which covers 8 domains and 10 languages. We also investigated effectiveness of two kinds of models: WFSA and Transformer-based encoder-decoders like mT5 and mBART. We found that mT5 provided us good XLM-Score, BLEU, METEOR and ROUGE-L of 76.2, 48.5, 40.3 and 59.4 respectively. This shows that the system is practically usable to generate candidate outlines which can be refined further by human editors to bootstrap generation of Wikipedia pages in low-resource languages.

References

1. Abhishek, T., Sagare, S., Singh, B., Sharma, A., Gupta, M., Varma, V.: XAlign: cross-lingual fact-to-text alignment and generation for low-resource languages. In: WebConf, pp. 171–175 (2022)

2. Banerjee, S., Mitra, P.: WikiWrite: generating Wikipedia articles automatically. In: IJCAI, pp. 2740–2746 (2016)
3. Bhatt, S.M., Agarwal, S., Gurjar, O., Gupta, M., Shrivastava, M.: TourismNLG: a multi-lingual generative benchmark for the tourism domain. In: ECIR (2023, to appear)
4. Briakou, E., Lu, D., Zhang, K., Tetreault, J.: Olá, bonjour, salve! xformal: a benchmark for multilingual formality style transfer. In: NAACL-HLT, pp. 3199–3216 (2021)
5. Chi, Z., et al.: MT6: multilingual pretrained text-to-text transformer with translation pairs (2021)
6. Chi, Z., Dong, L., Wei, F., Wang, W., Mao, X.L., Huang, H.: Cross-lingual NLG via pre-training. In: AAAI, vol. 34, pp. 7570–7577 (2020)
7. Ghalandari, D.G., Hokamp, C., Glover, J., Ifrim, G., et al.: A large-scale multi-document summarization dataset from the Wikipedia current events portal. In: ACL, pp. 1302–1308 (2020)
8. Giannakopoulos, G., et al.: MultiLing 2015: multilingual summarization of single and multi-documents, on-line Fora, and call-center conversations. In: SIG on Discourse and Dialogue, pp. 270–274 (2015)
9. Hayashi, H., Budania, P., Wang, P., Ackerson, C., Neervannan, R., Neubig, G.: WikiAsp: a dataset for multi-domain aspect-based summarization. TACL **9**, 211–225 (2021)
10. Hochreiter, S., Schmidhuber, J.: Long short-term memory. Neural Comput. **9**(8), 1735–1780 (1997)
11. Jhaveri, N., Gupta, M., Varma, V.: clstk: the cross-lingual summarization tool-kit. In: WSDM, pp. 766–769 (2019)
12. Lebret, R., Grangier, D., Auli, M.: Neural text generation from structured data with application to the biography domain. In: EMNLP, pp. 1203–1213 (2016)
13. Lewis, M., et al.: BART: denoising sequence-to-sequence pre-training for natural language generation, translation, and comprehension. In: ACL, pp. 7871–7880 (2020)
14. Liu, P.J., et al.: Generating Wikipedia by summarizing long sequences. In: ICLR (2018)
15. Liu, Y., et al.: Multilingual denoising pre-training for neural machine translation. TACL **8**, 726–742 (2020)
16. Maheshwari, H., et al.: DynamicToC: persona-based table of contents for consumption of long documents. In: NAACL-HLT, pp. 5133–5143 (2022)
17. Mei, H., Bansal, M., Walter, M.R.: What to talk about and how? Selective gen. using LSTMs with coarse-to-fine alignment. In: NAACL-HLT, pp. 720–730 (2016)
18. Mitra, R., Jain, R., Veerubhotla, A.S., Gupta, M.: Zero-shot multi-lingual interrogative question generation for "people also ask" at Bing. In: KDD, pp. 3414–3422 (2021)
19. Nema, P., Shetty, S., Jain, P., Laha, A., Sankaranarayanan, K., Khapra, M.M.: Generating descriptions from structured data using a bifocal attention mechanism and gated orthogonalization. In: NAACL-HLT, pp. 1539–1550 (2018)
20. Raffel, C., et al.: Exploring the limits of transfer learning with a unified text-to-text transformer. JMLR **21**(140), 1–67 (2020)
21. Ribeiro, L.F.R., Schmitt, M., Schütze, H., Gurevych, I.: Investigating pretrained language models for graph-to-text generation (2021)
22. Sagare, S., Abhishek, T., Singh, B., Sharma, A., Gupta, M., Varma, V.: XF2T: cross-lingual fact-to-text generation for low-resource languages. In: INLG (2023)

23. Sauper, C., Barzilay, R.: Automatically generating Wikipedia articles: a structure-aware approach. In: ACL-IJCNLP, pp. 208–216 (2009)
24. Shahidi, H., Li, M., Lin, J.: Two birds, one stone: a simple, unified model for text generation from structured and unstructured data. In: ACL, pp. 3864–3870 (2020)
25. Taunk, D., Sagare, S., Patil, A., Subramanian, S., Gupta, M., Varma, V.: XWiki-Gen: cross-lingual summarization for encyclopedic text generation in low resource languages. In: WebConf, pp. 1703–1713 (2023)
26. Tikhonov, P., Malykh, V.: WikiMulti: a corpus for cross-lingual summarization. arXiv:2204.11104 (2022)
27. Vougiouklis, P., et al.: Neural Wikipedian: generating textual summaries from knowledge base triples. J. Web Semant. **52**, 1–15 (2018)
28. Xue, L., et al: mT5: a massively multilingual pre-trained text-to-text transformer. In: NAACL-HLT, pp. 483–498 (2021)
29. Zhang, R., Guo, J., Fan, Y., Lan, Y., Cheng, X.: Outline generation: understanding the inherent content structure of documents. In: SIGIR, pp. 745–754 (2019)
30. Zhu, J., et al.: NCLS: neural cross-lingual summarization. In: EMNLP-IJCNLP, pp. 3054–3064 (2019)

Leveraging Secure Social Media Crowdsourcing for Gathering Firsthand Account in Conflict Zones

Abanisenioluwa Orojo[1(✉)], Pranish Bhagat[1], John Wilburn[2], Michael Donahoo[1], and Nishant Vishwamitra[2]

[1] Baylor University, Waco, USA
{abanisenioluwa_orojo1,pranish_bhagat1,jeff_donahoo}@baylor.edu
[2] University of Texas at San Antonio, San Antonio, USA
john.wilburn@my.utsa.edu, nishant.vishwamitra@utsa.edu

Abstract. The Russo-Ukrainian conflict underscores challenges in obtaining reliable firsthand accounts. Traditional methods such as satellite imagery and journalism fall short due to limited access to zones. Secure social media platforms such as Telegram offer safer communication from conflict zones but lack effective message grouping, hindering insight collection. The proposed framework aims to enhance firsthand account gathering by crowdsourcing secure social media data. We gathered 250,000 Telegram messages on the conflict and developed a language model-based framework to identify contextual groupings. Evaluation reveals 477 new groupings from 13 news sources, enriching firsthand information. This research emphasizes the significance of secure social media crowdsourcing in conflict zones, paving the way for future advancements.

Keywords: Crowdsourcing · Social Media · Conflict Zones · Firsthand Accounts · Secure Communication · Data Analysis · Natural Language Processing

1 Introduction

The intricate fabric of today's global societal landscape is increasingly dominated by geopolitical turmoil that dramatically affects societies on multiple levels. In this age of rapidly escalating conflicts, exemplified by situations such as the 2022 Russian invasion of Ukraine [6], gathering precise firsthand accounts from these volatile areas has never been more critical. As complexities grow, the need for nuanced, timely insight into conflict zones increases. Traditionally, firsthand intelligence gathering relied on satellite imagery, on-the-ground journalism, and diplomatic sources [1]. These methods offer a limited lens to view unfolding realities, as turbulent situations in conflict zones prohibit public surveying. They often provide a delayed, fragmented picture, lacking the full spectrum of human experiences within conflict zones. The digital age has ushered in new data sources

L. M. Aiello et al. (Eds.): ASONAM 2024, LNCS 15212, pp. 160–170, 2025.
https://doi.org/10.1007/978-3-031-78538-2_14

to enrich understanding of conflict-ridden regions [14]. Secure social media like Telegram has emerged as a powerful tool [12], offering people to safely communicate first-hand accounts, resulting in a stream of real-time, user-generated content reflecting diverse viewpoints. Studies show these platforms are increasingly used in conflict zones to communicate vital firsthand information. Yet, with vast data comes a challenge: extracting meaningful insights from complex data streams [4]. A major challenge to extracting meaningful insights from secure social media datastreams is the lack of organized communication patterns in groups where such information from the ground is shared. Relevant firsthand account reports are often lost among other conversations. Techniques that can capture these multiple interweaving conversations are needed to extract meaningful interactions and key themes [13]. We propose a framework to crowdsource firsthand account information from secure social media in conflict zones. Our framework uses TF-IDF [10] to extract key phrases about a conflict. It then organizes messages using a novel conversations extraction algorithm based on RoBERTa [8], identifying firsthand reports and separating them from irrelevant conversations. We address the following key research questions: **RQ1:** How can secure social media communication from conflict zones be organized to effectively augment existing open-source information? **RQ2:** Can the integration of crowdsourced information from secure social media applications enhance the richness and comprehensiveness of open-source data? We focus on the Russia-Ukraine conflict[1] and collect a dataset of 250,000 Telegram (*i.e.*, a social media platform popularly used for secure communication) messages[2]. Our framework addresses RQ1 by identifying conversations pertaining to firsthand information, outperforming baselines by 158.13%. For RQ2, we capture 477 new conversational groups with key insights on specific events, such as bombing of Ukrainian cultural heritage sites and health facilities.[3]

2 Related Works

Our exploration encompasses four domains within conflict research: social media's role, conflict data collection methodologies, empirical data use, and human rights violations detection via social media. Social media significantly influences modern conflicts, serving as a critical communication tool. It impacts conflicts by lowering communication costs, accelerating information dissemination, prompting strategic adaptation, and offering new data [15]. However, it also fuels conflict through disinformation and extremist recruitment. Peacebuilding efforts are incorporating digital perspectives to counteract these challenges.

Key strategies prioritize accuracy, consistency, and replicability [13]. The inclusion of "non-events" and intercoder reliability are crucial, especially for

[1] Our framework can be generically applied to any conflict. We use the Russia-Ukraine conflict to demonstrate our approach.

[2] Our dataset will be made publicly available.

[3] Complete Figures, Tables, and additional References can be found at: https://github.com/AKOrojo/ASONAM2024-Secure-Crowdsourcing.git.

social media data interpretation. High-quality, disaggregated data is fundamental for understanding conflict processes [5]. The advent of 'big data' has necessitated rethinking data aggregation methods. Challenges in collecting high-quality conflict data persist due to observation difficulties and varied information sources. Social media platforms have proven effective in uncovering evidence of atrocities in conflict zones. However, challenges include information overload, data reliability, verification, admissibility in legal contexts, and potential under-reporting due to technology access disparities.

Table 1. Samples of Articles and Discussed Topics

Name	Topics Discussed	Date
Ukrainian Cultural Heritage Potential Impact Summary	impacts on Ukrainian cultural heritage, climate and gastronomy	May 2022
Kyiv Falling into Darkness	Instability and decreased light production in Kyiv	Nov 2023

3 Methodology

3.1 Data Collection

Our data collection process involves three streams: satellite articles, Telegram messages, and a social media conversations dataset. The Yale Humanitarian Research Lab's Conflict Observatory provided satellite articles [7] covering the Russia-Ukraine conflict from May 2022 to 2023 (Table 1). Telegram messages were collected via Lyzem [9], totaling 258,101 messages from 2020 to Aug 2023, with 150,083 in 2022 and 107,351 in 2023. We also utilized Cornell University's ConvoKit Social Media Conversations Dataset [2] to train our LM algorithm, creating a dataset of message pairs labeled for conversational context. This comprehensive approach combines traditional reporting methods with secure social media data to provide a holistic view of the conflict.

3.2 Overview of Our Approach

Our framework comprises five main components: (1) Data Acquisition, which ingests Yale Conflict Observatory articles and Telegram messages; (2) Preprocessing & Keyword Extraction, involving data cleaning and TF-IDF application for keyword extraction; (3) Message Searching, using extracted keywords to identify relevant Telegram messages; (4) Conversation Modeling & Thread Identification, employing an LM-based model to identify conversation threads in message clusters; and (5) Topic Analysis, performing analysis on identified threads to extract firsthand accounts and key discussion topics. This approach integrates traditional conflict reporting with social media data analysis to provide a comprehensive view of the conflict landscape.

3.3 Preprocessing and Keyword Extraction

The initial stages involve collecting, cleansing, and structuring data to facilitate further analysis. This process prepares the data by removing extraneous information and organizing it efficiently. The keyword extraction was performed on the Articles from the yale conflict observatory. In the Keyword Extraction stage, we utilize the Term Frequency-Inverse Document Frequency (TF-IDF) method to identify key terms in the articles (Fig. 1).

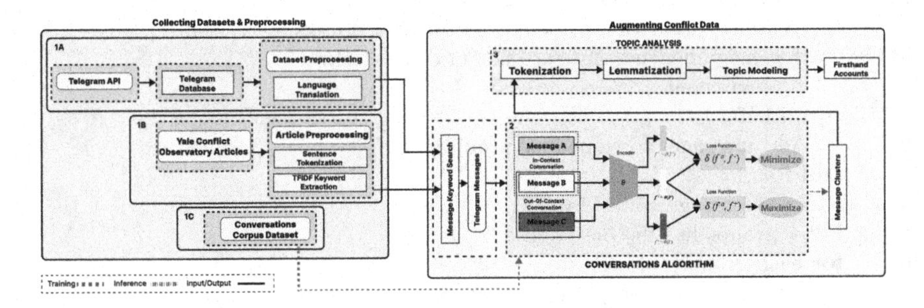

Fig. 1. Overview of our framework.

3.4 Conversation Thread Analysis

The Conversation Thread Analysis phase uses language models' natural language processing capabilities to analyze conflict communication. It starts with Message Search, using previously identified key phrases to find relevant Telegram messages. This targeted approach ensures collected data relates directly to the studied conflict and comes from active event discussions. After refining extracted keywords by removing special characters and duplicates, a curated set of distinct keywords retrieves a subset of 60,523 relevant messages from the Telegram dataset for detailed analysis.

Algorithm 1 modified from [16], depicts our conversations extraction process for the Telegram dataset subset. We use a fine-tuned RoBERTa-base model [3,8] for binary classification of sentence pairs as part of the same conversation thread. The model $f(s_{i1}, s_{i2}; \theta)$ is trained contrastively on sentence pairs $P = (s_{i1}, s_{i2})i = 1^n$ with labels $L = l_i i = 1^n$, where $l_i \in 0, 1$ indicates same-conversation membership. θ represents learnable model parameters. The final model achieved 74% accuracy on the classification task.

Our study optimizes parameters in a contrastive learning framework by minimizing Cross-Entropy loss, which evaluates binary classification accuracy by penalizing differences between predicted and actual labels. We use the AdamW optimizer, a variant of Adam, known for efficiently handling sparse gradients and adjusting learning rates based on gradient moments. Post-training, we generate token pairs from messages using a BertTokenizer. Our pre-trained model then

Algorithm 1. Message Conversation Grouping

Require: pre-trained model $\mathcal{M}$, input data set $D = \{d_1, d_2, ..., d_n\}$, minimum matches δ

1: Load pre-trained model $\mathcal{M}$
2: **for** each data d in D **do**
3: Load data into generic datastore DS
4: Initialize empty datastore DS_g for grouped messages
5: Initialize group id $g_{\text{id}} = 0$
6: **for** each pair of messages (m_i, m_j) in DS **do**
7: Extract embeddings for m_i and m_j
8: Generate prediction p_{ij} using $\mathcal{M}$ on embeddings of m_i and m_j
9: **if** p_{ij} signifies messages are not in context **then**
10: Increment $g_{\text{id}} = g_{\text{id}} + 1$
11: **end if**
12: Add messages to respective group in DS_g
13: **end for**
14: Format 'message_count' in DS_g
15: Filter groups in DS_g based on δ
16: **for** each group g in DS_g **do**
17: Save g to an appropriate output
18: **end for**
19: **end for**

predicts conversation membership for each pair by computing an output vector, transformed into probabilities via a softmax function. This approach ensures precise predictions and robust handling of diverse data inputs.

Where K is the number of possible classes (2: in the same conversation or not), we consider two sentences to be in the same conversation if the probability p_{ij} exceeds a chosen threshold θ. If p_{ij} indicates the messages are not in context, a new cluster is created by incrementing the cluster id $c_id = c_id + 1$. Messages are then added to their respective clusters in DF_c, our clustered messages datastore. This method dissects the vast Telegram message corpus into distinct conversations using refined keywords, akin to targeted crowdsourcing. Given the data volume and natural language complexity, this automated approach is more practical and effective than manual analysis.

We use a trained conversations detection model to filter and group messages into topically coherent conversation threads, iteratively refining our data organization. Our algorithm systematically examines each message's relationship with all others to determine conversation membership. Identified conversations are grouped, and the process repeats for each ungrouped message. After all messages are evaluated and grouped, we compare the hash of each group sorted by messages_id to identify and remove any duplicate groups. This LLM-driven approach ensures efficient and accurate conversation clustering, crucial for handling the complexity and volume of our data.

Our model accommodates non-transitive relationships between messages, reflecting natural conversational structures. Ambiguities in message clustering can lead to inconsistencies, as illustrated in this scenario with three messages (m1, m2, m3): (1) m1 and m2 are in the same conversation, (2) m2 and m3 are not, but (3) m1 and m3 are. To address such challenges, our algorithm allows message replication across groupings, ensuring no conflicts arise and enabling messages to exist in multiple groupings as context demands.

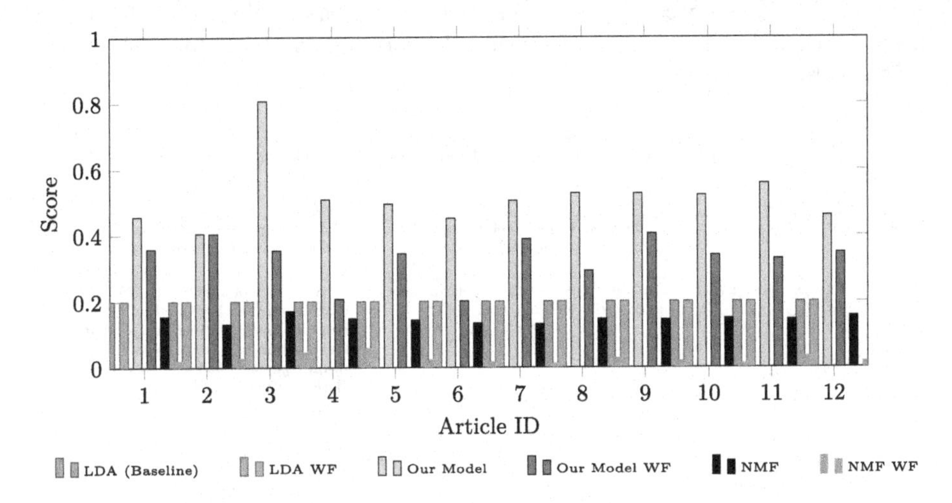

Fig. 2. Comparison of topic model scores for each article.

3.5 Clustering Conversations: Context vs. Topic

Our 'contextual topic clustering' approach combines context-based and topic-based clustering methods to overcome their individual limitations. We leverage the language model's classification for context clustering and reconstruct original conversation threads using message IDs and reply-to IDs for topic clustering. This method yields sub-corpora of related message groups connected by both context and topic. The resulting clusters contain messages from individuals discussing the same subject within the same conversation thread, ensuring both contextual relevance and semantic coherence. This approach provides a more comprehensive representation of conversations, capturing overall context and specific topics. It's valuable for applications like sentiment analysis, opinion mining, and trend detection, offering a granular understanding of the discourse within the dataset.

3.6 Augmenting Conflict Data

Our final process involves topic analysis on conversation clusters using Latent Dirichlet Allocation (LDA), a generative probabilistic model. For a corpus of D

documents, each document d with words w, and T topics with word distribution β_t, LDA finds: (1) topic distribution θ_d and (2) word topic assignment $z_{d,w}$. The generative process involves choosing $\theta_d \sim \text{Dirichlet}(\alpha)$ for each document, then for each word: $z_{d,w} \sim \text{Multinomial}(\theta_d)$ and $w_{d,w} \sim \text{Multinomial}(\beta_{z_{d,w}})$. Parameters α and β control topic and word mixtures. We apply a Coherence Model [11] to assess topic coherence, yielding a quantitative measure of semantic coherence. This process provides a structured understanding of conflict data, unveiling underlying themes and sentiments, forming the basis for our conclusions and recommendations.

Table 2. Evaluation of Framework & Articles

Top-Article Keyword	New Insights	Sample(s)
impact, climate, gastronomy	Findings on the extent of potential damage to Ukrainian cultural heritage	Ukraine **bombed** the museum and **civilians were targeted** and injured as well
vdv, defence, force, oblast, airborne	Exploring the causes and implications of the deteriorating situation in Kyiv	Drone footage of a tank of the 35th Ukrainian marine brigade **shelling Russian positions in close combat. Destruction of a large amount of** enemy equipment and manpower by Marines of the 36th Brigade

3.7 Implementation Details

We implement our LDA model using Python libraries NLTK, Gensim, and Pandas. The pipeline includes: text preprocessing (tokenization with NLTK's regular expression tokenizer and lemmatization with WordNet), converting text to bag-of-words representation using Gensim's dictionary class, training the LDA model (Gensim implementation with adjustable topic number), generating topics, and calculating coherence scores using Gensim's CoherenceModel. This score evaluates topic quality by measuring semantic similarity of high-scoring words within each topic. The implementation uses Gensim's internal handling of learning parameters like learning rate and epoch number, employing online stochastic inference for optimization and determining epochs based on model convergence.

4 Evaluation

This section assesses our framework's efficacy and validity on the Telegram dataset through both quantitative and qualitative analyses. We evaluate our model's conversation clustering performance using intrinsic dataset attributes, reconstructing conversations from message ID and reply-to ID. This method allows us to directly compare the true conversation structure with our model's predicted groupings, providing an accurate assessment of the model's performance. Our approach combines quantitative metrics with qualitative evaluation to ensure a comprehensive understanding of the framework's effectiveness.

Quantitative Analysis. In our initial evaluation, we compared three topic modeling techniques: (1) Latent Dirichlet Allocation (LDA), a standard topic modeling algorithm; (2) Our Custom Model, which uses NLTK for preprocessing and an LDA-based approach; and (3) Non-Negative Matrix Factorization (NMF), an alternative topic modeling method. This assessment focused on keyword choice for topic modeling and our model's performance relative to traditional techniques.

To assess the impact of our preprocessing and keyword extraction stages, we compared the performance of the three techniques with and without these initial framework phases, labeling results without these stages as "Without Framework" (WF). We evaluated performance based on each model's ability to derive meaningful topics from Telegram conversations, using topic coherence scores as the metric. Higher coherence scores indicate greater semantic similarity among top-ranking words in a topic, reflecting the model's proficiency in capturing meaningful topics. This comparison helped us understand the specific influence of our preprocessing and keyword extraction on topic analysis outcomes.

Figure 2 compares the three topic modeling techniques for each article, with and without our proposed framework. Our custom model, using NLTK for preprocessing and an LDA-based approach, achieved an average coherence score of 0.5219 across all articles, compared to 0.2 for standalone LDA and 0.1462 for NMF. Using the formula $\text{Improvement} = \frac{\text{Our Model Score} - \text{Baseline Score}}{\text{Baseline Score}} \times 100\%$, our model showed a 158.13% improvement over the standalone LDA baseline: $\frac{0.5219 - 0.2}{0.2} \times 100\% = 158.13\%$.

Our custom model's 158.13% improvement over standalone LDA underscores the effectiveness of our preprocessing techniques. The graph in Fig. 2 shows that excluding our framework's initial stages leads to comparable or diminished performance across all models. While standalone LDA sometimes lacked clarity and NMF showed inconsistencies, our NLTK-enhanced model excelled in providing nuanced insights into the Telegram dataset's geopolitical discourse. To optimize keyword count, we tested configurations ranging from 3 to 10 keywords, evaluating their impact using average coherence scores. This approach balanced comprehensive theme capture with avoiding redundancy. The results of this keyword count evaluation are presented in Table 3.

Table 3. Different Keyword Counts

Keywords	Average Score
3	0.4235
5	0.5219
10	0.4523

Table 4. Top Keywords per Article

Keywords
news, preservation, artifacts, restoration, und
light, darkness, instability, energy, soldier

Table 3 shows that using 4 keywords decreased the average coherence score by 15% (from 0.5219 to 0.4519) compared to 5 keywords, indicating missed crucial themes. Increasing beyond 5 keywords didn't significantly improve scores, with 6 keywords showing a marginal 2–3% decrease, suggesting potential noise introduction. The 5-keyword configuration emerged as optimal, achieving the highest average coherence score of 0.5219. This balance captured dataset nuances while maintaining thematic coherence and relevance, maximizing the effectiveness and clarity of our thematic analysis.

Qualitative Evaluation. Following our model's quantitative improvement, we conducted a qualitative evaluation to explore nuanced thematic interpretations within the dataset. We identified 'top topics' from conversation threads for each article, assigned by ID. This interpretative analysis, presented in Table 2, provides a multi-dimensional view of the dataset's subtler narratives and dominant discourses. Top keywords served as indicators of main narratives: terms like 'russian', 'ukraine', and 'war' highlighted geopolitical conflict; 'health', 'covid', and 'vaccine' indicated public health discussions; while 'looting', 'heritage', and 'damage' revealed concerns about cultural preservation amidst global unrest. This approach unraveled the complex thematic landscape within the conversations.

We extracted the top 5 keywords from topic modeling across each article's clusters (Table 4) for a more granular view of dominant narratives. Keywords like 'child', 'health', and 'covid' in Articles 5, 8, and 12 highlight the crisis's impact on children's well-being and public health. Terms such as 'looted', 'destroyed', and 'damage' in Articles 3, 4, and 10 underscore discussions about widespread destruction and cultural heritage issues. This combined quantitative and qualitative evaluation validates our framework's effectiveness and demonstrates its potential for extracting and analyzing meaningful topics from complex conversation datasets like Telegram, accurately identifying and interpreting conversation threads and their underlying themes.

5 Discussion

5.1 Beyond Ukraine and Future Works

The methodology and framework developed in this research, while tailored to the context of Ukraine, hold the potential for broad applicability in other regions and scenarios. The modular nature of our approach, which integrates data collection, message clustering, and contextual topic analysis, can be adapted to different datasets, languages, and cultural contexts. By refining search terms, adjusting the model parameters, or incorporating region-specific nuances, the approach can be generalized to study other conflict zones or areas of interest. Furthermore, the ethical considerations and biases identified in our research context provide valuable insights that can guide adaptations in other scenarios, ensuring rigorous and responsible research practices.

5.2 Limitations and Ethical Considerations

This research faces several limitations and ethical considerations. The use of specific search terms for data collection risks excluding relevant content, necessitating iterative refinement of queries and continuous data validation. Relying on Yale Humanitarian Research Lab's satellite articles, while reputable, may introduce inherent biases or focus areas that could influence result interpretation. Additionally, analyzing messages outside their original context risks misinterpretation or loss of nuanced meanings, highlighting the challenge of accurately understanding and presenting data in conflict communication research. Another limitation; Social media data may include misinformation or hate speech, partially mitigated by crowdsourcing also Keyword extraction sometimes yields irrelevant content, which could be addressed through human evaluation.

6 Conclusion

In this work, we introduced our framework for the crowdsourcing of secure social media conversations in conflict zones. We collected a dataset of Telegram posts to demonstrate the capability of our framework and draw new insights into the Russia-Ukraine conflict. Our framework outperforms baselines by 158.13% and captures 477 new conversational groups pertaining to key new insights on specific events in the ongoing conflict, such as the bombing of Ukrainian cultural heritage sites and health facilities. Our research showcases the utility of crowdsourcing firsthand account in conflict zones using secure social media conversations.

Acknowledgments. This research project and the preparation of this publication were funded in part by the NSF Grant No. 2245983.

References

1. College, N.W.: Types of intelligence collection - intelligence studies (2023). https://usnwc.libguides.com/intelligence/studies. Accessed 2 July 2023
2. Cornell University: Cornell ConvoKit: A Collection of Conversations from Wikipedia Talk Pages (2023). https://convokit.cornell.edu/documentation/awry.html. Accessed 11 July 2023
3. FacebookAI: roberta-base (2019). https://huggingface.co/FacebookAI/roberta-base
4. Ge, Q., Hao, M., Ding, F., et al.: Modelling armed conflict risk under climate change with machine learning and time-series data. Nat. Commun. **13**, 2839 (2022). https://doi.org/10.1038/s41467-022-30356-x
5. Gleditsch, K.S., Metternich, N.W., Ruggeri, A.: Data and progress in peace and conflict research. J. Peace Res. **51**(2), 301–314 (2014). http://www.jstor.org/stable/24557423
6. Institute, F.P.R.: Understanding Russia's invasion of Ukraine (2022). https://www.fpri.org/. Accessed 2 July 2023
7. Lab, Y.H.R.: Conflict observatory (2023). https://medicine.yale.edu/lab/khoshnood/. Accessed 10 July 2023

8. Liu, Y., et al.: Roberta: a robustly optimized BERT pretraining approach. arXiv preprint arXiv:1907.11692 (2019)
9. Lyzem: Lyzem - Privacy Friendly Search Engine (2023). https://lyzem.com/. Accessed 11 July 2023
10. Ramos, J.: Using TF-IDF to determine word relevance in document queries. In: Proceedings of the First Instructional Conference on Machine Learning, pp. 29–48. Citeseer (2003)
11. Rehurek, R.: Gensim: topic modelling for humans (2022). https://radimrehurek.com/gensim/models/coherencemodel.html. Accessed 13 June 2023
12. Research, C.P.: Telegram becomes a digital forefront in the conflict - news feeds from fighting zones (2023). https://blog.checkpoint.com/. Accessed 2 July 2023
13. Salehyan, I.: Best practices in the collection of conflict data. J. Peace Res. **52**(1), 105–109 (2015). http://www.jstor.org/stable/24557521
14. Tsovaltzi, D., Judele, R., Puhl, T., Weinberger, A.: Leveraging social networking sites for knowledge co-construction: positive effects of argumentation structure, but premature knowledge consolidation after individual preparation. Learn. Instr. **52**, 161–179 (2017). https://doi.org/10.1016/j.learninstruc.2017.06.004
15. Zeitzoff, T.: How social media is changing conflict. J. Conflict Resolut. **61**(9), 1970–1991 (2017). http://www.jstor.org/stable/26363973
16. Zhang, Y., Wang, Z., Shang, J.: ClusterLLM: large language models as a guide for text clustering. arXiv abs/2305.14871 (2023). https://doi.org/10.48550/arXiv.2305.14871

Applying the Ego Network Model to Cross-Target Stance Detection

Jack Tacchi[1,3](✉), Parisa Jamadi Khiabani[2], Arkaitz Zubiaga[2], Chiara Boldrini[1], and Andrea Passarella[1]

[1] Istituto di Informatica e Telematica, Consiglio Nazionale delle Ricerche, Pisa, Italy
jacktacchi@gmail.com

[2] School of Electronic Engineering and Computer Science, Queen Mary University of London, London, UK

[3] Scuola Normale Superiore, Pisa, Italy

Abstract. Understanding human interactions and social structures is an incredibly important task, especially in such an interconnected world. One task that facilitates this is Stance Detection, which predicts the opinion or attitude of a text towards a target entity. Traditionally, this has often been done mainly via the use of text-based approaches, however, recent work has produced a model (CT-TN) that leverages information about a user's social network to help predict their stance, outperforming certain cross-target text-based approaches. Unfortunately, the data required for such graph-based approaches is not always available. This paper proposes two novel tools for Stance Detection: the Ego Network Model (ENM) and the Signed Ego Network Model (SENM). These models are founded in anthropological and psychological studies and have been used within the context of social network analysis and related tasks (e.g., link prediction). Stance Detection predictions obtained using these features achieve a level of accuracy similar to the graph-based features used by CT-TN while requiring less and more easily obtainable data. In addition to this, the performances of the inner and outer circles of the ENM, representing stronger and weaker social ties, respectively are compared. Surprisingly, the outer circles, which contain more numerous but less intimate connections, are more useful for predicting stance.

Keywords: Stance Detection · Cross-Target · CT-TN · Ego Network · Signed Relationships · Social Media

1 Introduction

Humans have always been social animals. Everyday we interact countless times with one another and these interactions form the basis of our modern societies. It has even been argued that our ability to maintain larger social groups was the primary reason that the volume of our brains increased significantly in size [9].

J. Tacchi and P. J. Khiabani—Co-first authors with equal contribution and importance.

L. M. Aiello et al. (Eds.): ASONAM 2024, LNCS 15212, pp. 171–182, 2025.
https://doi.org/10.1007/978-3-031-78538-2_15

What's more, since the advent of the internet, humans have also been able to interact with each other regardless of geographical location. Because humans are now more interconnected than ever, understanding social connections and how they contribute to the spreading of ideas and opinions has never been more important.

The exponential expansion of social networks has introduced fresh obstacles in the realm of information retrieval [15]. One research task that deals directly with opinions on social networks is Stance Detection (SD), which aims to predict the stance of a given text towards a target entity [5]. Of course, being able to monitor the opinions of individuals or even overall trends in larger communities and populations can be extremely impactful. Especially, given its application to politics, where it can be used to quickly understand how people feel towards a given topic or to predict how they will vote.

SD has often viewed interactions in isolation, predicting a user's opinion towards a given entity purely based on what they have written (e.g. [27]). However, recent research has shown that considering an individual's surrounding social network can greatly improve the accuracy of SD [16], highlighting the influence of social connections on opinions. Specifically, a model called CT-TN (Cross-Target Text-Net), which uses predictions from a more traditional text-based model, RoBERTa [18], together with multiple network features from the X (formerly Twitter) social media platform: likes (a list of users whose posts have been liked by the target user), followers (the users who follow the target user) and friends (the users who are followed by the target user). This was done by creating an embedding for each feature and passing it through a classification model to obtain a prediction for each feature. The final prediction was then generated based on a majority vote of all the features (see Fig. 1). The CT-TN model outperformed other competitive models, such as CrossNET [28] and TGA-Net [1] in six different experimental conditions.

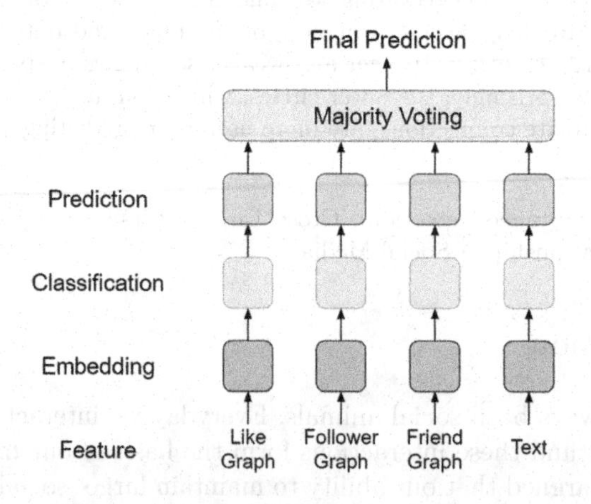

Fig. 1. Architecture of the CT-TN model.

However, the different network features that CT-TN requires are not always available. Indeed, the multiple different data sources required for CT-TN may be impossible or extremely costly to obtain in many situations. Therefore, it would be pertinent to investigate alternative approaches or features that are more parsimonious.

To this effect, in this work, we exploit models of users' personal networks, grounded on well-established findings in anthropology. In addition to providing solid quantitative models of the structure of users' social relationships, which can be obtained whenever users communicate publicly, thus minimising restrictions from the aforementioned shortcomings as much as possible. Indeed, given the aforementioned importance of understanding how humans communicate, it is unsurprising that this topic had been researched long before the internet. For instance, anthropological and psychological studies have found that the number of relationships that an individual is able to maintain is remarkably consistent across members of the same species [8]. The maximum maintainable group size is correlated with the proportional size of the neocortex part of said species' brain; strongly suggesting that a group's size is innately limited by the cognitive ability of the species within it [9]. What's more, these connections can be sorted by contact frequency into a series of concentric groups [10], with a smaller number of high-frequency connections at the centre and larger numbers of less-frequent connections towards the edges.

A representation of this structure, known as the Ego Network Model (ENM), places a target individual (the Ego, from which the model takes its name) at the centre and surrounds them by all of their connections (Alters) (see Fig. 2). For humans, the expected sizes of the concentric groups range from 5 (support clique), to 15 (sympathy group), then around 45–50 (affinity group) and finally 150 (active network) [10]. An additional inner circle of 1–2 connections has also been found in many online contexts [3]. Interestingly, these groups increase in size by a factor of around 3 each time, and this scaling factor has also been found for certain non-human animals, such as other primates and even birds [7].

The ENM has been studied not only offline but has also been used to reveal many novel insights about communications in online contexts [6]. For instance, the ENM has been used to discover differences between the behaviours of various types of users, including journalists [25] and members of different online communities [23]. It has also been shown that using ENM information can significantly improve "classical" tasks in Online Social Networks (OSNs), such as link prediction [26]. Moreover, recent extensions of ENM have been proposed to label links in the Ego Network with positive or negative signs, depending on whether the corresponding relationship between Ego and Alter displays a predominantly positive or negative sentiment [22]. This extension, called the Signed ENM (SENM) has been used to characterise, in detail, the prevailing polarity of social relationships in OSNs, in contrast to the offline world [24]. Therefore, the ENM and SENM could reasonably be expected to provide significant information pertinent to the task of SD. Indeed, Ego Networks go beyond providing mere structural information but can contribute semantically to a better knowledge of the underlying mechanisms of a network and, therefore, provide pertinent contextual insights for SD. In addition, the base ENM requires only a list of Ego-Alter pairs and the frequencies of their interaction; data which can be easily obtained just by monitoring public interactions.

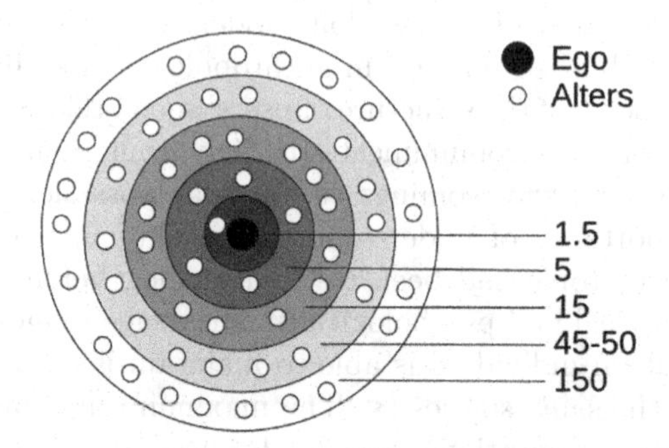

Fig. 2. The Ego Network Model, with the expected number of Alters of each circle (for humans).

1.1 Contributions

Given the potential restrictions of data availability discussed above, it is important to have a toolbox of diverse approaches to ensure that researchers have a viable means of tackling as many situations as possible. Indeed, the authors of the CT-TN model specifically advocate for exploring further network features to enhance SD [16]. In response to this, the current work proposes two features from adjacent network research areas for use within the task of SD: the ENM and the SENM. Furthermore, this paper demonstrates that these features perform better than a text-only approach, RoBERTa, and similarly, although slightly worse, than the cutting-edge CT-TN model. Thus, demonstrating that these novel features are viable alternatives for SD when not all the required features for CT-TN can be obtained. Finally, a comparison of the performances between the inner and outer circles provides evidence that a user's less intimate but more numerous outer connections have more of an impact on their stance.

2 Background

2.1 Stance Detection

The focus of most SD studies is to classify target dependency in one of four main ways: Target-Specific, Multi-Related Targets, Target-Independent and Cross-Target [2]. For this paper, we primarily concentrate on Cross-Target Stance Detection (CTSD), which is when a model is trained using data for one target entity (source) and then tested on a different, although related, target entity (destination). For example, a model trained using texts containing opinions towards

Joe Biden could be used to predict the stance of texts concerning another politician, such as Donald Trump or Bernie Sanders.

Approaches to CTSD, as well as to SD more generally, differ based on the text's context and the particular relationship being discussed. On social media platforms such as X, there's often a focus on discerning the author's stance (supportive, opposing, or neutral) towards a specific proposition or target [20]. Recent advances in SD encompass a range of linguistic features, such as word or character n-grams, dependency parse trees, and lexicons [21]. Moreover, there has been a shift towards end-to-end neural network methods that independently learn topics and opinions, integrating them via mechanisms such as memory networks or bidirectional conditional Long Short-Term Memory models (LSTMs) [4].

Past studies have primarily relied on the text of a post to gauge its stance, neglecting the valuable insights that other features within social media platforms could offer. However, the performance of the aforementioned CT-TN model demonstrates the importance of considering structural features of the surrounding social network. Thus, knowledge about a user's connections can reveal important insights about the user themself and, therefore, about the texts they author.

2.2 The Ego Network Model

As mentioned in Sect. 1, the ENM views a social network from the point of view of an individual Ego and organises their Alters around them based on their contact frequency, while the SENM extends this by adding the addition of signed relationships. Specifically, these signed relationships are obtained by analysing the sentiment of the interactions between each Ego and Alter, thus obtaining a list of sentiments for each relationship, which can then be used to infer an overall sentiment for the relationship as a whole. It has been observed that many different types of relationships start to have negative effects on the people involved once the ratio of negative interactions generated by that relationship passes a certain ratio: around 17% [12]. Therefore, this threshold is used to infer the sentiments of relationships based on their interactions.

While the Signed Ego Network provides an additional layer of information, it does require text for each interaction in a dataset. By contrast, the unsigned Ego Network requires only the frequency of interactions between each pair of users, without the need for any text.

3 Methodology

3.1 Performing Cross-Target Stance Detection

Feature Embeddings. In order to compute SD predictions, the data first need to be transformed into representations that are readable by a prediction model. For this, node2vec [13] was applied to each of the previously established graph-based features (likes, followers, friends) as well as the novel Ego Networks and Signed Ego Networks, which, although they are converted into the same vector-space representation, can be better thought of as proxy measures of the way humans function socially. node2vec is an unsupervised Deep Learning algorithm that uses a flexible, biased, random walk procedure to explore networks. The visited nodes can then be transformed into a vector space representation using a variety of methods, such as skip-grams or a continuous bag-of-words [13]. This is similar to how the word2vec algorithm [19] treats words (nodes) and sentences (walks). In addition to the graph-based features, text-based predictions were also used. These were generated using RoBERTa [19], an incredibly well-performing pretrained model that is used for many different natural language processing tasks. RoBERTa maps every token in a sentence to a vector representation in a continuous space. As mentioned in Sect. 1, the CT-TN model takes the predictions of each of these features, RoBERTa, likes, followers and friends, and obtains a final prediction using majority voting, where each feature's prediction acts as a vote for either "FAVOR" or "AGAINST". This allows for a thorough analysis of both textual and social network information, providing valuable insights for CTSD.

Aside from the aforementioned features, this paper also investigates two novel graph-based features: Ego Networks and Signed Ego Networks. These are also converted to a vector space representation using node2vec. Further details on how the Ego Networks are obtained are explained in Sect. 3.2.

Model Hyperparameters. Each of the features was used to train a neural network model with two hidden layers for classification task. For the text-based embedding, this was done using RoBERTa, a batch size of 128, a dropout of 0.2, a learning rate of 3e−5 (AdamW), and 40 epochs. For the graph-based embeddings, this was done using node2vec, with the same batch size and dropout, a learning rate of 1e−2 (SGD), and 100 epochs.

Experimental Settings. The CT-TN model and the individual RoBERTa feature predictions were used as baselines against which to test the two Ego Network features. These were all prepared using few-shot cross-target training, whereby the training data consisted of roughly 1,000 source target data points with 4 injections of destination target texts, increasing in size by increments of 100, from 100-shot to 400-shot (inclusive). For example, the Biden-Trump predictions were obtained by training on around 1,000 Biden texts with 100 Trump texts for the 100-shot condition, with 200 Trump tweets for the 200-shot condition, and so on. The stance predictions were then tested using between 500 and 800 data points (depending on the amount of remaining unseen data) that were solely related to the destination target (i.e. only using Trump-related texts for the aforementioned example). We conducted these experiments with five different random seeds: 24, 524, 1024, 1524, and 2024. Finally, we averaged the results from these five seeds for each shot size.

3.2 Computing Ego Networks

Computing Ego Networks. An Ego Network can be obtained for each active[1] user in the data by calculating the interaction frequencies of each of their relationships and then applying a clustering algorithm to them. One of the most popular methods for this is MeanShift [11], an unsupervised algorithm that automatically finds the most appropriate number of circles for an Ego [6], and so that is the algorithm that was employed for this work.

Additionally, the unsigned Ego Network feature was also separated into inner circles (1 and 2) and outer circles (3+), to better understand the importance of the different levels of the ENM for SD.

Generating Signed Connections. Once the unsigned Ego Networks have been obtained, it is relatively simple to generate the signed version. The interactions were grouped for each Ego-Alter pair and then a model, called Valence Aware Dictionary and sEntiment Reasoner (VADER) [14], was used to obtain a sentiment for each interaction. VADER is a very competitive sentiment analysis model that is specifically designed for performing sentiment analysis on English-language tweets. It is also the model used in the original SENM research paper [22].

Once sentiments had been obtained for the individual interactions, the psycho- logy-based threshold of 17% [12], mentioned in Sect. 2.2, was used to obtain a sentiment label for each relationship, resulting in signed Ego Networks.

3.3 Data

The data used in this study come from a well-established and publicly available set: P-Stance [17]. This dataset contains 21,574 English-language tweets collected during the 2020 U.S. presidential election. These tweets were specifically collected to be used for SD and each one is associated with one of 3 targets, Joe Biden, Donald Trump or Bernie Sanders, and a corresponding stance label, either "AGAINST" or "FAVOR". Hashtags, such as "#BidenForPresident" and "#NeverBernie" were used to both search for the tweets and to determine their target and stance. While the original dataset only included the text and stance of each tweet, the authors of the P-Stance dataset provided 9,307 tweet IDs upon request, allowing further data to be collected for each tweet, including information about the authoring users. In addition to providing the information required for computing the users' Ego Networks (see Sect. 3.2), this also made it possible to obtain, for each user, the remaining features required by CT-TN: likes, followers and friends.

[1] An active user is defined as a user, with a timeline of at least 6 months, that posts at least once every three days for half of the months that they are included in the data, based on previous research [6].

4 Results

The performances of the CT-TN model, RoBERTa, and the two Ego Network features can be seen in Fig. 3. Overall, they all perform very well, with most reaching a macro F1 score of above 0.7 before 400-shot for all target pairs, with CT-TN sometimes even going above 0.8. However, RoBERTa does not perform quite as well as the others, and only achieves macro F1 scores of around mid-0.6 for half of the target pairs.

Surprisingly, the signed and unsigned Ego Networks' F1 scores are very close, being within 0.01 of each other for 5 of the 6 target pairs (at 400-shot), and within 0.02 for the sixth pair (Sanders-Trump). This suggests that the additional information of signed connections does not provide a significant amount of information for the task of CTSD. Rather, it appears that the people we interact with regularly have an impact on our stances regardless of whether we have a negative or positive relationship with them.

Next, observing the outer circles, one can see that they perform similarly to, and often even outperform, the full Ego Network. However, they are less consistent, as displayed by the Biden-Trump and Sanders-Trump target pairs. By comparison, the inner circles perform slightly worse overall, with performances closer to those of RoBERTa. Since the outer circles contain weaker social relationships, it seems that weaker, but more numerous, ties are more informative than stronger, but less numerous, ones when it comes to stance prediction. This is rather surprising given that previous research on a similar network-based task, link prediction, found that the more intimate inner circles are better predictors of where new relations will form [26]. Thus, there seems to be a disconnect between how we form new connections and how we are influenced by them. Indeed, paired with the fact that the signed and unsigned ENMs performed similarly, it appears that the existence of a social connection may influence an individual regardless of any qualitative aspects, such as closeness or polarity.

The Ego Networks appear to perform slightly worse than the CT-TN model. However, as they only require interaction data, they could be used as a viable alternative whenever specific network features are not provided or obtainable for a given dataset. Moreover, as the signed and unsigned Ego Networks achieved similar performances, one could focus on employing the unsigned version, which would require even less data: only the frequencies of interactions, without the need for their texts.

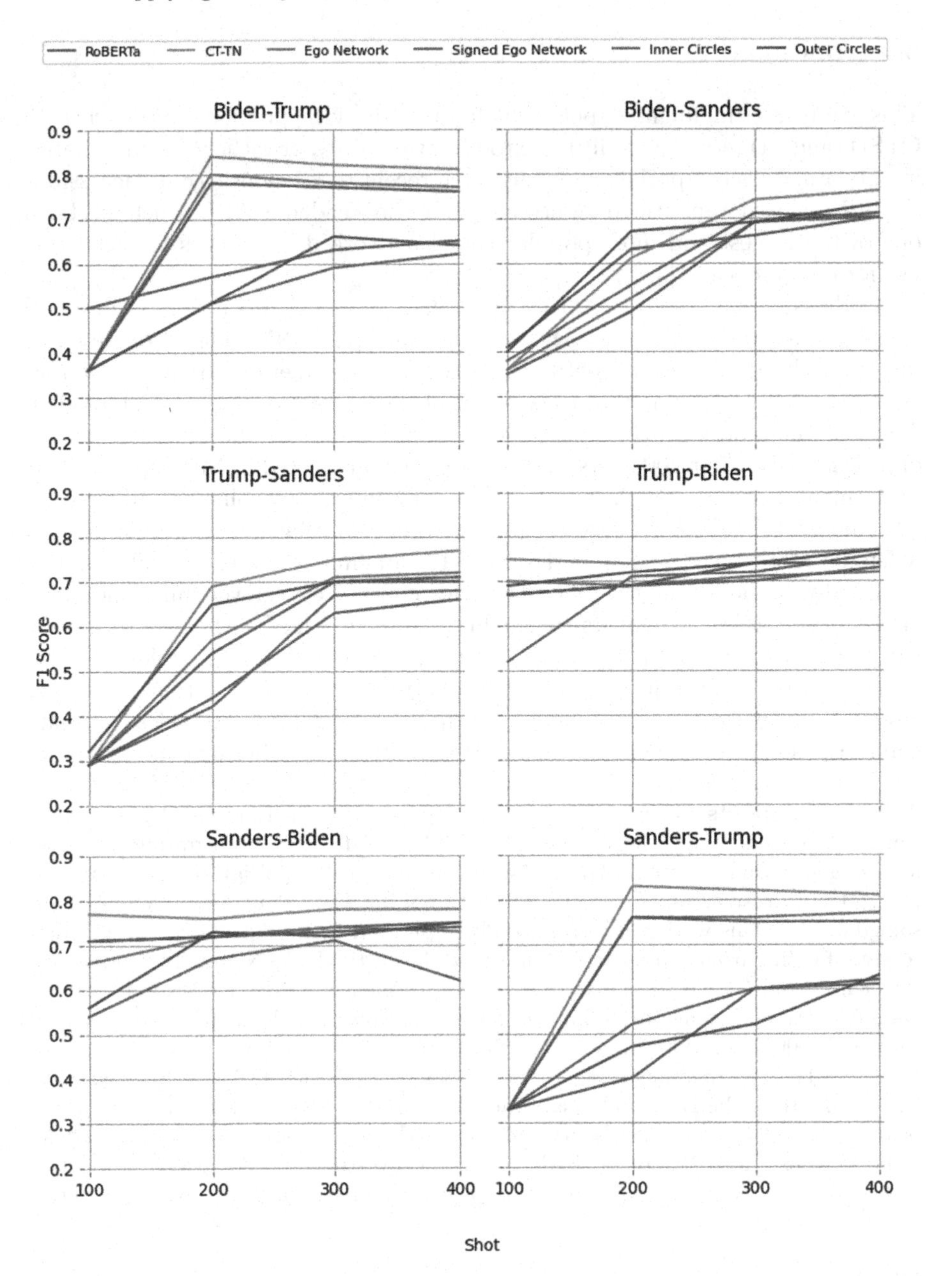

Fig. 3. Graphs displaying the performances of the individual text-based RoBERTa, the CT-TN model and the signed and unsigned Ego Networks features, as well as the unsigned inner and outer circle features. The shot number is displayed along the X axis and the averaged macro F1 scores along the Y axis.

5 Conclusion

This study has highlighted potential limitations with previous approaches to CTSD (and SD more generally) due to the availability, or rather the unavailability, of data. This is especially relevant given recent restrictions to data accessibility on X, most notably the discontinuation of the Academic API, which has been one of the largest and most popular social network data sources for academic use for over a decade [3,6,22,24].

Addressing this limitation, this paper has proposed two novel graph-based features that can be used for SD: the ENM and the SENM. The latter requires only text-based interactions between users and the former only requires interaction frequencies, meaning that the content of the interactions can remain hidden. These features perform consistently well for CTSD (achieving macro F1 scores of at least 0.71, after 400 shots, for all 6 target pairs tested in this paper). Their performances are only slightly worse than a previously established and cutting-edge model, CT-TN. Ego Networks present viable alternatives that could be used when not all the features required for CT-TN (or similar models) are obtainable.

Finally, by observing the different performances between the inner and outer circles of the ENM, it appears that, while the inclusion of both leads to a more consistent performance across different target pairs, the outer circles on their own can often perform just as well and sometimes better. By contrast, the inner circles do not perform as well, suggesting that the greater number of less intimate connections in the outer circles are more important for predicting a user's stance.

Acknowledgments. This work is partially supported by the European Union – Horizon 2020 Program under the scheme "INFRAIA-01-2018-2019 – Integrating Activities for Advanced Communities", Grant Agreement n.871042, "SoBigData++: European Integrated Infrastructure for Social Mining and Big Data Analytics" (http://www.sobigdata.eu). This work was also partially supported by SoBigData.it. SoBigData.it receives funding from European Union – NextGenerationEU – National Recovery and Resilience Plan (Piano Nazionale di Ripresa e Resilienza, PNRR) – Project: "SoBigData.it – Strengthening the Italian RI for Social Mining and Big Data Analytics" – Prot. IR0000013 – Avviso n. 3264 del 28/12/2021. C. Boldrini was also supported by PNRR - M4C2 - Investimento 1.4, Centro Nazionale CN00000013 - "ICSC - National Centre for HPC, Big Data and Quantum Computing" - Spoke 6, funded by the European Commission under the NextGeneration EU programme. A. Passarella and M. Conti were also supported by the PNRR - M4C2 - Investimento 1.3, Partenariato Esteso PE00000013 - "FAIR", funded by the European Commission under the NextGeneration EU programme.

Disclosure of Interests. The authors have no competing interests to declare that are relevant to the content of this article.

References

1. Allaway, E., McKeown, K.: Zero-shot stance detection: a dataset and model using generalized topic representations. arXiv preprint arXiv:2010.03640 (2020)

2. Alturayeif, N., Luqman, H., Ahmed, M.: A systematic review of machine learning techniques for stance detection and its applications. Neural Comput. Appl. **35**(7), 5113–5144 (2023)
3. Arnaboldi, V., Conti, M., Passarella, A., Pezzoni, F.: Ego networks in Twitter: an experimental analysis. In: 2013 Proceedings IEEE INFOCOM, pp. 3459–3464. IEEE (2013)
4. Augenstein, I., Rocktäschel, T., Vlachos, A., Bontcheva, K.: Stance detection with bidirectional conditional encoding. arXiv preprint arXiv:1606.05464 (2016)
5. Biber, D., Finegan, E.: Adverbial stance types in English. Discourse Process. **11**(1), 1–34 (1988)
6. Boldrini, C., Toprak, M., Conti, M., Passarella, A.: Twitter and the press: an ego-centred analysis. In: Companion Proceedings of the The Web Conference 2018, pp. 1471–1478 (2018)
7. Dunbar, R.I.M., et al.: Mind the Gap: or Why Humans Aren't Just Great Apes. Oxford University Press, Oxford (2014)
8. Dunbar, R.I.M.: Neocortex size as a constraint on group size in primates. J. Hum. Evol. **22**(6), 469–493 (1992)
9. Dunbar, R.I.M.: The social brain hypothesis. Evol. Anthropol. Issues News Rev. **6**(5), 178–190 (1998)
10. Dunbar, R.I.M., Spoors, M.: Social networks, support cliques, and kinship. Hum. Nat. **6**, 273–290 (1995)
11. Fukunaga, K., Hostetler, L.: The estimation of the gradient of a density function, with applications in pattern recognition. IEEE Trans. Inf. Theory **21**(1), 32–40 (1975)
12. Gottman, J., Gottman, J.M., Silver, N.: Why Marriages Succeed or Fail: and How You Can Make Yours Last. Simon and Schuster (1995)
13. Grover, A., Leskovec, J.: node2vec: scalable feature learning for networks. In: Proceedings of the 22nd ACM SIGKDD International Conference on Knowledge Discovery and Data Mining, pp. 855–864 (2016)
14. Hutto, C., Gilbert, E.: VADER: a parsimonious rule-based model for sentiment analysis of social media text. In: Proceedings of the International AAAI Conference on Web and Social Media, vol. 8, pp. 216–225 (2014)
15. Khiabani, P.J., Basiri, M.E., Rastegari, H.: An improved evidence-based aggregation method for sentiment analysis. J. Inf. Sci. **46**(3), 340–360 (2020)
16. Khiabani, P.J., Zubiaga, A.: Few-shot learning for cross-target stance detection by aggregating multimodal embeddings. IEEE Trans. Comput. Soc. Syst. (2023)
17. Li, Y., Sosea, T., Sawant, A., Nair, A.J., Inkpen, D., Caragea, C.: P-stance: a large dataset for stance detection in political domain. In: Findings of the Association for Computational Linguistics: ACL-IJCNLP 2021, pp. 2355–2365 (2021)
18. Liu, Y., et al.: Roberta: a robustly optimized BERT pretraining approach. arXiv preprint arXiv:1907.11692 (2019)
19. Mikolov, T., Chen, K., Corrado, G., Dean, J.: Efficient estimation of word representations in vector space. arXiv preprint arXiv:1301.3781 (2013)
20. Mohammad, S., Kiritchenko, S., Sobhani, P., Zhu, X., Cherry, C.: Semeval-2016 task 6: detecting stance in tweets. In: Proceedings of the 10th International Workshop on Semantic Evaluation (SemEval-2016), pp. 31–41 (2016)
21. Sun, S., Luo, C., Chen, J.: A review of natural language processing techniques for opinion mining systems. Inf. Fusion **36**, 10–25 (2017)
22. Tacchi, J., Boldrini, C., Passarella, A., Conti, M.: Signed ego network model and its application to Twitter. In: IEEE BigData 2022 (2022)

23. Tacchi, J., Boldrini, C., Passarella, A., Conti, M.: Cultural differences in signed ego networks on Twitter: an investigatory analysis. In: Companion Proceedings of the ACM Web Conference 2023, pp. 1039–1049 (2023)
24. Tacchi, J., Boldrini, C., Passarella, A., Conti, M.: On the joint effect of culture and discussion topics on X (Twitter) signed ego networks. arXiv preprint arXiv:2402.18235 (2024)
25. Toprak, M., Boldrini, C., Passarella, A., Conti, M.: Structural models of human social interactions in online smart communities: the case of region-based journalists on twitter. Online Soc. Netw. Media **30** (2021)
26. Toprak, M., Boldrini, C., Passarella, A., Conti, M.: Harnessing the power of ego network layers for link prediction in online social networks. IEEE Trans. Comput. Soc. Syst. **10**(1), 48–60 (2023)
27. Wei, P., Mao, W.: Modeling transferable topics for cross-target stance detection. In: Proceedings of the 42nd International ACM SIGIR Conference on Research and Development in Information Retrieval, pp. 1173–1176 (2019)
28. Xu, C., Paris, C., Nepal, S., Sparks, R.: Cross-target stance classification with self-attention networks. arXiv preprint arXiv:1805.06593 (2018)

FOCI: Fair Cross-Network Node Classification via Optimal Transport

Anna Stephens, Francisco Santos(✉), Pang-Ning Tan, and Abdol-Hossein Esfahanian

Michigan State University, East Lansing, MI 48825, USA
{steph496,santosf3,ptan,esfahanian}@msu.edu

Abstract. Graph neural networks (GNNs) have demonstrated remarkable success in addressing a variety of node classification problems. Cross-network node classification (CNNC) extends the GNN formulation to a multi-network setting, enabling the classification to be performed on an unlabeled target network. However, applying GNNs to a multi-network setting in practice is a challenge due to the possible presence of concept drift and the need to account for link biases in the graph data. In this paper we present `FOCI`, a powerful, model-agnostic approach for cross-network node classification that enables the GNN to overcome the concept drift issue while mitigating potential biases in the data. `FOCI` utilizes a fair Sinkhorn distance function with optimal transport to learn a fair yet effective feature embedding of the nodes in the source graph. We experimentally demonstrate the effectiveness of `FOCI` at addressing the CNNC task while simultaneously mitigating unfairness compared to other baseline methods.

1 Introduction

The proliferation of graph-based data in various application domains has motivated the need to develop more advanced techniques based on deep learning to harness the network data from multiple sources in order to improve the performance of node classification algorithms. Current graph neural network techniques would learn a feature embedding of the nodes from multiple graphs, which are then presented to a fully-connected network layer to perform the final classification. However, despite the notable advances in graph neural networks, there are still numerous challenges that must be addressed in order for the techniques to be successfully deployed to solve real-world problems.

First, existing techniques often assume that the graphs share similar distributional properties, thus enabling us to apply a model trained on one graph, *a.k.a.* the *source graph*, to the nodes in another, *a.k.a.* the *target graph*. Unfortunately, in practice, one would likely encounter some form of distributional shift, where the training data only captures the essence of a particular graph but fails to account for some unforeseen differences in another graph. Such type of concept drift is illustrated in Fig. 1, where the decision boundary induced from the

L. M. Aiello et al. (Eds.): ASONAM 2024, LNCS 15212, pp. 183–193, 2025.
https://doi.org/10.1007/978-3-031-78538-2_16

learned representation of a source graph (left) does not reflect the class separation of the nodes in the target graph (right). Second, incorporating fairness into node classification is another concern to prevent the model from generating biased prediction results. Graph neural networks (GNNs) are especially prone to fairness issues due to an artifact of the homophily principle, which states that similar nodes are likely to be connected to each other [14]. Past research on graph fairness have shown that neighborhood structure is often more dependent on the protected attributes than the classification labels [7,16]. As a consequence, GNNs have the potential to exacerbate unfairness as the learned embeddings may capture more information about a node's protected group than its class label [9], a phenomenon known as *link bias* [6].

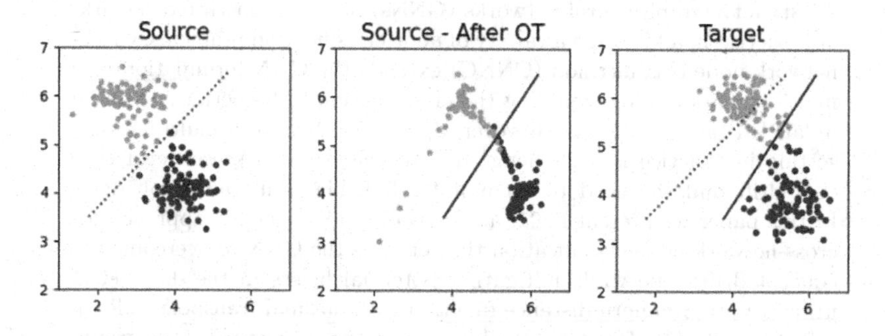

Fig. 1. An illustration of OT for domain adaptation, where the color represents class labels. The diagram on the left is the original source dataset and on the far right is the target dataset. The dotted line is the decision boundary of logistic regression trained on the source dataset. The middle diagram shows the transported source dataset using OT, with the solid line representing the decision boundary obtained by logistic regression.

To address the concept drift problem, domain adaptation and transfer learning methods have been developed. For graph data, transfer learning is typically studied under the guise of *cross-network node classification* (CNNC) problems [17,18,20], in which there is a source graph, with fully or mostly labeled nodes, and a target graph, which has either a few or no class labels. Existing CNNC methods are primarily model-dependent, tailored towards the specific neural architecture, and thus, limiting their flexibility. Optimal transport (OT) [5,8] is another widely-used transfer learning approach. Specifically, OT would learn a transportation plan that maps data points from a source domain to a target domain in a way that minimizes the total cost [3]. The primary advantage of employing OT lies in its model-agnostic nature, allowing it to be seamlessly integrated into any graph neural network model. The impact of OT becomes evident in the middle diagram of Fig. 1, which shows the decision boundary obtained after applying OT to the source dataset depicted in the left diagram. The modified decision boundary post-OT demonstrates superior alignment with the dataset in the target domain as shown in the right diagram. However, similar to CNNC, there has yet been any studies integrating fairness into OT formulation.

In this paper, we present a novel, model-agnostic framework called FOCI (**F**air Cr**O**ss-Network Node **C**lass**I**fication) that performs fair optimal transport while mitigating the concept drift issue in CNNC task. Specifically, FOCI considers the nodes' protected attribute information when transporting the source nodes to their corresponding target nodes when learning their feature embedding. We introduce a fair Sinkhorn distance measure for OT to encourage diversity in the mapping of the source nodes to target nodes. This strategy ensures that the learned features are oblivious to the protected attribute information. Finally, we also proposed FastFOCI, an extension of FOCI, which provides an improved run-time performance.

2 Problem Statement

Consider a set of graphs, $\mathcal{G}_1, \mathcal{G}_2, \cdots \mathcal{G}_n$, where each $\mathcal{G}_k(V_k, E_k, X_k, Y_k)$ is an attributed graph, with a node set V_k, edge set $E_k \subseteq V_k \times V_k$, node feature matrix X_k, and node label vector Y_k. Let $X_k = X_k^{(p)} \cup X_k^{(u)}$, where $X_k^{(p)}$ corresponds to the set of protected attributes (e.g., gender, race, or age group). Let $A^{(k)}$ denote the adjacency matrix representation of E_k, where $A_{ij}^{(k)} > 0$ if $(v_i, v_j) \in E_k$ and 0 otherwise. Assume the set of graphs are divided into two disjoint groups—source graph $\mathcal{G}_s(V_s, E_s, X_s, Y_s)$, where the node labels in Y_s are known, and a target graph, $\mathcal{G}_t(V_t, E_t, X_t, Y_t)$, where the node labels in Y_t are unknown. The adjacency matrices corresponding to the source and target graphs are denoted as $A_s \in \mathbb{R}^{n_s \times n_s}$ and $A_t \in \mathbb{R}^{n_t \times n_t}$, respectively, where n_s and n_t are the number of source and target nodes. Assuming the graphs have identical node features and class labels, therefore $X_s \in \mathbb{R}^{n_s \times d}$ and $X_t \in \mathbb{R}^{n_t \times d}$, $Y_s \in \{0, 1, \cdots, k-1\}^{n_s}$ and $Y_t \in \{0, 1, \cdots, k-1\}^{n_t}$, where k is the number of classes.

Given a source graph, $\mathcal{G}_s = (V_s, E_s, X_s, Y_s)$ and target graph, $\mathcal{G}_t = (V_t, E_t, X_t, Y_t)$, the goal of fair CNNC is to learn a function, $f : V \rightarrow \{0, 1, \cdots, k-1\}$, that accurately classifies the labeled nodes in Y_s and the unlabeled nodes in Y_t while minimizing the disparity in classification performance for different groups of the protected attributes.

3 Proposed FOCI Architecture

Figure 2 shows a high-level overview of FOCI, consisting of modules for representation learning, OT for domain adaptation, and fully connected layers for node classification.

3.1 Representation Learning and Pretraining

The GCN approach learns a feature embedding of the nodes in layer $l+1$, $H^{(l+1)}$, using the embedding from its previous layer $H^{(l)}$ as follows:

$$H^{(l+1)} = \sigma(\tilde{D}^{-\frac{1}{2}} \tilde{A} \tilde{D}^{-\frac{1}{2}} H^{(l)} W^{(l)}) \quad (1)$$

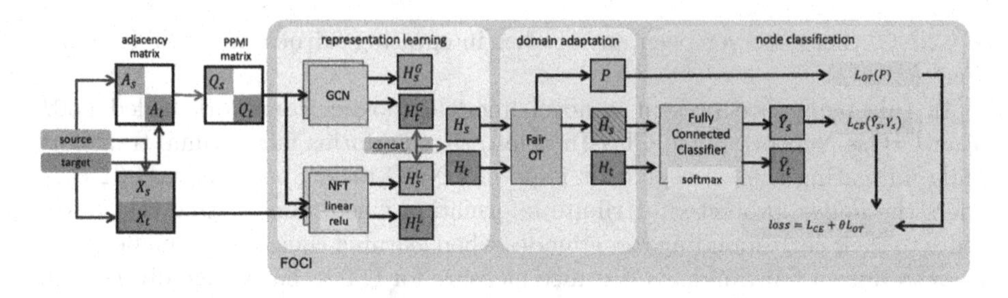

Fig. 2. A schematic illustration of the FOCI framework. The framework contains integrated modules for representation learning (using 2-layer GCN with PPMI matrix along with 2-layer NFT), fair OT for domain adaptation, and a fully connected node classification layer.

where $\tilde{A} = A + I$, $\tilde{D}_{ii} = \sum_j \tilde{A}_{ij}$, and $\sigma(\cdot)$ is the ReLU activation function.

`FOCI` modifies the GCN architecture [10] to learn a joint embedding of the nodes in the source and target networks. The modifications are needed for two reasons. First, due to oversmoothing effect [13], current GCN cannot be easily extended beyond 2 or 3 layers, thus throttling the effective neighborhood size of its message passing. Because of its limited neighborhood size and the link bias problem noted in the introduction, the learned representation may inadvertantly encode the protected attribute information [6]. To overcome this limitation, we expand the neighborhood utilized by the GCN through the use of the *positive pointwise mutual information* (PPMI) matrix [12]. A PPMI matrix measures the topological proximity of nodes within some K-steps within the network and has been used with other GNN approaches [2,17,20] to expand the neighborhood of interest. By using a PPMI matrix, the embedding learned will more likely capture information from nodes beyond those from the same protected group.

The representation learning module also learns a separate nonlinear embedding of the node features. This allows us to maintain a node embedding that is free of link bias for later use. We refer to this as the *node feature transformation* (NFT) layer, which consists of a fully-connected linear layer with ReLU activation function. We pre-train the network to classify only the labeled nodes in the source network. Specifically, `FOCI` uses two GCN layers with the PPMI matrix and two NFT layers. The learned node embeddings are concatenated as shown in Fig. 2 before being sent to a fully connected classification layer, which is trained to minimize the following cross-entropy loss: $\mathcal{L}_{CE} = -\frac{1}{n_s}\sum_{x_i \in X_s}\sum_{j=0}^{l} Y_{ij}\log(\hat{Y}_{ij})$.

3.2 Fair Optimal Transport

Our strategy to incorporate fairness into OT is by encouraging mappings between members of different protected groups. This heuristic helps to alleviate the link bias problem due to the inherent homophily effect in network data. Let $X_s^{(p)}$ and $X_t^{(p)}$ be the protected attributes of the nodes in the source and target networks, respectively. We first create two sparse matrices, $R \in [0,1]^{n_s \times n_t}$ and $S \in [0,1]^{n_s \times n_t}$, where

$$R_{ij} = \begin{cases} 1 & \text{if } X_{s,i}^{(p)} = X_{t,j}^{(p)} \\ 0 & \text{otherwise} \end{cases}, \quad S_{ij} = \begin{cases} 1 & \text{if } X_{s,i}^{(p)} \neq X_{t,j}^{(p)} \\ 0 & \text{otherwise} \end{cases}$$

Given a transportation plan matrix P, we compute the following fairness loss: $\ell_F = \frac{\langle P,R\rangle_F}{\langle R,R\rangle_F} - \frac{\langle P,S\rangle_F}{\langle S,S\rangle_F}$, where $\langle A,B\rangle_F = \sum_{ij} A_{ij}B_{ij}$ denotes the Frobenius inner product between two matrices. The smaller the fairness loss, the greater the emphasis is on transportation between samples in different protected groups. This enables us to incorporate fairness consideration into OT by introducing the following γ-fair Sinkhorn distance:

$$W_\gamma(\mu_s, \mu_t) = \langle P^\gamma, C\rangle \tag{2}$$

where $P^\gamma = \text{argmin}_{P \in U(\mu_s,\mu_t)} \mathcal{L}_{OT}$ and $\mathcal{L}_{OT} = \langle P, C\rangle - \lambda \text{ h(P)} + \gamma \left[\frac{\langle P,R\rangle}{\langle R,R\rangle} - \frac{\langle P,S\rangle}{\langle S,S\rangle}\right]$.

Theorem 1. *Given the γ-fair Sinkhorn distance in Eqn. (2), the solution for P^γ is*

$$P^\gamma = diag(u) K diag(v)$$

where $K = e^{-\frac{1}{\lambda}(C + \gamma(\frac{R}{n_1} - \frac{S}{n_2}))}$, $u = e^{-\frac{1}{2} - \frac{1}{\lambda}\alpha}$, $v = e^{-\frac{1}{2} - \frac{1}{\lambda}\beta}$.

Proof. The Lagrangian of the function in Eqn. (2) is

$$\mathcal{L} = \sum_{ij} \left[P_{ij}C_{ij} + \lambda P_{ij} \log P_{ij} + \gamma P_{ij}\left(\frac{R_{ij}}{n_1} - \frac{S_{ij}}{n_2}\right)\right] + \alpha^T(P1_d - \mu_s) + \beta^T(P^T 1_d - \mu_t)$$

where $n_1 = \langle R, R\rangle$ and $n_2 = \langle S, S\rangle$. The solution can be found by taking the derivative of the Lagrangian function with respect to P_{ij} and setting it to zero.

$$P_{ij} = e^{-\frac{1}{2} - \frac{1}{\lambda}\alpha_i} e^{-\frac{1}{\lambda}(c_{ij} + \gamma(\frac{r_{ij}}{n_1} - \frac{s_{ij}}{n_2}))} e^{-\frac{1}{2} - \frac{1}{\lambda}\beta_j}$$

The theorem follows by replacing the terms for K, u, and v into the above equation.

Theorem 1 enables us to compute P^γ using the modified fair Sinkhorn algorithm shown in Algorithm 1, which in turn, allows us to find a fairness-aware transportation plan. These changes can be seen in Algorithm 1 and have no significant impact on computational complexity of existing Sinkhorn algorithm [5]. Additionally, since we are still utilizing squared l_2 loss for our cost and uniform marginals, the transported source node representation by OT remains unchanged, i.e., $\hat{H}^s = n_s P^\lambda H_t$.

Algorithm 1. Fair Sinkhorn

Require: $H_s, H_t, X_s^{(p)}, X_t^{(p)}, \lambda, \gamma$
  $C = \text{computeCost}(H_s, H_t)$
  $\mu_s, \mu_t = \text{computeUniformMarginals}(H_s, H_t)$
  $R, S = \text{getMasks}(g_s, g_t)$
  $K = e^{-\frac{1}{\lambda}(C+\gamma(\frac{R}{n_1} - \frac{S}{n_2}))}$ {Modification}
  $u = \text{ones}(length(\mu_s))/length(\mu_s)$
  $\tilde{K} = \text{diag}(1/\mu_s)K$
  while u changes **do**
    $u = 1/(\tilde{K}(\mu_t/(K'u))$
  end while
  $v = \mu_t/(K'u)$
  $W = \text{sum}(u \odot ((K \odot C)v))$ {Distance Measure}
  $P = \text{diag}(u)K\text{diag}(v)$ {Transport Plan}
  $\hat{H}_s = n_s P H_t$

3.3 Cross-Network Node Classification

The fair OT module will generate a transported source node embedding matrix, $\hat{H}^s$. Next $\hat{H}_s$ and the un-transformed target embedding H_t are passed to a fully connected network for node classification. After pre-training, the entire network is trained end-to-end to optimize the following joint objective function, which is a combination of the cross-entropy loss and optimal transport loss.

$$\mathcal{L} = \mathcal{L}_{CE} + \theta \mathcal{L}_{OT} \tag{3}$$

where θ is a hyperparameter that controls the trade-off between the two losses.

3.4 FastFOCI

A hindrance to performance in FOCI is that the representation learning step must evaluate both the source and target graphs at the same time. The first major improvement implemented in FastFOCI is that the representation learning phase will consider the source data only. Additionally, in FastFOCI, the source graph will be fed to source GCN and NFT layers, GCN_s and NFT_s. In pretraining the output of these layers will be sent to a classifier and source predictions will be used to update the GCN_s, NFT_s, and classifier. GCN_s and NFT_s will be used to construct a source representation H_S while GCN_t and NFT_t is used to construct a target representation H_T. In this way we reduce the number of edges required by the GCN, which no longer requires full access to both source and target graphs at the same time. Finally, as the classifier does not need all node representations at the same time, we can improve the speed of the OT layer by inserting a sampling step after node representations are acquired and before OT occurs. Specifically, we randomly sample representations from both the source and target embeddings H_S and H_T described above and then pass those samples on to OT. We can then transport the sampled source representations to send on to the classifier.

Table 1. Breakdown of datasets used in experiments.

dataset	nodes	edges	$\#(Y=1)$	$\#(X^{(p)}=1)$
pokec-n	2933	16821	2145 (73%)	1722 (59%)
pokec-z	3285	22454	2137 (65%)	1844 (56%)
compas-0	2170	137473	1042 (48%)	1690 (78%)
compas-1	1434	148012	674 (47%)	1140 (79%)
Abide-large	804	93886	370 (46%)	685 (85%)
Abide-small	67	587	33 (49%)	42 (63%)
Credit-0	10468	75669	8029 (76%)	5462 (52%)
Credit-1	10169	29227	8384 (82%)	5360 (52%)

4 Experimental Evaluation

We consider the following datasets for our experiments (see Table 1 for more details):

- **Pokec** [11] is a popular social media platform in Slovakia. The node feature information is obtained from the user profile data while the link structure represents relationships between users. The classification task is to predict whether a user smokes with the user's sex as protected attribute. We use data from two geographical regions—Pokec-n and Pokec-z—to form the source and target networks.
- **Compas** [1] is a recidivism dataset in which each node corresponds to an incarcerated individual while an edge is formed by connecting individuals who were incarcerated during an overlapping period. The classification task is to predict whether an individual will re-offend again in the future with race as the protected attribute. The source and target networks are created by applying spectral clustering to split the network into 2 subgraphs, denoted as Compas-0 and Compas-1, respectively.
- **ABIDE** [4] is a popular dataset for studying Autism Spectrum Disorder (ASD). Simular to previous work [15], we construct a population graph by using phenotypic subject data as node features and resting-state fMRI similarity to construct the edges. Here the presence of ASD is node label and sex is the protected attribute. The dataset was split into two separate graphs according to their data collection sites. Given the need to have a large enough sample to train GNN models, ABIDE will only be trained with the larger graph as source and the smaller graph as target.
- **Credit** [19] is a financial dataset where the task is to predict whether an individual will default on a loan. The protected attribute is the age of the individual, and edges are formed between individuals with similar spending and payment patterns. Credit-0 was created by selecting the highest degree node and then repeatedly adding all immediate degree neighbors until the graph has over 10,000 nodes. Once the nodes in Credit-0 were removed, the process was repeated to obtain Credit-1.

The source code of FOCI can be found at https://github.com/ajoystephens/foci. We compared FOCI against the following CNNC baseline methods.

- **ACDNE** [17]: This approach uses an adversarial deep network embedding approach for domain adaptation. It learns a separate embedding from the node features and link structure before sending them to a discriminator.
- **ASN** [20]: This approach uses a series of 2-layer GCNs and GCN variational autoencoders (VAE) to learn the feature embedding. It addresses the domain adaptation issue using an adversarial discriminator.

Table 2. Comparison of FOCI and baselines ACDNE and ASN in terms of F1-Score (F1) and statically parity (SP). First and second place values are highlighted in gold and silver respectively.

Source	Target		ACDNE	ASN	FOCI
abide-large	abide-small	F1	0.944 +/- 0.038	**0.318 +/- 0.201**	0.970 +/- 0.008
		SP	**0.110 +/- 0.012**	0.008 +/- 0.008	0.102 +/- 0.012
pokec-n	pokec-z	F1	0.836 +/- 0.002	**0.748 +/- 0.022**	0.790 +/- 0.004
		SP	**0.080 +/- 0.015**	0.008 +/- 0.006	0.007 +/- 0.006
pokec-z	pokec-n	F1	0.836 +/- 0.003	**0.750 +/- 0.021**	0.784 +/- 0.008
		SP	**0.148 +/- 0.005**	0.012 +/- 0.009	0.056 +/- 0.016
compas-0	compas-1	F1	0.634 +/- 0.011	**0.028 +/- 0.058**	0.560 +/- 0.187
		SP	**0.172 +/- 0.007**	0.001 +/- 0.002	0.122 +/- 0.041
compas-1	compas-0	F1	0.604 +/- 0.008	**0.318 +/- 0.201**	0.638 +/- 0.009
		SP	**0.170 +/- 0.013**	0.008 +/- 0.008	0.163 +/- 0.011

For every method, we performed 10-fold cross fold validation on the source dataset to select their best hyperparameters. For the baseline methods, the range of their hyperparameter values include those reported in their authors' published papers and source code. We repeated our experiment 10 times with different random seeds and recorded the average F1-score and statistical parity values.

4.1 Experimental Results

The results comparing FOCI to other CNNC approaches are shown in Table 2. Here the goal is to achieve high node classification results, as measured by F1-score, while simultaneously improving fairness, as measured by the statistical parity measure:

$$SP = |P(\hat{Y} = 1|X^{(p)} = 0) - P(\hat{Y} = 1|X^{(p)} = 1)|$$

The results in Table 2 suggest FOCI is consistently one of the best methods in both F1-Score and statistical parity. ACDNE has the best F1-Score in most

Table 3. Comparison of FOCI and FastFOCI and in terms of F1-Score, statically parity (SP), and epoch time in seconds. Best values are **bold**.

Source	Target	Method	F1-Score	SP	Epoch Time (s)
pokec-n	pokec-z	FOCI	**0.790 +/- 0.004**	**0.007 +/- 0.006**	0.569 +/- 0.067
		FastFOCI	0.773 +/- 0.041	0.013 +/- 0.014	**0.117 +/- 0.042**
pokec-z	pokec-n	FOCI	**0.784 +/- 0.008**	0.056 +/- 0.016	0.627 +/- 0.065
		FastFOCI	0.756 +/- 0.071	**0.047 +/- 0.027**	**0.110 +/- 0.029**
compas-0	compas-1	FOCI	0.560 +/- 0.187	**0.122 +/- 0.041**	0.277 +/- 0.024
		FastFOCI	**0.599 +/- 0.085**	0.158 +/- 0.027	**0.108 +/- 0.022**
compas-1	compas-0	FOCI	**0.638 +/- 0.009**	**0.163 +/- 0.011**	0.177 +/- 0.021
		FastFOCI	0.622 +/- 0.037	0.169 +/- 0.026	**0.084 +/- 0.023**
credit-0	credit-1	FastFOCI	**0.883 +/- 0.001**	**0.014 +/- 0.001**	**0.123 +/- 0.089**
credit-1	credit-0	FastFOCI	**0.868 +/- 0.009**	**0.006 +/- 0.004**	**0.145 +/- 0.195**

cases, but the worst statistical parity. ASN has often improved statistical parity but this coincides with reduced F1-Score. FOCI, however, manages to maintain a high F1-Score while reducing statistical parity, balancing the fairness/utility trade off. Note that the credit dataset is not considered in Table 2 as methods aside from FastFOCI were unable to process the large credit dataset.

Table 3 compares results from FOCI and FastFOCI in terms of F1-score, statistical parity, and run time. FOCI results for the credit datasets are excluded from this table because the FOCI method was unable to process the credit dataset. FOCI's optimal transport layer encountered numerical errors when attempting to transport the entire source and target credit datasets. Observe that FastFOCI provides significant run time improvements over FOCI, but with a slight decline in model utility and fairness. This suggests that FastFOCI may be the best option with larger datasets, but that FOCI may be the better choice if dataset size and runtime are not significant concerns.

4.2 Ablation Study

We investigate the impact of the hyperparameter γ on the OT transport plan and model outcome. Here, we restrict our discussion to the Pokec datasets and the FOCI approach. First, we vary γ and evaluate the resulting transport plan matrix P^γ. In these and all other experiments involving OT we held $\lambda = 0.03$. Figure 3 shows these results by plotting γ against the mean value of p_{ij} where nodes i and j do or do not share protected groups respectively. Note that mean values for p_{ij} are very small as $P^\gamma \in \mathbb{R}^{m \times n}$. Next we examine the impact of γ on statistical parity and F1-score. Here, γ was varied within the range that produces an impact on P^γ in Fig. 3, while all other hyperparameters were kept the same. The results shown in Fig. 4 suggest that the statistical parity and F1-scores do not vary significantly within the given range.

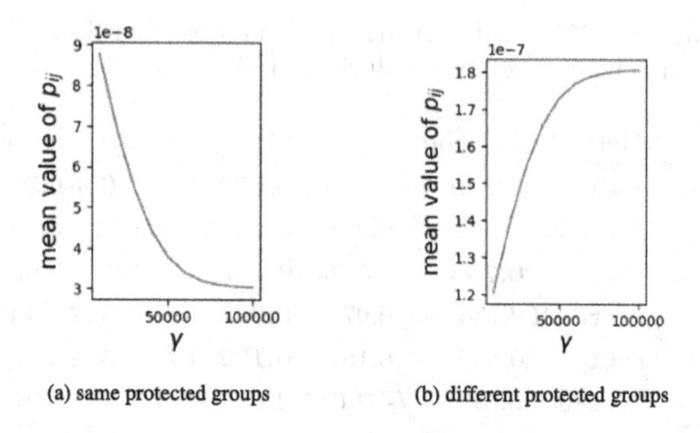

(a) same protected groups (b) different protected groups

Fig. 3. Impact of γ on transport plan matrix when mapping from pokec-n to pokec-z. As γ increases transport plan values decrease between nodes which share a protected group and increase for nodes which are in different protected groups.

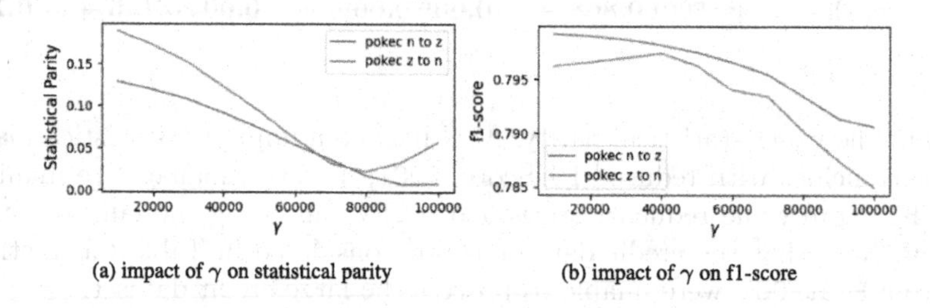

(a) impact of γ on statistical parity (b) impact of γ on f1-score

Fig. 4. Impact of γ on model outcomes.

5 Conclusion

This paper presents a framework called FOCI for fair cross-network node classification and a framework called FastFOCI which offers improved efficiency. Both frameworks use a novel fair Sinkhorn distance measure to encourage mapping between members of different protected groups in the source and target networks. We have experimentally shown that the proposed fair Sinkhorn distance helps to mitigate unfairness while maintaining high accuracy, comparable to other state-of-the-art CNNC baselines.

Acknowledgment. This material is based upon work supported by NSF under grant #IIS-1939368 and #IIS-2006633. Any opinion, findings, and conclusions or recommendations expressed in this material are those of the author(s) and do not necessarily reflect the views of the National Science Foundation.

References

1. Angwin, J., Larson, J., Mattu, S., Kirchner, L.: Machine bias. ProPublica (2016)
2. Cao, S., Lu, W., Xu, Q.: Deep neural networks for learning graph representations. In: Proceedings of the AAAI Conference on Artificial Intelligence, vol. 30 (2016)

3. Courty, N., Flamary, R., Tuia, D.: Domain adaptation with regularized optimal transport. In: Proceedings of ECML PKDD, pp. 274–289 (2014)
4. Craddock, C., et al.: Towards automated analysis of connectomes: the configurable pipeline for the analysis of connectomes (C-PAC). Front. Neuroinform. **42**, 10–3389 (2013)
5. Cuturi, M.: Sinkhorn distances: lightspeed computation of optimal transport. In: Advances in Neural Information Processing Systems, vol. 26 (2013)
6. Dai, E., Wang, S.: Say no to the discrimination: learning fair graph neural networks with limited sensitive attribute information. In: Proceedings of the 14th ACM International Conference on Web Search and Data Mining, pp. 680–688 (2021)
7. Dong, Y., Lizardo, O., Chawla, N.V.: Do the young live in a "smaller world" than the old? age-specific degrees of separation in a large-scale mobile communication network. arXiv preprint arXiv:1606.07556 (2016)
8. Kantorovitch, L.: On the translocation of masses. Manage. Sci. **5**(1), 1–4 (1958). https://doi.org/10.1287/mnsc.5.1.1
9. Karimi, F., Génois, M., Wagner, C., Singer, P., Strohmaier, M.: Homophily influences ranking of minorities in social networks. Sci. Rep. **8**(1), 11077 (2018)
10. Kipf, T.N., Welling, M.: Semi-supervised classification with graph convolutional networks. arXiv preprint arXiv:1609.02907 (2016)
11. Leskovec, J., Krevl, A.: SNAP Datasets: Stanford large network dataset collection. http://snap.stanford.edu/data (2014)
12. Levy, O., Goldberg, Y.: Neural word embedding as implicit matrix factorization. In: Advances in Neural Information Processing Systems, vol. 27 (2014)
13. Li, Q., Han, Z., Wu, X.M.: Deeper insights into graph convolutional networks for semi-supervised learning (2018)
14. McPherson, M., Smith-Lovin, L., Cook, J.M.: Birds of a feather: homophily in social networks. Ann. Rev. Sociol. **27**(1), 415–444 (2001)
15. Parisot, S., et al.: Spectral graph convolutions for population-based disease prediction. In: Proceedings of MICCAI, pp. 177–185 (2017)
16. Rahman, T., Surma, B., Backes, M., Zhang, Y.: Fairwalk: Towards fair graph embedding. In: Proceedings of IJCAI, pp. 3289–3295 (2019)
17. Shen, X., Dai, Q., Chung, F.l., Lu, W., Choi, K.S.: Adversarial deep network embedding for cross-network node classification. In: Proceedings of AAAI Conference on Artificial Intelligence, vol. 34, pp. 2991–2999 (2020)
18. Shen, X., Dai, Q., Mao, S., Chung, F.l., Choi, K.S.: Network together: Node classification via cross-network deep network embedding. IEEE Trans. Neural Netw. Learn. Syst. **32**(5), 1935–1948 (2020)
19. Yeh, I.C., Lien, C.H.: The comparisons of data mining techniques for the predictive accuracy of probability of default of credit card clients. Expert Syst. Appl. **36**(2), 2473–2480 (2009)
20. Zhang, X., Du, Y., Xie, R., Wang, C.: Adversarial separation network for cross-network node classification. In: Proceedings of the 29th ACM International Conference on Information & Knowledge Management, pp. 2618–2626 (2021)

Fast Flocking of Protesters on Street Networks

Guillaume Moinard(✉) and Matthieu Latapy

Sorbonne Université, CNRS, LIP6, 75005 Paris, France
{guillaume.moinard,matthieu.latapy}@lip6.fr

Abstract. We propose a simple model of protesters scattered throughout a city who want to gather into large and mobile groups. This model relies on random walkers on a street network that follow tactics built from a set of basic rules. Our goal is to identify the most important rules for fast flocking of walkers. We explore a wide set of tactics and show the central importance of a specific rule based on alignment. Other rules alone perform poorly, but our experiments show that combining alignment with them enhances flocking.

Keywords: street networks · protests · gathering · flocking · tactics

1 Introduction

Consider the following scenario. Protesters are scattered throughout a city and want to gather into groups large enough to perform significant actions. They face forces that may break up groups, block some places or streets and seize any communication devices protesters may be carrying. As a consequence, protesters only have access to local information on people and streets around them. Furthermore, formed protester groups must keep moving to avoid containment by adversary forces.

In this scenario, protesters need a distributed and as simple as possible protocol, that utilises local information exclusively and ensures the rapid formation of significantly large, mobile, and robust groups. We illustrate these objectives in Fig. 1. Our goal in this paper is to identify the key building blocks for such protocols.

To do so, we model the city as a network of streets and intersections in Sect. 2. We then assume that protesters are biased random walkers on such a network. In Sect. 3, we present the set of basic rules that compose *tactics* walkers can use to move. We define, in Sect. 4, what flocking features groups must achieve and run extensive experiments to measure how effective each tactic is regarding those.

2 Framework

We need a framework to simulate displacements of protesters in a city. We model cities as undirected graphs we call street networks. Protesters are then biased

L. M. Aiello et al. (Eds.): ASONAM 2024, LNCS 15212, pp. 194–203, 2025.
https://doi.org/10.1007/978-3-031-78538-2_17

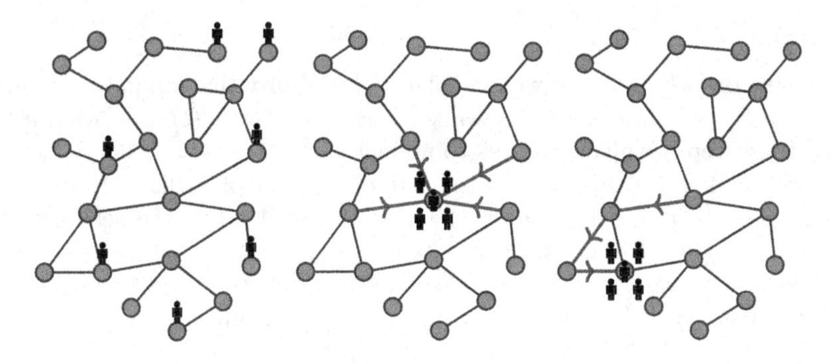

Fig. 1. Walkers scattered on a street network (left) must rapidly gather (center) and subsequently flock (right).

random walkers on this network. They can move from node to node with simple rules we introduce in next Section.

2.1 Street Networks

In order to model real-world cities, we leverage OpenStreetMap [5] data and the OSMnx library [3]. For a given city, we use this library with its default settings to extract the graph $G = (V, E)$ defined as follows: the nodes in V represent street intersections in this city and the links in $E \subseteq V \times V$ represent pieces of streets between them. We take the undirected graph G, meaning there is no distinction between (u, v) and (v, u) in E. In addition, we denote by $N(v) = \{u, (u, v) \in E\}$ the set of neighbors of any node v in V.

We performed experiments on a wide ranges of large worldwide cities of diverse sizes and structures. This led to no significant difference on obtained results. We therefore use a typical instance, namely Paris, to present our work in this paper. This street network has 9 602 nodes, 14 974 links, leading to average degree 3.1. Its diameter is 83 hops and its average distance is 39.4 hops. The average street length is 99 m, and the average distance is 5552 m.

The links of a street network generally represent street segments of very heterogeneous lengths [11]. Then, moves from a node to another one may have very different duration. In order to model this, we use a classical discretization procedure [12] that consists in splitting each link of the street network into pieces connected by evenly spaced nodes. In the obtained graph, each link represents a street slice of length close to a step δ. Then, the walkers defined in previous section consistently make a move of length approximately δ at each time step.

In this paper, we use δ equal to 10 m, leading to a network of $N = 145000$ nodes and $M = 300736$ links. It gives a sufficient precision in our context, and experiments with other reasonable values of δ displayed no significant difference.

2.2 Walkers

Given a network $G = (V, E)$, we consider a set W of walkers numbered from 1 to $|W|$. We denote the location of walker i at time t by $x_i(t) = v$, with $v \in V$. At each time step t, walker i moves to node $x_i(t+1) \in N(v)$.

For any link (u, v) in E, we also define the *flux* of walkers from u to v as $J_{u\to v}(t) = |\{i, x_i(t-1) = u, x_i(t) = v\}|$. We call *group* the set of walkers at a given node v at a given time t: $g_v(t) = \{i, x_i(t) = v\}$. We denote by $n_v(t) = |g_v(t)|$ the number of walkers located at node v at time t and by $g(t) = |\{g_v(t), v \in V, g_v(t) \neq \emptyset\}|$ the number of non-empty groups at step t.

3 Walker Model

At each time step t, each walker i moves from location $x_i(t)$ to location $x_i(t+1)$ in $N(x_i(t))$. This section presents how we choose the new location $x_i(t+1)$.

3.1 Available Information

We assume that walkers have very limited capabilities: they have no access to the actual location of other walkers, even on neighbor nodes, and they have no long-term memory and no communication protocol. Instead, we assume they only have access to estimates of aggregated observables, such as the number of walkers on neighbor nodes and the flow intensity on surrounding links.

More formally, walker i may use the following information:

- its previous location $x_i(t-1)$,
- the number $n_v(t)$ of walkers on node v for all v in $N(x_i(t)) \cup \{x_i(t)\}$,
- the flux of walkers arriving and leaving its current location $v = x_i(t)$, *i.e.* $J_{u\to v}(t)$ and $J_{v\to u}(t)$ for all u in $N(v)$.

Thanks to the information above, each walker i knows it previous location $u = x_i(t-1)$, its current location $v = x_i(t)$, and it has access to various criteria to decide its next location $w = x_i(t+1)$. Then, it needs a way to derive walking rules from criteria, and a way to combine these walking rules into a tactic.

3.2 Criteria

A criterion is a parameter from which we construct walking rules. We consider the following set of criteria $\mathcal{C}$.

- *Random.* The walker makes no difference between all possible neighbors: the criterion has value 1 for each of them.
- *Propulsion.* The walker never goes back to its previous location u: the criterion has value 0 for u and value 1 for other neighbors of v.
- *Attraction.* The walker preferably moves to nodes where there are already many walkers: the criterion is equal to the number of walkers $n_w(t)$.
- *Follow.* The walker preferably follows the most popular moves of other walkers: the criterion has value $J_{v\to w}(t)$.
- *Alignment.* The walker takes into account the net flux in both directions: the criterion is equal to $\Delta J_{v\to w}(t) = J_{v\to w}(t) - J_{w\to v}(t)$.

3.3 Walking Rules

Let us consider a criterion $C_{u,v,w}(t)$. We define the corresponding walking rule using the classical *logit rule*. It gives the probability $\omega^C_{u,v,w}(t)$ that walker i moves from v to w, given the fact that it arrived at v from u:

$$\omega^C_{u,v,w}(t) = \frac{e^{\beta \cdot C_{u,v,w}(t)}}{\sum_{z \in N(v)} e^{\beta \cdot C_{u,v,z}(t)}} \tag{1}$$

Parameter $\beta \geq 0$ is the intensity of choice: it quantifies the influence of the criterion on walker choices. If $\beta = 0$, the criterion has no influence and walkers make purely random choices. If $\beta \to \infty$ walkers necessarily choose a neighbor among the ones that maximize the criterion.

3.4 Tactics

A *tactic* is a linear combination of walking rules that defines the probability $\pi_{u,v,w}(t)$ to move from v to one of its neighbors w when coming from u:

$$\pi_{u,v,w}(t) = \sum_{C \in \mathcal{C}} \alpha_C \cdot \omega^C_{u,v,w}(t) \tag{2}$$

where α_C is the coefficient of criterion C, with $\sum_{C \in \mathcal{C}} \alpha_C = 1$. Therefore, a tactic is defined by a set of criteria and their coefficients. We call *strict tactic* one that has all its coefficients, except one, set to zero and then always follows the same criterion.

In practice, at each time step, each walker selects a criterion C with probability α_C. Then, it computes its probability $\omega^C_{u,v,w}$ to go to each neighbor node w and selects its new location accordingly.

As transition probabilities at time t depend on previous non-deterministic moves, formal analysis of such processes is generally out of reach [4]. We will therefore explore possible tactics using simulations.

3.5 Baseline

In addition to the tactics above, we consider a reference baseline that easily achieves flocking thanks to collective decisions. This means that, at each time step, all walkers at a given node make the same choice. We then obtain reference results that we expect our walker models to reach or outperform, even though they are unable to make collective decisions.

More formally, for each node v at time t, a unique random neighbor u of v is chosen and all walkers i, such that $x_i(t) = v$, move to $u = x_i(t+1)$. This is equivalent to purely random walks of groups until they meet another group.

Since we consider non-bipartite connected graphs, it is well known that all groups will eventually merge. The number of needed steps is called the coalescence time [6]. Even though this number may be prohibitive, this means that the baseline successfully produces large groups. In addition, groups are mobile since, once formed, they perform purely random walks. As a consequence, this baseline makes a relevant reference for the success of flocking walkers.

4 Experiments

We seek to design good tactics for our walkers. They must produce flocking faster than the baseline presented in Sect. 3.5. Moreover, we want to obtain short time convergence, as our walkers model the action of protesters that can not walk forever. In this Section we define metrics to evaluate how well walkers flock. We then run simulations for a reasonable duration and evaluate which tactics have performed the best in this time interval.

4.1 Flocking Metrics

We characterize *flocking* as a gathering of walkers exploring the network. With notations of Sect. 2.2, this leads to the two following *score* definitions for a given run of a given tactic.

Definition 1. (cluster, gathering score $\rho(t)$) *A cluster is a maximal connected sub-graph with walkers on all its nodes. The gathering score $\rho(t)$ is the average number of walkers in clusters.*

For example, when every agents are in the same cluster, $\rho(t) = |W|$. If instead all agents are in different ones, it equals 1.

Definition 2. (mobility score $\mu(t)$) *The mobility $\mu_i(t)$ of walker i is the number of distinct nodes a walker has already visited at time t: $\mu_i(t) = |\{v, \exists t' \leq t, x_i(t') = v\}|$. The mobility score $\mu(t)$ is the average walker mobility.*

Notice that the mobility score is monotonically non-decreasing with time: $\mu(t+1) \geq \mu(t)$. In addition, if all walkers move to a node they already visited then $\mu(t+1) = \mu(t)$.

We are interested in tactics with high gathering and mobility scores. Indeed, the high gathering score ensures that walkers form significant groups. In addition, the high mobility score implies that walkers continue to move. However, the fact that the gathering score remains high shows that walkers stay grouped. This means that the tactic successfully achieves flocking.

In order to gain more insight on group structure and dynamics, we introduce an additional metric.

Definition 3. (sprawling score $\sigma(t)$) *The sprawling score $\sigma(t)$ is the average number of nodes in clusters.*

If the sprawling score is 1, all groups are isolated from each other: whenever walkers are at a node, there is no walker on neighbor nodes. If instead the sprawling score is high, walkers form large clusters of neighbor groups. Its largest possible value is the total number of nodes in the network, meaning that there are walkers on each node.

We are interested in tactics with low sprawling score, meaning that they succeed in merging neighbor groups.

4.2 Extensive Exploration of Tactics

In this Section, we follow an extensive method to explore the wide set of possible tactics on an entire city network:

- we consider the Paris street network discretized with parameter $\delta = 10$ meters, leading to a network of $N = 145000$ nodes and $M = 300736$ links,
- we consider N walkers initially distributed uniformly at random on nodes,
- we perform 1000 time steps, thus considering reasonably short walks of approximately 10 kilometers,
- we set β large enough to ensure each walking rule strictly follows its criterion,
- we finally consider all tactics obtained as combinations of α_C parameter values from 0 to 1 by steps of 0.1 such that $\sum_{C \in \mathcal{C}} \alpha_C = 1$.

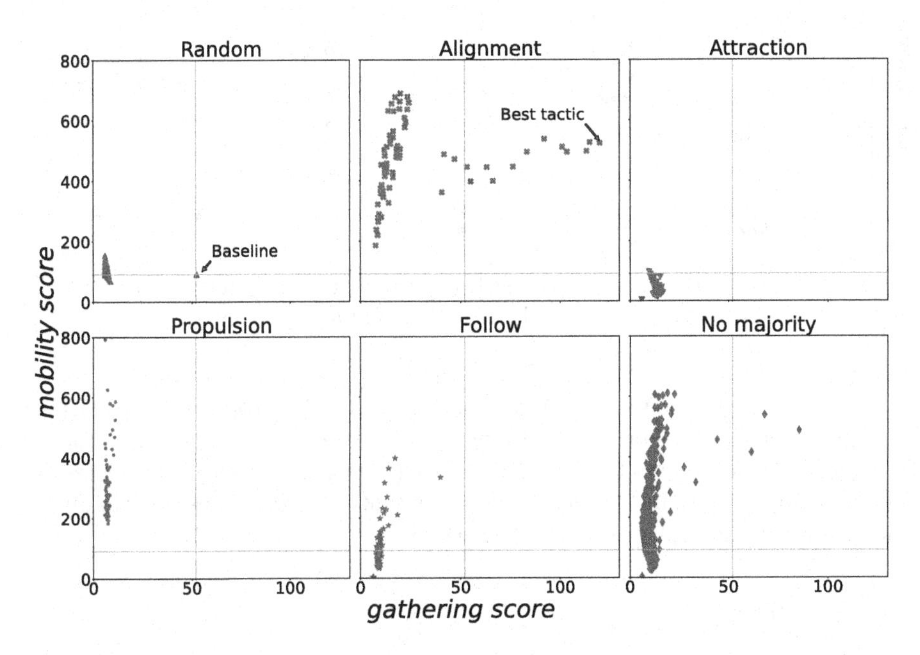

Fig. 2. *Gathering and mobility of all tactics.* Each dot corresponds to the average last step value from ten runs of a tactic. The horizontal axis gives the gathering score, the vertical one gives the mobility score. From left to right and from top to bottom: tactics based mostly on the Random, Alignment, Attraction, Propulsion and Follow rule, respectively. On each of these plots, we indicate the strict tactic, that exclusively uses the corresponding rule. The bottom-right plot corresponds to tactics with no prevailing rule.

With this setup, we obtain 1001 different tactics. We run 10 simulations of each tactic and we plot the average mobility and gathering scores in Fig. 2. In these plots, each dot corresponds to a tactic defined by a set of α_C parameter values. We split these tactics into six plots: we display a set of tactics on the same plot if they all have $\alpha_C > 0.5$ for the same criterion C, and we display on the last plot the set of all other tactics.

We also display in each plot of Fig. 2 a vertical and an horizontal line that indicate baseline results. Then, the tactics achieving the best flocking performances are the ones in the upper right corner: they obtain groups bigger and are more mobile that the baseline.

We pay particular attention to strict tactics, which performances are spotted by red dots in Fig. 2. We also highlight what we identify as the best tactic regarding our scores. It is the tactic with the highest gathering score among tactics that outperform our baseline. We display the evolution of mobility and gathering scores over time for these tactics in Fig. 3.

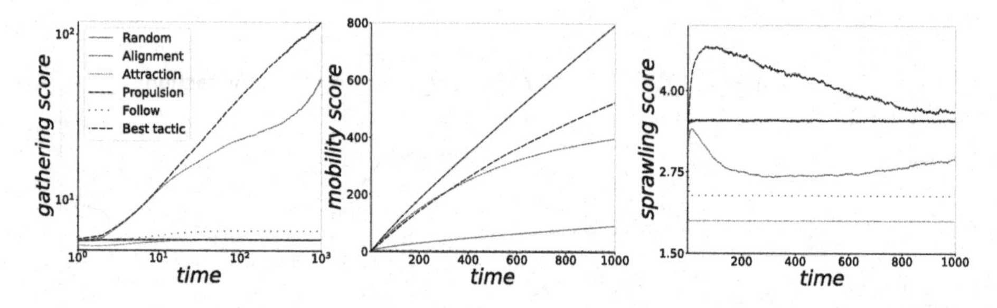

Fig. 3. Plots showing the evolution of gathering, mobility and sprawling scores for the strict tactics and the best tactic. Gathering score is in log-log scale.

Figure 2 clearly shows that Alignment-based tactics (top center plot) outperform others. All other sets of tactics perform poorly, except a few tactics for which no rule weights up more than 50% (bottom right plot). These tactics actually also use Alignment rule, to a lesser extent. This identifies the Alignment rule as a key building block for flocking tactics.

4.3 Best Tactics

We now focus on the two main tactics that achieve flocking: the Alignment strict tactic and the best tactic (the one that corresponds to the rightmost dot on Fig. 2). Figure 3 displays their scores over time in green and purple colors, respectively.

The plots show that the mobility score of both tactics first rapidly grow, and that this growth significantly decreases over time. Even if these tactics are not as good as the Propulsion strict tactic regarding mobility, they have comparable performances for this metric.

Notice that the Propulsion strict tactic has very low gathering scores (Fig. 3, left), which makes it an irrelevant tactic despite its mobility score. Instead, both the Alignment strict tactic and the best static quickly reach excellent gathering scores. The best tactic significantly outperforms the Alignment strict tactic and has a linear gathering score plot in log-log scale. This means that its evolution has a polynomial growth (of exponent below 1), indicating a fast growth, but also that the evolution of group size tends to flatten over time. This is due to groups reaching a state where all clusters of groups are in distant regions of the network. Then, it takes longer for groups to meet other groups, merge, and grow in size.

Finally, Fig. 3 also displays the sprawling score for all considered tactics. We observe that the best tactic produces a greater sprawling score than Alignment. The sprawling, for those two tactics, is due to groups following each others when they detect another group on a neighbor node, without necessarily merging with it.

With the Alignment strict tactic, the sprawling of groups first very quickly increases, then decreases and stabilizes. This is because this tactic forms groups immediately at the beginning, mostly as lines of walkers following each others. The sprawling reduces as groups reach intersections and split, until the aggregation and splitting dynamics reach an equilibrium.

For the best tactic, groups aggregate into lines for a longer time period, resulting into a much higher sprawling score. It then slowly linearly decreases until the end of the run. This is because the Attraction rule, when chosen in the best tactic, will make the front group wait for the groups behind it, leading to less sprawled clusters.

As explained in Sect. 4.1, an efficient tactic should have a low sprawling score. The sprawling of the baseline is 1 (up to the third digit), thanks to the collective decision. This is the optimum value.

Our walkers do not have access to collective decision. When a cluster of groups arrives at an intersection, at the end of a street, it can split into multiple groups. In the case of a lone group (a cluster with no sprawling) it will split into multiple groups, with equal number of walkers on average. This reduces the gathering score.

4.4 Interpretation

Recall that the best tactic corresponds to the rightmost dot among those in the upper right corner of plots in Fig. 2. It is defined by the following parameters: $\{\alpha_{follow} = 0.1, \alpha_{align} = 0.8, \alpha_{attr} = 0.1\}$ and a null weight for other rules. This tactic produces groups of 121 walkers on average, and walkers explore on average 540 distinct nodes during their 1000 steps. These scores are more than twice and five times more than what the baseline gets, respectively.

Figure 4 illustrates the behavior of this tactic. First, Alignment imposes walkers to move forward, may they be alone or part of a cluster, as shown in the first configuration of Fig. 4. Indeed, a walker i alone at location $x_i(t) = u$ and $x_i(t-1) = v$ will measure a negative flux $\Delta J_{u \to v}(t) = -1$ at time t, while it will be $\Delta J_{u \to w}(t) = 0$ for all $w \neq v$. This implies the walker never goes back. This effect is left unchanged with multiple walkers in a cluster.

Second, this same rule guarantees that, if two groups cross path, they then merge in a single cluster in which all walkers will follow the same path. Indeed

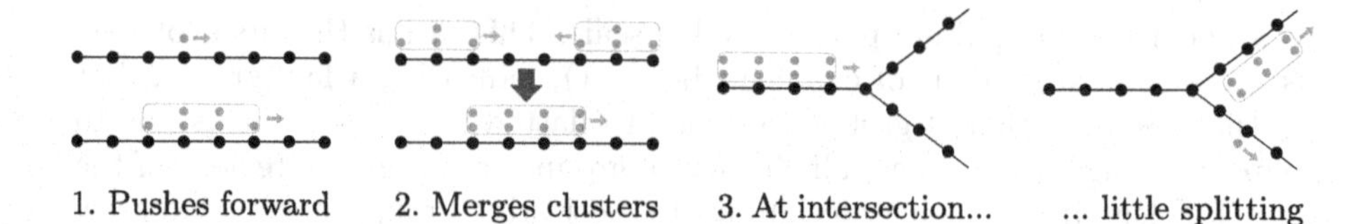

Fig. 4. Schematic configurations that walkers, groups and clusters of groups achieve with the best tactics.

when a group u cross path with a smaller group v, we have $x_i(t+1) = x_j(t)$ and $x_j(t+1) = x_i(t)$ for all $i \in g_u(t)$ and all $j \in g_v(t)$. The net flux is then $\Delta J_{u \to v}(t) = n_u(t) - n_v(t) > 0$ for all walkers, which drive them all in the same direction: the smaller group goes back towards v (where it came from) while the trajectory of the largest group is left unchanged.

However, this rule alone is not perfectly effective at avoiding splitting. The walkers in the first group in a cluster will not always all choose the same node at an intersection, as illustrated in the two bottom pictures of Fig. 4, as the different possible nodes all have a flux equal to 0. In this context, the Attraction rule allows walkers that split in the least chosen direction at an intersection to go backward and avoid loosing sight of the cluster.

Finally, the Follow rule improves the gathering. Indeed, clusters of groups tend to sprawl when walkers use the Alignment rule. In such chain of groups following each others, the Follow rule allows walkers in the front group to move backwards, merging with the group behind them, while it forces walkers in other groups to move forward to catch up the leading group.

This equilibrium between those three rules gives an outcome where groups flock very efficiently.

5 Related Work

Our approach is different from usual protest studies based on thresholds [9] or agent-based [1,8,10] model. Indeed, these works focus on how people decide to participate to a protest; they do not deal with protester mobility.

In distributed systems, computer scientists proposed solutions to gathering problems where walkers follow a common distributed algorithm to meet on any connected graph [2,13].

Flocking is a collective dynamic where groups of walkers move spontaneously in the same *direction* [7,15,16]. It has been largely studied in free space. Few articles explore flocking when trajectories of walkers are network constrained. For example, in [14] the authors implement rules similar to ours to study the possible outcomes of their combinations on a line. They get results very similar to ours in a single street.

Still, to our knowledge, we are the first to experiment rules for flocking that we can apply on any network.

Acknowledgments. We warmly thank Ivan Mulot-Radojcic for his help with implementation and documentation (Supplementary material: https://k-avi.github.io/protesting_on_graphs). This work is funded in part by the CNRS through the MITI interdisciplinary programs.

Disclosure of Interests. The authors have no competing interests to declare that are relevant to the content of this article.

References

1. Agamennone, M.: Riots and Uprisings: Modelling Conflict between Centralised and Decentralised Systems. Ph.D. thesis, King's College London (2021)
2. Bhagat, S., Pelc, A.: How to meet at a node of any connected graph. In: 36th International Symposium on Distributed Computing (DISC 2022). Schloss Dagstuhl-Leibniz-Zentrum für Informatik (2022)
3. Boeing, G.: OSMnx: New methods for acquiring, constructing, analyzing, and visualizing complex street networks. Comput. Environ. Urban Syst. **65**, 126–139. https://doi.org/10.1016/j.compenvurbsys.2017.05.004, https://linkinghub.elsevier.com/retrieve/pii/S0198971516303970
4. Brémaud, P.: Markov Chains: Gibbs fields, Monte Carlo simulation, and queues, vol. 31. Springer Science & Business Media (2001). https://doi.org/10.1007/978-3-030-45982-6
5. Coast, S.: OpenStreetMap (2004). https://www.openstreetmap.org/
6. Cooper, C., Elsässer, R., Ono, H., Radzik, T.: Coalescing random walks and voting on connected graphs. SIAM J. Discret. Math. **27**(4), 1748–1758 (2013). https://doi.org/10.1137/120900368
7. Cucker, F., Smale, S.: Emergent behavior in flocks. IEEE Trans. Autom. Control **52**(5), 852–862 (2007). https://doi.org/10.1109/tac.2007.895842
8. Epstein, J.M.: Modeling civil violence: an agent-based computational approach. Proc. Natl. Acad. Sci. **99**, 7243–7250 (2002)
9. Granovetter, M.: Threshold models of collective behavior. Am. J. Sociol. **83**(6), 1420–1443 (1978)
10. Lemos, C., Coelho, H., Lopes, R.J., et al.: Agent-based modeling of social conflict, civil violence and revolution: state-of-the-art-review and further prospects. EUMAS, pp. 124–138 (2013)
11. Masucci, A.P., Smith, D., Crooks, A., Batty, M.: Random planar graphs and the London street network. Eur. Phys. J. B **71**, 259–271 (2009)
12. Neri, I., Kern, N., Parmeggiani, A.: Totally asymmetric simple exclusion process on networks. Phys. Rev. Lett. **107**(6), 068702 (2011)
13. Pelc, A.: Deterministic rendezvous in networks: a comprehensive survey. Networks **59**(3), 331–347 (2012)
14. Raymond, J., Evans, M.: Flocking regimes in a simple lattice model. Phys. Rev. E **73**(3), 036112 (2006)
15. Reynolds, C.W.: Flocks, herds and schools: a distributed behavioral model. In: Proceedings of the 14th Annual Conference on Computer Graphics and Interactive Techniques, pp. 25–34 (1987)
16. Vicsek, T., Czirók, A., Ben-Jacob, E., Cohen, I., Shochet, O.: Novel type of phase transition in a system of self-driven particles. Phys. Rev. Lett. **75**(6), 1226 (1995)

Unraveling the Italian and English Telegram Conspiracy Spheres Through Message Forwarding

Lorenzo Alvisi[1,2], Serena Tardelli[2(✉)], and Maurizio Tesconi[2]

[1] IMT School for Advanced Studies, Lucca, Italy
lorenzo.alvisi@imtlucca.it

[2] Institute of Informatics and Telematics, National Research Council, Pisa, Italy
{serena.tardelli,maurizio.tesconi}@iit.cnr.it

Abstract. Telegram has grown into a significant platform for news and information sharing, favored for its anonymity and minimal moderation. This openness, however, makes it vulnerable to misinformation and conspiracy theories. In this study, we explore the dynamics of conspiratorial narrative dissemination within Telegram, focusing on Italian and English landscapes. In particular, we leverage the mechanism of message forwarding within Telegram and collect two extensive datasets through snowball strategy. We adopt a network-based approach and build the Italian and English Telegram networks to reveal their respective communities. By employing topic modeling, we uncover distinct narratives and dynamics of misinformation spread. Results highlight differences between Italian and English conspiracy landscapes, with Italian discourse involving assorted conspiracy theories and alternative news sources intertwined with legitimate news sources, whereas English discourse is characterized by a more focused approach on specific narratives. Finally, we show that our methodology exhibits robustness across initial seed selections, suggesting broader applicability. This study contributes to understanding information and misinformation spread on Italian and English Telegram ecosystems through the mechanism of message forwarding.

Keywords: Telegram · Message forwarding · Linked chats · Conspiracy · Network · Communities

1 Introduction

Telegram has grown popular as a significant hub for news and information thanks to its commitment to anonymity, low moderation, and privacy. Yet, the very features that attract users also open doors for misinformation and conspiracy theories to spread on topics such as the infodemic, pandemic, and other societal issues [6,16,20] In fact, the platform has also facilitated ideology radicalization, coordination of attacks, mobilizing protests, and the promotion of other conspiratorial narratives, thus playing a crucial role in influencing public discourse and

L. M. Aiello et al. (Eds.): ASONAM 2024, LNCS 15212, pp. 204–213, 2025.
https://doi.org/10.1007/978-3-031-78538-2_18

impacting democratic processes [12,25]. Analyzing how these phenomena organize and characterize is crucial for understanding the direction of public discourse and the factors influencing it. This understanding is vital not only for making online environments safer but also for grasping potential offline developments.

In this study, we analyze the spread of conspiratorial narratives within Telegram communities through message forwarding, specifically within Italian and English language landscapes. Message forwarding on Telegram involves sharing a message from one chat directly into another, serving as a critical mechanism for distributing content across different user groups. We hypothesize that forwarded messages not only distribute content but also signal homophily, that is shared interests and beliefs, among community members, similar to how the diffusion of invite links has been studied in the past [1,17]. Specifically, we first collect data from Telegram by leveraging message forwarding. Starting from selected initial chats as seeds, we perform iterative, snowball sampling and expand the data by retrieving new chats, including channels, groups – often overlooked in existing literature, and messages. For the first time, we also incorporate linked chats, which are two-tiered structures consisting of channels linked to their respective groups. We collect two large datasets from January to February, 2024: the Italian dataset comprises more than 1K chats and 3.4M messages, while, the English dataset consists of more than 600 chats and 5M messages. We build two Telegram networks based on message forwarding, identify key communities and characterize conspiratorial narratives within Telegram communities, focusing on both English and Italian spheres, shedding light on Italian Telegram dynamics not extensively explored in existing literature. We show that the Italian landscape of conspiracy theories forms a network involving religious groups, Russian influences, anti-vaccination proponents, and news source of varying reliability. In contrast, the English landscape appears more tied to structured conspiracies, involving ties with cryptocurrency scams. Finally, we validate our method by showing that our findings do not depend on the initial selection of seeds, suggesting the robustness and broad applicability of our methodology.

2 Related Works

2.1 Telegram Data Collection Methods

Several studies relied on message forwarding to collect data from Telegram. For example, the authors in [15] aimed to create the largest collection of English Telegram channels, spanning a wide range of diverse topics, with their analysis primarily centered on dataset statistics. In contrast, research in [24] analyzed communities by building user networks from forwarded messages, and exploring the narratives within. Similarly, research in [4] and [3] followed a snowball characterized specific English-speaking Telegram communities of channels. Our study, however, expands on this foundation by incorporating not just channels but also groups into our analysis. Specifically, we uniquely consider the *linked chat* feature on Telegram, where a channel is directly connected to a group. To the best of our knowledge, this is the first research effort to include this duality feature in literature. Other studies adopted snowball approaches on Telegram, focusing on different elements like mentions or invite links [17,22]. Lastly, other studies employed different data collection strategies, such as gathering messages from an initial set of seeds without employing a snowballing approach [2] or leveraging invite links. These studies primarily aim to illustrate the unfolding of specific events, like instances of toxicity or fraud schemes.

2.2 Conspiracy in Italian and English Telegram Discussions

Conspiracy theories have been identified and analyzed across various platforms, thriving in numerous online environments [5,8,10], including Telegram. The majority of the research on Telegram has focused on conspiracy theories within English-speaking discussions, including studies on the far-right and the QAnon movements [12]. Notably, the QAnon conspiracy, in particular, has been linked to a wide range of conspiratorial narratives, highlighting its broad influence [24]. On the other hand, the realm of conspiracy theories within Italian-speaking Telegram communities remains largely unexplored. The Italian conspiracy ecosystem on Telegram came to the spotlight during the COVID-19 pandemic [23], as protest movements gained significant social momentum, leading to widespread protests [19], sometimes with ties to the Italian alt-right, a phenomenon also observed in other European countries [25]. Other studies focused into the Italian QAnon disinformation infrastructure [18], highlighting the closed nature of these communities within the Italian sphere, similarly to English-speaking environments [24]. Despite these insights, a comprehensive understanding of the broader conspiracy landscape in Italy remains unexplored. Our study seeks to fill this gap by examining the connections between various conspiracy narratives in Italian-speaking Telegram communities, and comparing them with English-speaking communities.

3 Methodology

3.1 Telegram Terminology

Telegram offers a variety of chat types. *Channels* are unidirectional chats where typically only administrators broadcast content to an audience that cannot interact directly. *Groups* are chat rooms where all members have permission to share contents by default and interact with each other. *Supergroups* are a variation of groups, differentiated mainly by administrative powers and member limits. However, for our study, we treat the latest as equivalent to regular groups. A notable feature in Telegram is the ability for channel admins to link a channel to a corresponding group, creating a two-tiered structure known as *linked chat*. In this structure, a channel enables any user, whether a follower or not, to reply directly to each post. Simultaneously, the associated group houses these conversational threads and operates as a standard group. This composite structure allows unrestricted interaction on the channel's posts and fosters broader discussion within the group. For the scope of our paper, we consider public channels, groups, and linked chats. We use the term *chat* interchangeably to refer to all three types. As mentioned, we highlight a key Telegram feature, that is the ability for users to share posts and messages from one chat to another via *message forwarding*. This feature preserves the original chat's information, effectively creating a bridge between chats and facilitating the discovery and retrieval of connected content.

3.2 Dataset

We retrieve two distinct Telegram datasets pertaining to conspiracy discussions in Italian and English using the following approach. We employ a snowball technique focused on message forwarding. For the first time, we expand this technique to include groups and linked chats. We start by retrieving seed chats known for conspiracy content on *tgstats.com*, a platform providing a categorized catalog of existing Telegram chats. For the Italian data, we focus on terms associated with pandemic conspiracy theories, identifying 43 Italian chats related to conspiracies as seeds. For the English seeds, we search for keywords associated with the QAnon conspiracy, resulting in 20 seed chats. We start from two different conspiracy theories to anchor our study in the specific cultural and linguistic contexts, ensuring a focus on the conspiracy sphere and exploring how these conspiracies expand and evolve in these settings. We leverage Telegram APIs to collect the messages: starting with seed chats at iteration 0, we parse messages to identify forwarded messages, following them to retrieve new chats that meet our language criteria, either Italian or English, determined by the most frequently detected language. Our data collection concludes after iteration 2. The final datasets cover the period from January to February, 2024: the Italian dataset includes 1,346 chats and 3.4M messages; the English dataset comprises 634 chats and 5M messages. Notably, when examining the distribution of users and comments per chat, we find that the log-number of users and messages follows a Gaussian distribution. This contrasts with the typical heavy-tailed distribution of conversational trees found in previous research [2]. This difference might suggest that linked chats, similar to chat rooms, behave differently from traditional social media feeds. Alternatively, our snowball sampling might be missing smaller, less influential chats.

3.3 Network Construction and Community Detection

The message forwarding mechanism enables us to construct a directed weighted graph $\mathcal{G} = (\mathcal{N}, \mathcal{E})$, where $\mathcal{N}$ represents the set of nodes and $\mathcal{E}$ the set of edges. In this graph, nodes correspond to chats, which include unlinked channels, unlinked groups, and linked chats. For any two nodes $u, v \in \mathcal{N}$, the weight of the edge $w_{e_{u,v}} \in \mathcal{E}$ is determined by the number of messages forwarded from chat u to chat v. To prevent loops, forwards from a chat to itself, including within linked chats, are excluded. This exclusion is crucial as, in linked chats, each message from the channel is automatically forwarded to the associated group to form conversational trees. The Italian network consists of 1,346 nodes and 35,802 edges, and the English network comprises 634 nodes and 24,546 edges. We employed community detection within our graph using the Louvain algorithm tailored for directed graphs [7], focusing only on communities with more than 10 chats.

4 Results

4.1 Uncovering Narratives

Here, we present summary information for each community, alongside their main narratives. We uncover the main topics by leveraging topic modeling techniques, channel information, and by examining TF-IDF weighted hashtags used by each community. To perform topic modeling, we adopted a state-of-the-art algorithm known as Anchored Correlation Explanation (CorEx) [9]. Unlike traditional methods like Latent Dirichlet Allocation (LDA), CorEx identifies hidden topics within a collection of documents without assuming any particular data generating model. Given that our networks consists of chat platforms and their messages, we trained separate models for each community using the chat messages as corpora. We set the expected number of topics to 10, as additional topics were adding negligible correlation to the learned models. Finally, we ranked the obtained topics according to the fraction of the total correlation that they explain. Results are discussed as follows.

Italian Narratives. The Italian-speaking communities are presented as follows:

- `Freedom`: This community is centered around concepts of liberal democracy and dissent, discussing geopolitical topics, democracy, governance, and control-related issues.
- `Warfare`: A community concerned with international warfare, particularly focusing on the Ukrainian conflict and Russian propaganda.
- `ConspiracyMix`: A community about various conspiracy theories involving government actions, health-related topics such as the pandemic, and foreign political figures.
- `ConspiracyMix2`: Similar to ConspiracyMix, this community spans across conspiracy theories, touching on warfare, vaccines, farmers' protests, and QAnon.
- `NewsSource`: A community that encompasses a spectrum of information sources ranging from conspiracy theory-driven outlets to reputable journalistic sources (e.g., "IlSole24Ore"). This convergence reflects the dynamics of conspiratorial contexts, where genuine information is often filtered through a conspiratorial lens [14].
- `Politics`: A political community discussing economic issues, government policies and European affairs.
- `AltNews`: A community focused on counter-information and alternative news sources, focusing on issues of censorship, globalism, and societal control.

- `Fight`: A community engaged in civil struggles, emphasizing the importance of truth, freedom, and action in the face of societal challenges.
- `Novax`: A community characterized by dissent against vaccinations, and other health related studies.
- `Religious`: A community centered on Italian religious values, discussing Jesus, sacraments, and other themes of rebirth, envy, exorcism, and healing.
- `Spiritual`: A community centered on spiritual topics, such as spiritual awakening and meditation.

These communities discuss conspiracy theories with alternative information challenging mainstream narratives to news source offering more traditional views. In addition, conspiracy narrative ties to religiosity, alternative health, and conspiratorial thinking, as observed in literature for English-speaking groups [11,24]. Exploring these groups gives us insight into the Italian conspiracy ecosystem on Telegram, a subject that is relatively unexplored in existing literature.

English Narratives. To provide valuable comparative insights into conspiracy theories in different cultural contexts, we present the English-speaking communities as follows:

- `QAnonCrypto`: A community where conspiracy discussions are hijacked by the cryptocurrency world.
- `Warfare`: A community similar to its Italian counterpart, focusing on the Ukrainian conflict, military issues, and other war rhetoric.
- `QAnonHealth`: A community where QAnon conspiracy theories intersect with health concerns, discussing food, cancer, and parasites, along with other medical aspects.
- `CHScams`: A community that relies on conspiracy theory discussions to promote financial scams and fraudulent activities in Chinese language.
- `QAnon`: This community focuses on pure QAnon conspiracy theories, involving topics such as child abuse, government control, and political figures.
- `ConspiracyMix`: This community discusses various conspiracy theories, with a focus on legal issues, while also touching the cryptocurrency sphere.
- `Covid`: A community centered around discussions of COVID-19, vaccine skepticism, and related health and governmental issues.
- `OldSchoolConsp`: A community focused on traditional conspiracy topics such as UFOs, aliens, the paranormal, and discussions of time and consciousness.

The English-speaking communities exhibit a marked tendency towards insularity, as QAnon is a very closed community [4]. Indeed, many communities, although primarily connected with QAnon themes, show a distinct emphasis on topics such as cryptocurrency, health, or governmental affairs, unified by an underling QAnon narrative. This phenomenon of thematic variations within a singular ideological framework is indicative of the QAnon community's cohesiveness. Indeed, prior work has observed an increasing association of QAnon with religiosity, alternative health, wellness philosophies, and affective states promoting conspiratorial thinking [24].

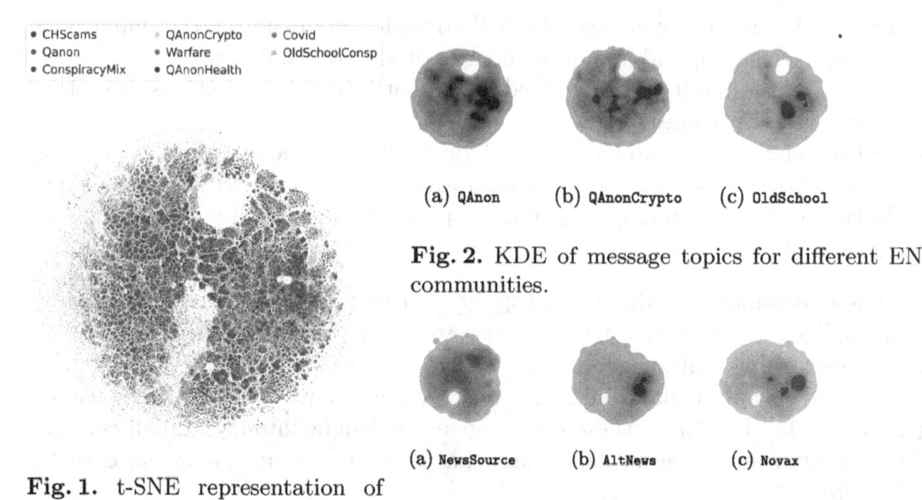

Fig. 1. t-SNE representation of message distribution by topic in the EN Dataset.

(a) `QAnon` (b) `QAnonCrypto` (c) `OldSchool`

Fig. 2. KDE of message topics for different EN communities.

(a) `NewsSource` (b) `AltNews` (c) `Novax`

Fig. 3. KDE of message topics for different IT communities.

4.2 T-SNE for Context Analysis

To provide a comprehensive representation of the topics discussed within our datasets, we represent all messages using t-Distributed Stochastic Neighbor Embedding (t-SNE) [13]. Spatial proximity in the t-SNE map can suggest how topics fit into the larger conversation on conspiracies. We build the t-SNE visualization on topics identified by the CorEx algorithm. In particular, we developed two distinct models, one for Italian and one for English to analyze the entire corpus of messages. We opted to identify 50 topics to further our understanding of the context dynamics inside the clusters. This is particularly important because Telegram chats often cover a broad range of topics rather than focusing on a single subject [12]. We obtain and $n \times m$ matrix where n and m are respectively the number of messages and the number of topics we wanted to detect. Each value $v_{i,j}$ represents the correlation between the ith message and the jth topic. We lower the dimensionality of our matrix using the tSNE and plot all messages in a two-dimensional space, coloring them according to the community of origin to show how clusters are closely related or share similar discussions. Figure 1 presents the results on the English dataset. The varying distributions of the messages across communities highlight the differences in discussion in terms of quantity, focus, and framework, even among similar communities. This spatial arrangement underlines the nuanced interactions between these communities. For example, we can observe the proximity of the `QAnonCrypto` community to the `QAnon` and the `QAnonHealth` communities, suggesting that crypto topics tend to piggyback engage with QAnon-related discussions. Figure 2 better presents the differences in distributions through Kernel Density Estimation (KDE) of the messages, where areas of higher density indicate a higher likelihood of encountering messages related to specific topics. For instance, in Fig. 2a, the distribution of messages in chats of the `QAnon` community is notably widespread, suggesting correlations with many different topics, similarly to `QAnonCrypto`. This suggests that some communities on Telegram tend to discuss a broad array of topics, they each enrich the discourse with their unique frameworks and worldviews. In contrast, more specialized communities like `OldSchoolConsp` (Fig. 2d) are localized to very specific areas. We conduct the same analysis for the Italian dataset. Due to space constraints, we highlight only some notable patterns. We observe distinct patterns between the `NewsSource` (Fig. 3a) and `AltNews` (Fig. 3b) communities, which both cover alternative news topics. However, `NewsSource` also includes legitimate news sources, resulting in messages that show dual density peaks, possibly indicating interdependence, whereas `AltNews` messages display a single density peak, reflecting a more homogeneous topic focus.

5 Validation

To assess the robustness of our findings, we aim to determine if starting from different seeds results in the same chat composition in our dataset. We focus on the Italian dataset and create a counterpart validation dataset using the snowballing process, this time starting from a distinct set of 28 seeds that were not among the original 43 Italian seeds used in the initial data collection. These new seeds are sourced from the *butac.it* blacklist, a list of Italian disinformation Telegram channels. The collected dataset includes 1,591 chats active from February to March, 2024. We stopped the collection after two iterations of the process to maintain consistency with the original methodological framework. We determine if the chats retrieved in the validation dataset match those in our original dataset, by examining the overlap between the Italian datasets and the validation dataset. We find that 80% of the chats in the validation dataset are also present in our original dataset, suggesting that our results would remain robust even with a different set of seeds. Moreover, chats excluded from the original dataset have lower averages in size, in-degree, and out-degree, suggesting that the missing chats have less influence within the dataset. These results show that the insights derived from our network analysis are not overly dependent on the initial seeds used to construct the dataset.

6 Conclusions

In this study, we analyzed online Italian and English conspiracy-related Telegram communities through the lens of message forwarding, aiming to uncover the dynamics of conspiracy theory discussions in different speaking contexts. Using snowball sampling, we collected two extensive datasets encompassing Telegram channels, groups, linked chats, and messages shared over from January to February, 2024. We uncovered trends of thematic diversity within a cohesive ideological framework, as trends similarly observed in literature for English-speaking groups [24]. In addition, the presence of news sources and alternative news outlets shows a dynamic interplay in the legitimization of conspiracy theories, highlighting the intricate balance between mainstream credibility and counter-narratives. This enriches our understanding of the Italian conspiracy ecosystem on Telegram, a relatively uncharted territory in existing literature. Finally, we tested our methodology's robustness against variations in initial dataset seeds, showing the reliability of our insights and broader applicability. As the diffusion of misinformation cannot be fully captured through a static analysis, future work should incorporate temporal analyses to uncover temporal dynamics on Telegram [21]. This research contributes new perspectives on misinformation spread, paving the way for further exploration of conspiracy discourse, especially in the under-explored Italian context, and misinformation diffusion on Telegram.

Acknowledgments. This work was partly supported by SoBigData.it which receives funding from European Union – NextGenerationEU – National Recovery and Resilience Plan (Piano Nazionale di Ripresa e Resilienza, PNRR) – Project: "SoBigData.it – Strengthening the Italian RI for Social Mining and Big Data Analytics" – Prot. IR0000013 – Avviso n. 3264 del 28/12/2021; and by project SERICS (PE00000014) under the NRRP MUR program funded by the EU – NGEU.

Disclosure of Interests. The authors have no competing interests to declare that are relevant to the content of this article.

References

1. Anderson, A., Huttenlocher, D., Kleinberg, J., Leskovec, J., Tiwari, M.: Global diffusion via cascading invitations: structure, growth, and homophily. In: Proceedings of the 24th international conference on World Wide Web, pp. 66–76 (2015)
2. Avalle, M., et al.: Persistent interaction patterns across social media platforms and over time. Nature (2024)
3. Baumgartner, J., Zannettou, S., Squire, M., Blackburn, J.: The pushshift telegram dataset. In: Proceedings of the International AAAI Conference on Web and Social Media, vol. 14, pp. 840–847 (2020)
4. Bovet, A., Grindrod, P.: Organization and evolution of the UK far-right network on telegram. Appl. Network Sci. **7**(1), 76 (2022). https://doi.org/10.1007/s41109-022-00513-8
5. Calamusa, A., et al.: Twitter monitoring evidence of Covid-19 infodemic in Italy. Eur. J. Public Health **30**(Supplement_5), ckaa165–066 (2020)
6. Curley, C., Siapera, E., Carthy, J.: Covid-19 protesters and the far right on telegram: co-conspirators or accidental bedfellows? Soc. Media+ Soc. **8**(4), 20563051221129187 (2022)
7. Dugué, N., Perez, A.: Direction matters in complex networks: a theoretical and applied study for greedy modularity optimization. Phys. A **603**, 127798 (2022)
8. Engel, K., Hua, Y., Zeng, T., Naaman, M.: Characterizing reddit participation of users who engage in the QAnon conspiracy theories. Proc. ACM Hum.-Comput. Interact. **6**(CSCW1), 1–22 (2022)
9. Gallagher, R.J., Reing, K., Kale, D., Ver Steeg, G.: Anchored correlation explanation: topic modeling with minimal domain knowledge. Trans. Assoc. Comput. Linguist. **5**, 529–542 (2017)

10. Gambini, M., Tardelli, S., Tesconi, M.: The anatomy of conspiracy theorists: unveiling traits using a comprehensive twitter dataset. Comput. Commun. **217**, 25–40 (2024)
11. Greer, K., Beene, S.: When belief becomes research: conspiracist communities on the social web. Front. Commun. **9**, 1345973 (2024)
12. Hoseini, M., Melo, P., Benevenuto, F., Feldmann, A., Zannettou, S.: On the globalization of the QAnon conspiracy theory through telegram. In: Proceedings of the 15th ACM Web Science Conference 2023, pp. 75–85 (2023)
13. van der Maaten, L., Hinton, G.: Visualizing data using T-SNE. J. Mach. Learn. Res. **9**(86), 2579–2605 (2008). http://jmlr.org/papers/v9/vandermaaten08a.html
14. Mahl, D., Schäfer, M.S., Zeng, J.: Conspiracy theories in online environments: an interdisciplinary literature review and agenda for future research. New Media Soc., 14614448221075759 (2022)
15. Morgia, M.L., Mei, A., Mongardini, A.M.: TGDataset: a collection of over one hundred thousand telegram channels (2023)
16. Ng, L.H.X., Loke, J.Y.: Analyzing public opinion and misinformation in a COVID-19 telegram group chat. IEEE Internet Comput. **25**(2), 84–91 (2020)
17. Nizzoli, L., Tardelli, S., Avvenuti, M., Cresci, S., Tesconi, M., Ferrara, E.: Charting the landscape of online cryptocurrency manipulation. IEEE Access **8**, 113230–113245 (2020)
18. Pasquetto, I.V., Olivieri, A.F., Tacchetti, L., Riotta, G., Spada, A.: Disinformation as infrastructure: Making and maintaining the QAnon conspiracy on Italian digital media. Proc. ACM Hum.-Comput. Interact. **6**(CSCW1), 1–31 (2022)
19. Spitale, G., Biller-Andorno, N., Germani, F.: Concerns around opposition to the green pass in Italy: social listening analysis by using a mixed methods approach. J. Med. Internet Res. **24**(2), e34385 (2022)
20. Tardelli, S., et al.: Cyber intelligence and social media analytics: current research trends and challenges. In: Proceedings of the 2nd CINI National Conference on Artificial Intelligence (Ital-IA 2022) (2022)
21. Tardelli, S., et al.: Temporal dynamics of coordinated online behavior: stability, archetypes, and influence. arXiv preprint arXiv:2301.06774 (2023)
22. Urman, A., Katz, S.: What they do in the shadows: examining the far-right networks on telegram. Inf. Commun. Soc. **25**(7), 904–923 (2022)
23. Vergani, M., Martinez Arranz, A., Scrivens, R., Orellana, L.: Hate speech in a telegram conspiracy channel during the first year of the Covid-19 pandemic. Soc. Media+ Soc. **8**(4), 20563051221138758 (2022)
24. Willaert, T.: A computational analysis of telegram's narrative affordances. PLoS ONE **18**(11), e0293508 (2023)
25. Zehring, M., Domahidi, E.: German corona protest mobilizers on telegram and their relations to the far right: a network and topic analysis. Soc. Media+ Soc. **9**(1), 20563051231155106 (2023)

Masking the Bias: From Echo Chambers to Large Scale Aspect-Based Sentiment Analysis

Yeonjung Lee[1](✉), Yusuf Mücahit Çetinkaya[1,2](✉), Emre Külah[2], Hakkı Toroslu[2], and Hasan Davulcu[1]

[1] Arizona State University, Tempe, AZ, USA
ylee197@asu.edu
[2] Middle East Technical University, Ankara, Türkiye
ycetinka@asu.edu

Abstract. Aspect-based sentiment analysis (ABSA) is a natural language processing (NLP) task, ascribing precise sentiment linkages to specific entities and issues in text data. This paper addresses critical shortcomings in current ABSA methods, particularly the issues of limited aspects, training set biases, and lack of comprehensive stance-coded datasets. First, we develop a scalable MaskedABSA approach that masks aspect terms in training sentences to enable unbiased sentiment inference from the context alone. We show that the proposed method surpasses the state-of-the-art solutions in accuracy for the aspect term sentiment classification task, as verified by the SemEval datasets. Furthermore, we tackle the perennial challenges of limited training resources and the prohibitive costs of manual annotation in ABSA dataset creation by introducing an innovative weak supervision technique capitalizing on the inherent community clustering properties found within social media datasets. We utilize community detection algorithms to partition a share network into polarized groups with homogeneous adversarial stances, allowing large-scale aspect-based sentiment analysis dataset curation without labor intensive manual labeling. Our methodology is also validated using a real-world polarized dataset comprising diverse aspects and stances to showcase its efficacy and scalability.

Keywords: aspect-based sentiment analysis · social networks · weakly labeled data · social media

1 Introduction

Aspect-based sentiment analysis (ABSA) is a subset of sentiment analysis that not only discerns the sentiment within a given text but also links it to specific entities or aspects. This precision enhances understanding of stance, enabling scholars to derive actionable insights by pinpointing the sentiments on specific entities or issues. As a result, ABSA has become a critical tool in data analytics and artificial intelligence (AI) applications.

L. M. Aiello et al. (Eds.): ASONAM 2024, LNCS 15212, pp. 214–225, 2025.
https://doi.org/10.1007/978-3-031-78538-2_19

While current ABSA methodologies have proven valuable, they require reevaluation to effectively handle the complexities and vast diversity present within large real-world datasets. One of the primary challenges of ABSA is training set bias, a phenomenon where the data used to train an algorithm fails to comprehensively represent the diversity of contexts and opinions in the full spectrum of the target populations. This bias can skew the algorithm's performance, often to the detriment of minority stances on less represented views, thereby compromising ABSA systems' efficacy and fairness.

In our model, we simply *mask* the terms corresponding to the aspects we seek to analyze. This technique allows our model to focus on the context alone to infer the sentiment directed towards the masked aspect without being biased by the aspect term itself. By doing so, we refine the model's capacity to more precisely gauge sentiments towards both previously seen and unseen (new) aspects, thereby addressing one of the critical limitations of current ABSA models.

The construction of labeled datasets for NLP tasks presents challenges due to the costly nature of labor-intensive manual coding. The high costs associated with ABSA dataset development results in a scarcity of publicly available datasets. ABSA differs from traditional sentiment analysis by marking both specific aspects present in every sentence and labeling the sentiment corresponding to each aspect. This process necessitates understanding the domain-dependent key aspects and the contextual interplay between aspects and intended sentiments, which cannot be easily automated. This requirement for manual labeling significantly contributes to the high costs associated with ABSA dataset development and the lack of publicly available datasets. The limited number and size of available datasets poses challenges for reliably training, testing and comparing models, especially transformer-based ones with millions of parameters. Researchers currently try to determine a small model's accuracy by averaging outcomes across multiple runs. However, larger datasets would facilitate more accurate evaluations [4].

Data analysis on social media platforms like X (formerly Twitter) is especially relevant due to such platforms' pervasive roles in enabling viral information cascades that impact public discourse. Recent studies [3,8,20] show that social media users pay attention to opinions they agree with. It indicates a strong correlation between biases in the content people both produce and consume. In other words, *echo chambers* reinforcing certain biased stances and a tailored media experience that eliminates opposing viewpoints and differing voices are very real on social media. Our approach employs community detection algorithms on retweet (or share) networks to reveal camps and their more homogeneous subcommunities. This method allows us to use partisan 'barrier-bound' users within a camp to rapidly and cost-effectively produce stance annotated weakly-labeled ABSA datasets.

For example, recent studies on political discourse on social media [3,20] confirm that social media users are exposed mainly to opinions that agree with their own, and partisan users enjoy higher centrality and content endorsement. These findings indicate a strong correlation between biases in the content people both produce and consume. Our approach employs community detection algorithms on the retweet network, revealing camps and their more homogeneous sub-communities. Furthermore, by distinguishing between 'barrier-bound' partisan users, who only interact within their own camps vs. 'barrier-crossing' users [2,11], who attract positive engagement across camps, allows us to use 'barrier-bound' users and their shared core stances sub-community by sub-community to rapidly and cost-effectively produce weakly-labeled ABSA datasets. We summarize our key contributions as follows:

- We foster large-scale, cost-effective, weakly labeled ABSA datasets by harnessing homogeneous stances of partisan users in social media communities.
- We unveil MaskedABSA, a model that enhances aspect-based sentiment analysis by masking aspect terms in sentences. This improves accuracy in gauging sentiments toward both known and previously unseen aspects.
- We evaluate the MaskedABSA model's performance under imbalanced perspective conditions, particularly focusing on accuracies for detecting underrepresented minority stances to bolster the fairness of ABSA systems.
- We make experimental datasets and models publicly available via GitHub at tweetpie/masked-absa.

The remainder of this article is structured as follows: Sect. 2 explores previous research relevant to our topic. Section 3 details the procedures involved in preparing the MaskedABSA and the weakly labeled dataset. Section 4 presents various experiments and evaluations. Section 5 offers concluding remarks and future works.

2 Related Works

ABSA has witnessed significant advancements, mainly through deep learning-based approaches and large-scale pre-trained models. E2E-ABSA [12] utilizes BERT for token classification with an additional layer for aspect polarity determination, yielding notable results. Dual-MRC [13] adopts a unique architecture featuring two adjacent BERT models, one each for extracting Aspect Terms (AT) and Opinion Terms (OT), which are combined to produce the outputs. Similarly, Span-ASTE [22] employs a span-level representation approach coupled with BERT to generate Aspect Terms and their corresponding stance pairs. Conversely, BART-Aspect-Based Sentiment Analysis Model (BART-ABSA) [23] is a sequence-to-sequence model developed to address ABSA subtasks, leveraging the BART model to tackle the Aspect-Category-Opinion-Sentiment (ACOS) subtasks. BART-ABSA [9] has demonstrated notable performance improvements through modifications in data pre-processing and model architecture, capitalizing on BART's pre-trained capabilities.

InstructABSA [18] diverges from the BERT architectures by employing pre-trained models and a sequence-to-sequence transformer by utilizing instruction queries. Our study uses the T5 transformer to train Masked Sentiment Classifier models. Introduced by C. Raffel et al. [10], the T5 Transformer demonstrates remarkable performance, particularly in question-answering tasks, compared to BERTBASE [6]. Our approach involved training the T5 model using a teacher forcing method, where the model is provided input sequences and their corresponding target sequences during training, resulting in enhanced performance in ABSA tasks. Disentangled Linguistic Graph Model (DLGM) [15], and ABSA-DeBERTa [14], enrich BERT-based models' ability to capture intricate relationships between aspects and sentiments.

3 Methodology

3.1 Masked Aspect Sentiment Classification (MASC)

ABSA is a task that provides a polarity of each aspect term in the sentence. Therefore, Zhang et al. [24] define it as a comprehensive opinion summary at the aspect level. The process of aspect-based sentiment analysis comprises two distinct components: Aspect Term Extraction (ATE) and Aspect Term Sentiment Classification (ATSC). First, in the Aspect Term Extraction stage, the model identifies and extracts aspect terms from a provided sentence. Next, in the Aspect Sentiment classification phase, the model predicts the stance associated with identified aspect terms. As a result, the model returns aspect term and stance pairs. This task is illustrated in Fig. 1.

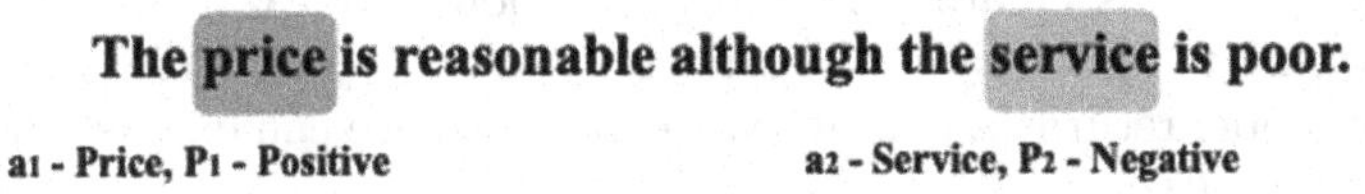

The price is reasonable although the service is poor.

a1 - Price, P1 - Positive **a2 - Service, P2 - Negative**

Fig. 1. Example of aspect-based sentiment annotation.

We formally define aspect-based sentiment analysis process as follows. Consider a sentence S_i expressed as:

$$S_i = \{w_1, w_2, w_3, \ldots, w_n\} \tag{1}$$

where each w_j represents a word in the sentence. Within the sentence S_i, we identify multiple aspect terms denoted by A_i:

$$A_i = \{a_1, a_2, \ldots, a_k\},\ where\ k <= n \tag{2}$$

The process of detecting and extracting these aspect terms is termed Aspect Term Extraction (ATE). Subsequently, each aspect term is associated with a polarity P_i:

$$P_i = \{p_1, p_2, \ldots, p_k\},\ where\ k <= n \tag{3}$$

Each polarity p_x is categorized as one of the following values:

$$p_x = \{pos, neg, neutral, conflict, none\} \tag{4}$$

For the sample sentence in Fig. 1, the model extracts aspects a_1 (price) and a_2 (service). It then determines the polarities as p_1 (positive) and p_2 (negative), respectively. As a result, the model outputs pairs such as (a_1, p_1) and (a_2, p_2), returning *(price, positive)* and *(service, negative)*.

During the Aspect Term Sentiment Classification task, we observe that current models tend to overemphasize the mere presence of the aspect term and the majority bias related to the term, neglecting the broader context of the sentence when determining sentiment. This tendency may not be readily apparent when utilizing SemEval datasets, as these are hand-crafted where training and test sets exhibit similar distributions. However, when deploying real-world datasets exhibiting highly skewed or missing labels for specific aspects, the vulnerability of current ABSA models becomes more evident. These models tend to memorize the associated word and repeatedly predict the sentiment label previously associated with it. To mitigate this issue, we introduce the MaskedABSA method. This technique involves replacing aspect terms within a sentence with a placeholder token, "[MASK]", thereby redirecting the model's focus towards the context of the sentence.

The procedure modifies the original sentence from $S_i = \{w_1, w_2, w_3, \ldots, w_n\}$ to $S'_i = \{w_1, a_1, w_3, \ldots, a_k, \ldots, w_n\}$ considering both the input S_i and aspect terms A_i. For the Mask method, we construct masked sentences such as $M_{i_1} = \{w_1, \text{[MASK]}, w_3, \ldots, w_n\}$ and $M_{i_k} = \{w_1, w_2, w_3, \ldots, \text{[MASK]}, \ldots, w_n\}$ from S'_i, where each aspect term a_k is replaced by "[MASK]" and for each aspect a new sample M_{i_k} is created. The model, denoted as F_{MASC}, is then tasked with predicting the polarity of the term represented by "[MASK]" in each sentence. The model returns a polarity value p_k. Thus, we can express the function of the model as $F_{MASC}(M_{i_k}) = p_k$.

3.2 Weakly Labeled Dataset

One effective strategy to address both the scarcity of ABSA datasets and the challenges of bias involves using weak supervision combined with social network structures, such as retweet networks [5]. Analyzing the patterns of retweets can infer the sentiment and aspect orientations prevalent within specific user groups or topics without requiring detailed sentence-by-sentence manual annotation. This approach allows for creating large-scale, weakly labeled datasets that can be refined using machine learning models and further validated through targeted reviews.

A well-defined codebook, offering guidelines for detecting and annotating stances on key aspects according to their distinct community orientations, significantly enhances the efficiency and consistency of the weakly labeled data production process.

Several researchers have utilized the Louvain community detection algorithm on social media networks [1,7,19], to extract non-overlapping communities. It was also observed that [25] users within the same Louvain community tend to share not only common interests but similar viewpoints and stances on popular issues and entities. Our approach harnesses per community homogeneous viewpoints, captured in a codebook, to generate vast amounts of weakly labeled datasets for ABSA tasks.

U.S. Race Relations Dataset. Our dataset, derived from a real-world case, contains a collection of 9,084,824 tweets from January 1, 2022 to May 31, 2023, related to race relations in the U.S. The dataset involves 296,540 users and includes a retweet network with 738,012 edges, depicted in Fig. 2. To collect the dataset, we used keywords related to U.S. race relations, specifically focusing on five topics - Black Lives Matter (BLM), All Lives Matter (ALM), Issues, Races, and Religion. Weakly labeled data is stance annotated data, leveraging shared community biases as recorded on the codebook'.

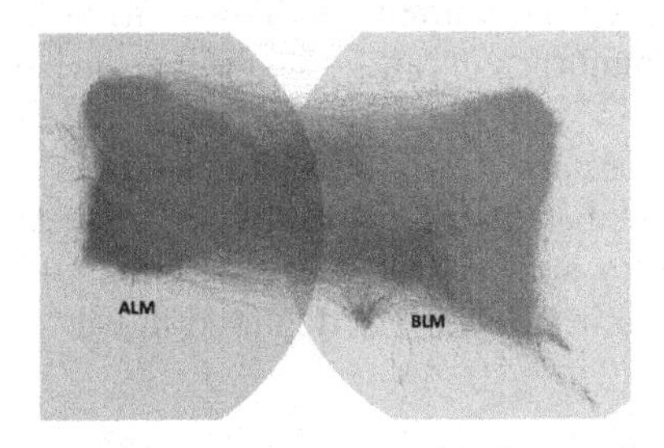

Fig. 2. Visualization of polarized camps in U.S. Race Relations

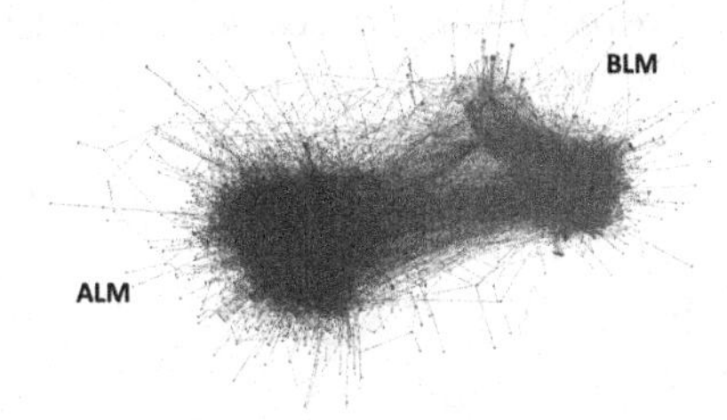

Fig. 3. Only Barrier-crossing users' network from U.S. Race Relations dataset

The primary advantages of producing weakly labeled datasets using our approach are its cost-efficiency and scalability, making it an appealing option in scenarios where acquiring comprehensive, fully labeled datasets is economically or logistically not feasible.

Camp Labeling of Users. The retweet network graph shows a polarized network comprising two camps with seven sub-communities. Each sub-community is depicted in a different color in Fig. 2.

An experimentally determined edge weight threshold is applied to partition the network into a pair of polarized camps. These camps are labeled by a panel of experts as either pro-BLM or pro-ALM, based on an analysis of their most viral 1,000 tweets, examined community by community. In social media analytics, users' communication patterns can be categorized as either barrier-bound or barrier-crossing. Barrier-bound partisan users interact only within their camps, whereas barrier-crossing users attract positive engagement from both camps. The barrier-crossing users' network shown in Fig. 3. Previous studies [11, 17] have shown that contents amplified by barrier-bound users tend to reinforce existing biases within their groups, creating echo chambers that further polarize opinions. Our approach leverages the consensus that exists among barrier-bound users who exhibit clear partisan biases to develop a codebook and rapidly generate vast amounts of weakly labeled training datasets for MaskedABSA models.

Bias and Sentiment Coding. We extracted all noun phrases and named entities from our corpus using weak labeling for significant aspect terms. Our codebook encompasses the most frequent and discriminating unigrams, bi-grams, and other n-grams, including all unambiguous bigrams such as "mental health" that could be overshadowed by more generic high-frequency unigrams like "health."

Subsequently, three domain experts independently coded the noun phrases and named entities for their shared stances within the BLM and ALM camps. These experts then convened in a panel to discuss and reconcile all their differences toward achieving a consensus on the codebook entries. From the top 3,000 phrases identified, 1,789 are determined to be non-ambiguous. The sentiment column reveals 496 phrases with negative connotations and 257 with positive ones. In terms of bias, 234 phrases are coded as "anti" and 160 as "pro" within the BLM camp; for the ALM camp, 177 phrases are "anti" and 90 are "pro".

The resulting dataset comprises weakly labeled tweets aimed at dissecting sentiment and ideological biases linked to specific camps and their sub-communities. Annotations comprise both aspect and stance as depicted in Fig. 4.

Increased [funding/PRO] for [community programs/PRO] will [repair/PRO] the systemic injustices of [racial profiling/ANTI]

Fig. 4. An example of a stance annotated sentence from the BLM camp.

Preparing U.S. Race Relations and Politics Datasets. From 62,196 samples, we separated two distinct sets of data focused on a pair of different themes: one on race-related issues and the other on politics-related issues and entities. Initially, we identified 1,206 terms associated with race-related topics and 593 terms with politics-related topics. We then created two separate collections of tweets: one matching terms related to race relations and another matching terms related to U.S. politics. We ensured these datasets did not overlap; any tweet that included terms from both categories was left out. After organizing these groups, we balanced the training and testing datasets for race relations and politics themes. We focused on ensuring fair representation from the All Lives Matter (ALM) and Black Lives Matter (BLM) camps. This approach led to the identification of 6,760 tweets for the politics dataset and 84,518 tweets for the race relations dataset.

4 Experiments and Evaluations

In our experiments, we tested our MaskedABSA approach using the SemEval Dataset and real-world datasets from Twitter.

4.1 SemEval Datasets

For the SemEval datasets, we used four training datasets: Laptop14, Restaurant14, Restaurant15, and Restaurant16. These datasets each include training and test subsets, allowing us to conduct our experimental evaluations comprehensively.

The data illustrated in Table 1 and Tabel 2 clearly demonstrate that the MaskedABSA model surpasses alternative models in terms of performance across all datasets, showing exceptional robustness even in cross-domain testing scenarios.

Table 1. ATSC accuracy results (%) on SemEval datasets compared with MaskedABSA models.

Model	Laptop14	Res.14	Res.15	Res.16
BART-ABSA-ALSC [23]	76.76	*87.29*	76.49	89.10
Dual-MRC [13]	75.97	82.04	75.08	-
InstructABSA2 [18]	*81.56*	85.17	*84.50*	*89.43*
MaskedABSA	**86.24**	**87.65**	**91.53**	**94.83**

Table 2. Cross-domain ATSC results (%) on SemEval datasets comparing with the top model in the literature.

Train	Test	Model	Accuracy
Restaurant 14	Laptop14	InstructABSA	82.44
		MaskedABSA	**84.79**
Laptop 14	Restaurant 14	InstructABSA	80.56
		MaskedABSA	**85.32**
Restaurant 15	Hotel 15	InstructABSA	89.74
		MaskedABSA	**91.37**

4.2 U.S. Race Relations and Politics Datasets

To ensure our dataset is similar in size and scope to the SemEval datasets, we randomly chose 3,000 tweets for training and 1,000 tweets for testing. While selecting these tweets for training and testing, we focused on maintaining a balance between the ALM and BLM camps. Due to the nature of Twitter, where negative messages often spread more effectively [16,21], our dataset ended up having more negative content than positive. The details and key metrics of our dataset are as shown in Table 3.

Table 3. Race and Politics datasets distribution.

Races Dataset	Positive	Negative	Total	Politics Dataset	Positive	Negative	Total
Train	885	2115	3000	Train	873	2122	2995
Test	309	699	1000	Test	257	751	1008

We started by developing two models, one trained on the U.S. race relations dataset and the other on the U.S. politics dataset. The performance of each model was measured, with the race relations model achieving an F1 score of 71.92 and the politics model scoring 72.10, as shown in Table 4.

For cross-domain validation tests, we tested the performance of each model using the dataset from the other domain. The results showed only a small drops in their corresponding F1 scores, only 2.68 points for the U.S. politics dataset and 2.70 points for the U.S. race relations dataset. These experiments indicate that the MaskedABSA model maintains strong cross-domain performance, as shown in Table 4.

Table 4. Cross-domain test results on the U.S. Race Relations and Politics datasets using MaskedABSA.

Train	Single		Cross		Differences
	Test	F1	Test	F1	
Race	Race	71.92	Politics	69.24	−2.68
Politics	Politics	72.10	Race	69.40	−2.70

4.3 Bias Test

Aspect terms bias is a phenomenon where a model shows favoritism toward specific stances for specific terms. To tackle this problem, we developed the masking method, where context is forced to determine the stance instead of the aspect itself. To evaluate the efficacy of our approach, we conducted a bias test analyzing the accuracy of stance detection corresponding to messages carrying minority category biases. For instance, if the word "gun control" appears ten times in the dataset with eight occurrences labeled as anti/negative and two as pro/positive, we aimed to determine the accuracy of ABSA on the minority stance category tweets. In this test, we evaluated the performance of the InstructABSA model in comparison to the MaskedABSA model. The results demonstrated that our MaskedABSA model achieved an average accuracy of 45.6% and 57.7% in the politics and race-related models, significantly surpassing the InstructABSA's accuracies of 2% and 8% only, respectively. This experiment reveals a fundamental limitation of state-of-the-art ABSA models: a susceptibility to memorizing majority category stance labels for aspects, even in contexts that unambiguously signal a minority category stance. This highlights a critical weakness in their ability to accurately discern and represent the perspectives of less prevalent groups.

4.4 New Terms Test

To assess the stance detection accuracies for the previously unseen, new aspect terms issue, we designed an experiment to gauge the effectiveness of the MaskedABSA. We analyzed the model's performance with new aspect terms across U.S. race relations and politics datasets. Our approach involves replacing the queried aspect with a new term that does not appear in the training set. As a result, the aspect terms used in the test datasets were completely new to the models, having not been introduced during their training. This testing procedure was replicated for both the race relations and politics datasets to measure the performance of the InstructABSA model. Despite the experimental conditions, the MaskedABSA model's accuracy remained the same, as it conceals the queried aspect term during prediction, rendering the novelty of the term inconsequential. Additionally, we applied this methodology to the SemEval dataset, introducing new terms to challenge the model further. The results, documented in Table 5, clearly illustrate the MaskedABSA model's enhanced ability to handle new terms compared to the other models.

Table 5. Accuracy with replacements of new terms.

	Laptop 14	Res. 14	Res. 15	Res. 16	Race	Politics
Instruct (%)	50.00	71.23	55.08	64.40	53.00	61.84
Masked (%)	**86.24**	**87.65**	**91.53**	**94.83**	**71.92**	**72.10**

5 Conclusions and Future Work

In this study, we addressed the challenge of training set bias in ABSA models by introducing the MaskedABSA model, designed to focus on the contextual information of the text rather than merely memorizing majority term sentiments. Additionally, our study presents a methodology for constructing weakly labeled ABSA datasets by utilizing social network connectivity and a codebook comprising frequent terms. To ensure the robustness and generalizability of our approach, we rigorously evaluated it using SemEval benchmark datasets and real-world datasets on U.S. race relations and politics.

We conducted comprehensive tests to assess the model's effectiveness in handling previously unseen new aspect terms. Our findings demonstrated the superiority of the MaskedABSA model over the current state-of-the-art (SOTA) models. Rigorous cross-domain testing showed the model's versatility and robustness, showing only a marginal performance drop when applied to diverse real-world domains. The bias test confirmed our model's superior accuracy in mitigating aspect term bias, even for minority-held stances.

ABSA can be a powerful tool in detecting and countering mis/disinformation and fake news while navigating the complexities of breaking news. By dissecting the sentiment surrounding various aspects of a breaking news story, ABSA can identify inconsistencies in the language use to raise a red flag when the reported stances are inconsistent across credible mainstream vs. alternative media sources.

Furthermore, ABSA empowers users to explore the multifaceted landscape of breaking news by uncovering diverse perspectives from competing factions. By pinpointing communities engaged in discussions on social media and analyzing the sentiments within those groups, individuals can gain a deeper understanding of the varied viewpoints surrounding an event. This allows them to form well-rounded, informed opinions based on a broader spectrum of perspectives, transcending the limitations of their own echo chambers.

Acknowledgment. The work of Yusuf Mücahit Çetinkaya was supported by TUBITAK-2214-A.

References

1. Blondel, V.D., Guillaume, J.L., Lambiotte, R., Lefebvre, E.: Fast unfolding of communities in large networks. J. Stat. Mech: Theory Exp. **2008**(10), P10008 (2008)
2. Buhrmester, M.D., Cowan, M.A., Whitehouse, H.: What motivates barrier-crossing leadership? New England J. Public Policy **34**(2), 7 (2022)
3. Çetinkaya, Y.M., Toroslu, İH., Davulcu, H.: Developing a twitter bot that can join a discussion using state-of-the-art architectures. Soc. Netw. Anal. Min. **10**, 1–21 (2020)
4. Chebolu, S.U.S., Dernoncourt, F., Lipka, N., Solorio, T.: A review of datasets for aspect-based sentiment analysis. In: International Joint Conference on Natural Language Processing (2022)
5. Cherepnalkoski, D., Mozetič, I.: Retweet networks of the European parliament: evaluation of the community structure. Appl. Network Sci. **1**, 1–20 (2016)
6. Devlin, J., Chang, M.W., Lee, K., Toutanova, K.: BERT: pre-training of deep bidirectional transformers for language understanding. arXiv preprint arXiv:1810.04805 (2018)
7. Elmas, T., Randl, M., Attia, Y.: # teamfollowback: Detection & analysis of follow back accounts on social media. In: Proceedings of the International AAAI Conference on Web and Social Media, vol. 18, pp. 381–393 (2024)
8. Garimella, K., De Francisci Morales, G., Gionis, A., Mathioudakis, M.: Political discourse on social media: echo chambers, gatekeepers, and the price of bipartisanship. In: Proceedings of the 2018 World Wide Web Conference, pp. 913–922 (2018)
9. Hoang, C.D., Dinh, Q.V., Tran, N.H.: Aspect-category-opinion-sentiment extraction using generative transformer model. In: 2022 RIVF International Conference on Computing and Communication Technologies (RIVF), pp. 1–6 (2022)

10. Khashabi, D., Chaturvedi, S., Roth, M., Upadhyay, S., Roth, D.: Looking Beyond the Surface: a challenge set for reading comprehension over multiple sentences. In: Proceedings of the Annual Conference of the North American Chapter of the Association for Computational Linguistics (NAACL) (2018)
11. Lee, Y., Ozer, M., Corman, S.R., Davulcu, H.: Identifying behavioral factors leading to differential polarization effects of adversarial botnets. ACM SIGAPP Appl. Comput. Rev. **23**(2), 44–56 (2023)
12. Li, X., Bing, L., Zhang, W., Lam, W.: Exploiting BERT for end-to-end aspect-based sentiment analysis. In: Proceedings of the 5th Workshop on Noisy User-generated Text (W-NUT 2019)
13. Mao, Y., Shen, Y., Yu, C., Cai, L.: A joint training dual-MRC framework for aspect based sentiment analysis. In: Proceedings of the AAAI Conference on Artificial Intelligence, vol. 35, pp. 13543–13551 (2021)
14. Marcacini, R.M., Silva, E.: Aspect-based sentiment analysis using BERT with disentangled attention. In: LatinX in AI at International Conference on Machine Learning 2021 (2021)
15. Mei, X., Zhou, Y., Zhu, C., Wu, M., Li, M., Pan, S.: A disentangled linguistic graph model for explainable aspect-based sentiment analysis. Knowl.-Based Syst. **260**, 110150 (2023)
16. Mousavi, M., Davulcu, H., Ahmadi, M., Axelrod, R., Davis, R., Atran, S.: Effective messaging on social media: what makes online content go viral? In: Proceedings of the ACM Web Conference 2022, pp. 2957–2966 (2022)
17. Rao, A., Morstatter, F., Lerman, K.: Retweets amplify the echo chamber effect. In: Proceedings of the International Conference on Advances in Social Networks Analysis and Mining. pp. 30–37 (2023)
18. Scaria, K., Gupta, H., Goyal, S., Sawant, S.A., Mishra, S., Baral, C.: InstructABSA: instruction learning for aspect based sentiment analysis. arXiv preprint arXiv:2302.08624 (2023)
19. Seckin, O.C., Atalay, A., Otenen, E., Duygu, U., Varol, O.: Mechanisms driving online vaccine debate during the Covid-19 pandemic. Soc. Media+ Soc. **10**(1), 20563051241229657 (2024)
20. Torres-Lugo, C., Yang, K.C., Menczer, F.: The manufacture of partisan echo chambers by follow train abuse on twitter. In: Proceedings of the International AAAI Conference on Web and Social Media, vol. 16, pp. 1017–1028 (2022)
21. Tsugawa, S., Ohsaki, H.: Negative messages spread rapidly and widely on social media. In: Proceedings of the 2015 ACM on Conference on Online Social Networks, pp. 151–160 (2015)
22. Xu, L., Chia, Y.K., Bing, L.: Learning span-level interactions for aspect sentiment triplet extraction. arXiv preprint arXiv:2107.12214 (2021)
23. Yan, H., Dai, J., Qiu, X., Zhang, Z., et al.: A unified generative framework for aspect-based sentiment analysis. arXiv preprint arXiv:2106.04300 (2021)
24. Zhang, W., Li, X., Deng, Y., Bing, L., Lam, W.: A survey on aspect-based sentiment analysis: tasks, methods, and challenges. IEEE Trans. Knowl. Data Eng. (2022)
25. Zhang, Y., Wu, Y., Yang, Q.: Community discovery in twitter based on user interests. J. Comput. Inf. Syst. **8**(3), 991–1000 (2012)

Computational Analysis of Communicative Acts for Understanding Crisis News Comment Discourses

Henna Paakki[1(✉)] and Faeze Ghorbanpour[2]

[1] Aalto University, Espoo, Finland
henna.paakki@aalto.fi
[2] Ludwig Maximilian University of Munich, Munich, Germany
faeze.ghorbanpour@lmu.de

Abstract. Social media analyses using computational methods are becoming increasingly important, especially for crisis communication and social media monitoring. We seek to investigate the validity and utility of computationally analyzing communicative acts in social media crisis news comment discourses. We implement an applied act classifier for a novel online context by utilizing few-shot learning and a small manually annotated dataset. To illustrate the usefulness of analyzing communicative acts, we show that how people use acts in comments notably changes across the crisis timeline. In contrast to classic crisis research, early social media crisis responses in our study do not show a heightened use of acts oriented to discursive struggle only, but instead resolution oriented acts are most common at first. In further analysis, we show that computational analysis of acts can complement traditional content analyses to reveal more specific insights on the functions and goals of comments. Our study paves the way for more fine-grained approaches to understanding social media discourses and crisis responses, offering potential new tools for crisis management.

Keywords: Crisis Discourses · Crisis Narratives · Communicative Acts · Few-shot Learning · Applied NLP · Social Media

1 Introduction

With the increasing use of social media as a communication tool and source for news, the capability for researchers, communication experts, and journalists to analyze the interactions taking place has become essential for understanding emerging narratives, perspectives, interactions between networks, and how opinions are formed. We argue that analyzing what acts social media comments represent can allow insights into how social media responses or reactions to crises evolve and how online participants position themselves toward emerging issues.

Computational modeling of communicative acts has not so far been widely applied to analyze social media interactions in crisis settings. We argue that

L. M. Aiello et al. (Eds.): ASONAM 2024, LNCS 15212, pp. 226–242, 2025.
https://doi.org/10.1007/978-3-031-78538-2_20

such analysis can reveal what functions online comments have, allowing more fine-grained understandings of their position to the crisis in contrast to e.g. classic topic modeling. There can be significant differences in comments' functions although their contents might be quite similar, e.g.: 'What are politicians planning to do?' vs. 'Politicians are planning to do nothing!'. In this applied study, we explore if the analysis of communicative acts can be used to uncover broader changes in audience reactions to crisis news during a long-lasting crisis.

The identification of communicative acts is an important step towards a deeper analysis of social media comments. Natural Language Processing (NLP) can help unveil how acts are used in online turns to mobilize reactions or appeal to others [21,35]. Although dialogue acts found in telephone and synchronous chat conversations have been extensively researched [5,16,22,26,34], how communicative acts appear and evolve in asynchronous conversations, e.g. in relation to crisis discourses, has not been researched much as of yet, despite its viability for investigating the dynamics of online discussion [38,44].

Social media is relevant during crises as a communication medium, because key information travels fast through social networks [28]. For this reason, this paper focuses on online reactions to COVID-19 crisis news on the Indian NDTV and Canadian CTV channels on YouTube. We are interested in discovering whether differing positions can be found in comments to crisis news at different stages of the crisis, and whether they show an early discursive struggle as outlined in crisis research [41], seeking to answer three research questions (RQs):

- RQ1: Are there significant differences along a crisis timeline in terms of what type of acts comments represent?
- RQ2: Is there a discursive struggle visible in the comments at the initial phase of the crisis?
- RQ3: Can analysis of communicative acts using NLP be applied to complement content analyses to gain deeper insights into reactions to crisis news?

As there are no pre-existing models or datasets for act classification for asynchronous data such as ours, we form a theory [9,25,38] and data-driven annotation scheme , and train a classifier. Then, we use act-tagged data to conduct a time-series analysis of acts used in crisis news comments during three significant phases of the COVID-19 crisis: the beginning phase (Phase 1), the vaccination phase (Phase 2), and the prolonged phase (Phase 3). We analyze how act-taking changes along the crisis timeline and use acts to complement a further content analysis of central topics of the comments.

2 Communicative Acts in Asynchronous Conversations

Research concentrating on applying NLP to the analysis of social media has often focused on content-based approaches like topic modeling, Bag-of-Words or word or sentence vectors (e.g. [37]), or analyses of key accounts sharing content [20]. Social media comments have specific functions in relation to the content they are responding to – whether a comment is, e.g., stating information or asking for it considerably influences how its content should be interpreted. Acts represent what the main functions of a comment are within a conversation, i.e. what it does and how it relates to other actions taken in previous comments [9,25]. Previous studies show that the concept of communicative act [25] can be used to analyze e.g. conversational flow, functions of comments, accountability, positioning, intentionality [21,35], and conversational breakdowns [38].

Dialogue acts and intent recognition have been central interests to developing bots' and service agents' capabilities to interact with humans [19], and many datasets and annotation standards have been developed for computational identification of acts in synchronous chats [16] and telephone conversations [22]. These, however, reflect a context different from asynchronous conversation. We focus on communicative acts found important in earlier qualitative studies of asynchronous online conversations, especially from the point of view of political and crisis-related online discussion [3,25,38] Based on the theoretical foundations of acts [42], previous CMC research [38] and typologies of common acts [9,25], we form an annotation scheme for acts that can account for the characteristics of asynchronous conversations related to crises and political discussions involving discursive struggle (see Table 1).

3 Crisis News and Social Media Discourse

People are increasingly using social media for news [48]. It is often an important tool for news actors, organizations, and crisis communicators for rapidly disseminating important information about the crisis [28], and for public for getting real-time updates on the crisis event(s), sense-making and mobilization [28]. Societal crises create an empty discursive space, requiring explanation regarding the crisis, related risks, and required mitigation actions [41]. This also makes such events vulnerable to manipulation [45], which is why there is need for more fine-grained computational approaches to understanding crisis discourses online.

Researchers examined various aspects of social media discourses: besides investigating the content of messages, analyses of actor networks and behaviors can offer deeper insights into online interactions [20]. However, studies have often relied on content-based analyses [37,45]. Not many applied studies have looked deeper into conversational behaviors like acts used in social media comments, although these insights can offer valuable information on citizen perspectives, positions, and conversation dynamics [38]. Such analyses can arguably provide more insight into how crisis narratives and discourses evolve on social media, and how the public is positioned to the ongoing crisis.

Acts are central, because crises are socially constructed by using acts in comments [29]. At the surface level, contributions to sense-making in news comments are formulated as communicative acts, which reflect how people position themselves to crisis news and what the function of each comment is [14,25,42]. Crisis discourses are seen to first involve an empty discursive space, where discourses struggle for visibility and legitimacy to explain the event [30,41]. This is followed by a resolution of the struggle, one main narrative taking governance [41]. Theories in linguistics also stress that discursive struggle is essential for establishing a main narrative on events, but some divide this into several phases, including orientation–complication–evaluation–resolution [32], in contrast to crisis research [41], which mainly highlights the struggle and its resolution. We take inspiration from these theories in our empirical analysis.

To sum, we see investigating changes in the use of communicative acts that contribute to crisis sense-making as an opportunity to better understand shifts in perspectives to the ongoing crisis. Furthermore, we expect that this enables the identification of critical points of discursive struggle.

4 Data

We are interested in crisis news comments due to their role in shaping people's perspectives on a crisis, e.g. dissenting comments on news on social media may persuade people to orient toward the news source negatively [48]. We are especially interested in how news audience discourses around the crisis develop during a long-lasting crisis. YouTube news channels' crisis news videos and their comments offer an opportunity to investigate how reactions to the news change over time. We analyze comments on NDTV News' and CTV News' YouTube channel videos related to the COVID-19 crisis in India and Canada. The channels and their comments are mostly in English. We concentrate here only on English content, excluding all non-English content (mostly in Hindi in NDTV data), by using Fasttext [31]. The data were collected using scraper scripts in Python and the Google YouTube API. The data include comments to videos from *Phase 1*: the beginning phase (1/2020–8/2020), *Phase 2*: vaccination phase (02/2021–08/2021), and *Phase 3*: the prolonged phase of the crisis (11/2021–02/2022).

NDTV was selected as it is among the most followed English-language public news providers in India and one of the most trusted [36]. The channel allows viewer comments, and has a wide viewership. We were interested in the Indian context as trust in news has been reported to be low [36], and because Global South perspectives have not been sufficiently represented in research. To contrast this data to another English-speaking context, we compare the NDTV data to similar CTV data. CTV News was selected due to being among the most trusted news providers in their region, quite highly trusted overall, being mostly neutral in political ideology and considered to provide highly factual reporting[1]. It was an interesting context to focus on due to some of the events that occurred during the Pandemic in Canada, e.g. the Convoy movement. We wished to compare stretches of time during the long-lasting crisis that generated a lot of interest. To identify these, we investigated peaks in Google Trends on people's search habits related to the Pandemic, using search terms including "Corona, Coronavirus, COVID-19, COVID-19 vaccine, Pandemic". This allowed us to identify three phases where interest in the crisis peaked, as illustrated in Fig. 1.

NDTV data has 216,455 comments to 1,942 videos in total. This includes 70,520 comments and 641 videos in Phase 1; 112,510 comments and 658 videos in Phase 2; 33,425 comments and 643 videos in Phase 3. Approximately 71%

[1] https://mediabiasfactcheck.com/ctv-news/.

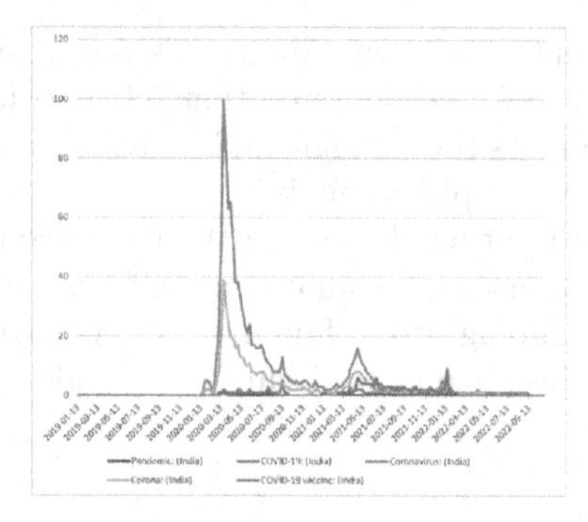

Fig. 1. Pandemic search term Google Trends from Dec. 2019 to Feb. 2022.

– 84% (depending on Phase) of comments were in English. 2% only contained emojis. The CTV set has 195,316 comments and 855 videos in total. These include 40,360 comments and 346 videos in Phase 1; 66,566 comments and 265 videos in Phase 2; and 88,390 comments and 244 videos in Phase 3.

5 Annotation Procedure

To our knowledge, there exist very few asynchronous conversation datasets manually labeled for acts. Existing models we found (e.g., [46]) have not included some key acts for analyzing crisis discourses: although challenges are included in [5], there exist no datasets with all relevant acts we need , e.g. challenges and denials [3,38]. Based on our own experiments, classifiers trained on instant chat conversations [16] do not generalize well on asynchronous data. Thus, we chose to annotate a small dataset to train a model to fit our empirical case. Our annotation scheme is based on earlier research on acts found relevant in (crisis-related) asynchronous discussion [25,38]. The data for manual annotation was selected randomly from the whole dataset, using Python scripts, making sure that each comment was used only once. We also ensured that the annotation data would have an equal distribution of comments from each phase.

According to research on digital conversation, a forum post may contain several acts [21]. This way, participants can achieve more by writing several points at one go [44]. Thus, we decided to prepare both single and multilabel annotations for the data. We assigned an order of importance and a confidence score from 1 to 5 to each comment. These were used in FlairNLP model training [24]. In annotation guidelines, we followed the descriptions of acts in Table 1[2].

We conducted the annotation iteratively: we, the authors, acted as the two annotators, continuing to annotate practice data until we had reached over 80% agreement. We independently annotated the same set of comments (first batch 96 comments), compared and negotiated our annotations and resolved all conflicts,

[2] See the annotation guidelines in detail on our GitHub page: https://github.com/Aalto-CRAI-CIS/Acts-in-crisis-narratives.

Table 1. Summary of acts used in our final action labeling scheme.

First act	Description
Question	requests information, expecting an answer providing or confirming the information.
Statement	asserts a fact or claim; provides information, evaluation/opinion announces information; answers a question (informatively).
Challenge	contests the epistemic basis of prior claims, actions, or feelings; or conveys a negative complaint, attributing responsibility for an event to a person or a group (accusation).
Request	proffers a service for the speaker, expecting acceptance/rejection. This class also contains proposals, both being directives.
Appreciation	includes various forms of positive evaluative reactions and thanking that can invite a response but do not require it.
Responding act	**Description**
Acceptance	marks the degree to which a speaker accepts a proposal, plan, opinion, or statement; or admission admitting that *e.g.* a statement, proposal or accusation is justified.
Denial	marks the degree to which a speaker resists, denies or rejects some previous proposal, plan, opinion, or statement; or contests some information or an accusation.
Apology	Excuse or apology for speaker's previous actions or statements.

analyzing especially difficult cases to improve our guidelines and scheme. Then we proceeded to batches 2–3, following the enhanced guidelines and the same process. After batch 3, we had achieved a sufficient percentage agreement (80%).

For the final set, we calculated inter-annotator agreement scores to evaluate our annotation, using Krippendorf's alpha with the MASI metric to calculate the distance between annotations [2], and Fleiss Kappa (multikappa for multi-label) [15], using NLTK's agreement metrics package [4]. For the single-label task – considering only the first annotation – we achieved 0.75 Fleiss Kappa (substantial agreement). Krippendorf's alpha for the single-label task was 0.75, at the upper bound for acceptable agreement. For multi-label annotation, we achieved a Krippendorf's alpha of 0.64, and a Fleiss' Kappa (multikappa) of 0.60. We considered the overall percentage agreement and inter-annotator scores sufficient for our analysis study, and thus we moved on to labeling our dataset. A total of 676 comments were included in the manually annotated dataset. We used 60% for training, 20% for development, and 20% for testing. As for the multi-label option in annotation, analyzing the final annotated dataset, we found that most comments were assigned two labels. 64.2% of the comments were annotated with two action labels, 35.8% with only one action, and 6.7% with three actions. Four labels or more were very rare for our data.

We compared the effects of using different numbers of classes: original set of 13 acts – including accusation, announcement, rejection, admission, and evaluation along with actions in Table 1 – to compare if merging classes that had similar functions in our data would affect performance. We compared these to models trained with 9 acts (with accusations as a separate class), and models trained with 8 acts in Table 1.

6 ML Models and Analysis Methods

A common problem with applied analyses is that pre-existing models or labeled datasets do not fit the context of the case at hand. This can often become an obstacle to producing a high-performing model that suits the empirical context, and can be implemented rapidly enough to investigate constantly evolving crisis discourses. For this reason, we utilize few-shot learning to train our applied model. Few-shot learning is a viable solution for applied models, allowing the training of a model with novel classes by using a small, manually labeled dataset. In few-shot learning, a model is trained with a small set of labeled data to direct predictions in the novel applied task [43]. At the time of inference, the predictions of the model are based on a few manually labeled examples. Adapting pre-trained models for novel tasks utilizing few-shot learning has become possible with the publication of the latest large Transformer-based LMs like BERT [13] and GPT models [6], as many LMs have learned a variety of tasks implicitly in their training on immense text datasets, allowing generalization beyond their original uses. The common approach of fine-tuning a general language model like BERT for a novel task (*e.g.* RoBERTa [47]) has been shown to be outperformed by some more recent few-shot learning approaches [24].

We were interested in state-of-the-art models that would support multiple languages, multi-label classification, easy adaptation for applied use, and light use of resources. There is an increasing amount of models that allow multi-label few-shot learning [24,26,43]. FlairNLP [24], PET [40], Fastforward few-shot learning[3], and SetFit [43] and adapter methods [27] met our requirements.

FlairNLP proposes utilizing novel annotated examples, but also the information contained in the names of the novel classes (*e.g.* politics) [24]. It also allows the use of confidence scores in training and achieves high reported performance trained only with a low number of examples, supporting multiple low-resource languages [24]. Our chosen FlairNLP model uses the TARS Classifier 'tars-base' [24]. PET or iPET utilizes Pattern-Exploiting Training, which is a semi-supervised training procedure that reformulates input examples as cloze-style phrases [40]. Fastforward pipeline relies on a relatively simple algorithm using BERT and words2vec to embed the texts. SetFit is an efficient state-of-the-art framework for few-shot fine-tuning of Sentence Transformers [39] without the use of prompts that can require a lot of manual labor. It can achieve high accuracies using only a small set of manually labeled data [24]. For SetFit, we use paraphrase-Mpnet-base-v2 [39] as a base model. We also experimented with

[3] https://github.com/fastforwardlabs/few-shot-text-classification.

other models, including the Albert-small-v2 model [39] , but their performance remained lower. Lastly, PEFT methods deliver performance similar to that of pre-training and full fine-tuning in downstream tasks while using fewer resources. Various PEFT techniques include prompt tuning, Low-Rank Adaptation (LoRA) [27] , and adapters [33]. Adapters, which insert small tunable modules into each transformer layer, have proven effective across different domains [33]. Adopting a similar approach to Mahabadi et al. [33], we train an adapter model with T5[4], a state-of-the-art transfer learning model [10]. We used Optuna [1] for hyperparameter optimization for all models.

In an initial test, SetFit and the adapter model had the best performance, FlairNLP third. We decided to focus mostly on SetFit and to drop PET and Fastforward due to the laboriousness of the prompt-use required in training, and low performance. As shown in Table 2, the T5-base adapter model (Adapt.T5) performs best, overall. For multilabel models macro-F1 best represents model performance, also better reflecting performance in our case where there is a class imbalance in data distribution. The best model performs sufficiently well on all our action classes based on closer examination. We resolved to use the T5 multilabel model as the most suitable option due to good Macro-F1 performance, and the fact that comments often included more than one action.

Table 2. Model performances in 5-fold cross-validation with single and multilabel classification using different numbers of acts

Model	Act set	Single label			Multilabel		
		Macro-F1	*Micro-F1*	*Acc.*	*Macro-F1*	*Micro-F1*	*Acc.*
Adapt.T5	8-act	**0.81**	**0.76**	**0.76**	**0.83**	**0.78**	**0.78**
Adapt.T5	9-act	0.70	0.74	0.74	0.76	0.72	0.72
Adapt.T5	13-act	0.48	0.74	0.74	0.49	0.68	0.68
SetFit	8-act	0.69	0.69	0.70	0.71	0.75	0.41
SetFit	9-act	0.67	0.68	0.69	0.64	0.70	0.37
SetFit	13-act	0.59	0.57	0.59	0.54	0.61	0.26
FlairNLP	8-act	0.67	0.59	0.64	0.70	0.72	0.27
FlairNLP	9-act	0.66	0.61	0.65	0.68	0.69	0.37
FlairNLP	13-act	0.54	0.58	0.57	0.58	0.59	0.18

To answer our RQs, we tagged our data with acts using the best ML model – the multilabel T5-base adapter model (Adapt.T5) with 8 acts. We extracted comments from the data pertaining to the three Phases of crisis on a week-by-week basis. We analyzed the frequencies of acts taken during each week divided

[4] Hyperparameters: learning rate= $3e-4$, AdamW with linear learning rate, schedule with warm-up, batch-size 32, epoch number 200, without early stopping.

by the total number of acts taken that week. We then plotted the time series analysis according to each act group's difference from group mean during each phase. We further investigated significant peaks in each act group, calculating an outlier threshold of $mean(G) + 2 \times SD(G)$ to indicate peaks that should be studied further. We excluded any weekly data that had less than 100 comments. To answer RQ3, we further analyzed significant peaks in the use of specific acts, using BertTopic for topic modeling [23].

7 Results

As indicated by our RQs, we aimed to show that communicative acts can provide meaningful insights into social media crisis discourses: showing significant changes in which acts would be most prevalent at different points in time (RQ1), confirming if early crisis involves discursive struggle (RQ2), and illustrating how different acts might involve different topics (RQ3). To answer RQ1, we wished

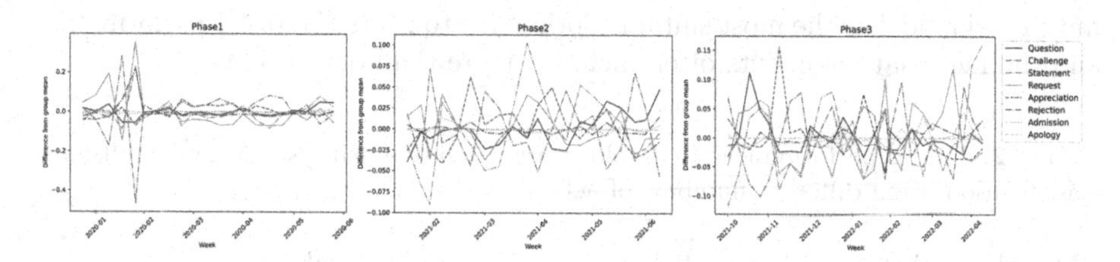

Fig. 2. Temporal differences in the use of acts during Phases 1, 2 and 3 in NDTV comments.

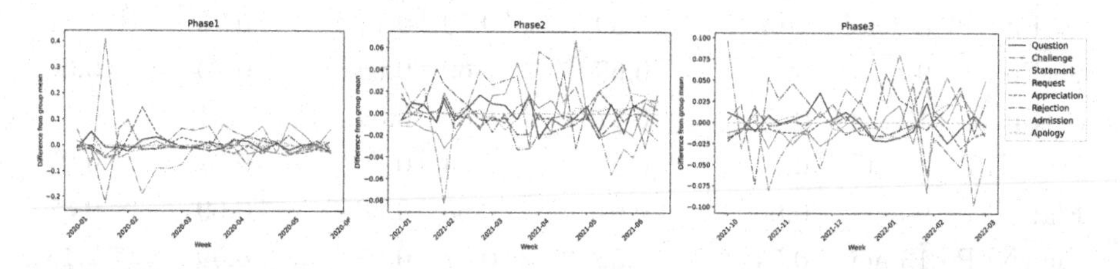

Fig. 3. Temporal differences in the use of acts during Phases 1, 2 and 3 in CTV comments.

to see if the relative use of acts would differ along the crisis timeline during three significant time phases, comparing the results from NDTV and CTV comments. Comparing the Phases in Fig. 2 for the NDTV data, there are notable differences in how various acts appear at different times. For example, during Phase 1 in NDTV data, notable peaks where differences from group means are above

Table 3. Peaks in acts according to Phase and week in NDTV and CTV data.

Channel	Peaking acts		
	Phase 1, 2020	Phase 2, 2021	Phase 3, 2021–22
NDTV	challenges 17/02–24/02	challenges 24/02–01/03	questions 17–24/11,-21
	requests 25/02–02/03	acceptances 01–07/03	apprec. 09–16/12,-21
	denials 25/02–02/03	apprec. 16–23/03	acceptances 09–16/04,-22
	statements 25/02–02/03	statements 17–24/04	requests 01–08/05,-22
		denials 01–08/07	acceptances 09–16/05,-22
		questions 01–08/07	statem. 25/05–01/06,-22
			denials 25/05–01/06,-22
CTV	questions 01–8/02	denials 01–07/03	statements 01–08/11,-21
	acceptances 01–08/02	requests 25/04–01/05	challenges 25/11–01/12,-21
	statements 09–16/02	requests 01–08/06	questions 25/12–01/01,-22
	apprec. 01–08/03	apprec. 17–25/06	apprec. 01–08/02,-22
	denials 01–08/05	challenges 01–08/07	apprec. 25/02–01/03,-22
	denials 25/05–01/06	acceptances 09–16/07	denials 01–08/04,-22
	requests 25/05–01/06		

the significance level include challenging and denying acts – however, requests are also highly prevalent. For CTV comments, questions and statements and more positively inclined acts (acceptances and appreciations) are prevalent during early Phase 1, but denials and requests toward the end of Phase 1. We can find significant peaks in different acts at different points in time during the crisis. These are also different when comparing NDTV and CTV comments.

To sum, our results indicate a positive answer to RQ1: there are significant differences in which acts comments to crisis news portray at different points in time during the crisis. The results also show notable differences when comparing NDTV and CTV datasets.

To answer RQ2, we analyzed whether the early crisis involved a discursive struggle. We consider discursive struggle to emerge through challenges (or accusations) and denials (e.g. of some information).

We see a peak in resolution-oriented requests in the NDTV news comments at the beginning of the Pandemic. There is also a peak in challenging acts, and denials, which is more in line with crisis theory. In CTV data, statements seem most prevalent in the early stages of Phase 1. Denials peak later, toward the end of the Phase. This indicates, concerning RQ1, that the beginning of the crisis produces somewhat of a different audience response than expected in classic crisis theory. It does not necessarily involve a significant amount of acts contributing to discursive struggle, but the audience may produce a significant amount of requests (or proposals) in response to the crisis – e.g. requesting authorities to act or otherwise aiming at crisis resolution. On the other hand, as seen in CTV data, (informative) statements about the crisis might be most common. Acts showing discursive struggle are also common in Phase 2 and Phase 3, which shows that several points in time during the crisis create discursive struggle over issues related to the crisis. In response to RQ2, our results show that audience responses to crisis differ somewhat from the classic theoretical view of how crisis discourses evolve during the early beginning phase of the crisis.

Table 4. 12 most common topics in each of the first two peaking acts during Phase 1, NDTV above and CTV topics below.

NDTV	Requests Jan-March/-20	Count	Challenges Jan-March/-20
805	should,you,demand,action	2064	is,it,virus,spread,not
302	virus,coronavirus	374	news,ndtv,channel,fake,media
85	china,send,him,chinese,yoga	260	you,stupid,idiot,fool,education
79	help,poor,can,people,food	199	she,teacher,be,why,mother
57	kerala,state,india	149	testing,cases,kits,symptoms
46	politics,you,party	140	india,indians,indian,people
43	god,pray,save,let,jesus	98	he,phfi,pm,guy
38	flights,international,all,airports	96	muslims,muslim,allah,islam
34	test,testing,negative,korea	91	god,spiritual,creator,garbage
30	stay,home,safe,guys,please,indoor	87	government,country,state,govt
27	tourist,ban,stop,traveling,karnataka	72	doctor,doctors,he,patients
25	mask,wear,n95,cover,surgical	70	die,selfish,parents,life,murder punish
CTV	**Questions Jan-March/-20**	**Count**	**Statements Jan-March/-20**
84	the,virus,china,we,this	161	trudeau,canada,canadians,is
71	canada,is,that,canadians	146	he,guy,him,this,is,was,like
65	this,why,video,who,high	128	doctors,medical,italy,thank
56	he,his,guy,this,does,what	114	lol,booshit,project,sadki,afieya
27	italy,doctor,would,treat,so,who	88	virus,coronavirus,china,corona
24	money,go,get,home,work,stay	75	masks,death,italy,pneumonia
24	lungs,be,you,with,smokers	68	too,late,little,ago,week,time
13	she,script,curious,write,adam	57	roxham,road,open,close,illegal
12	masks,mask,wearing,wear,why	48	she,just,did,fire,not,lad
-	-	46	my,asthma,temperature,cough oxygen
-	-	46	china,world,sg,now,government
-	-	42	home,pay,money,work,rent,food

To answer RQ3, we investigated if communicative acts could complement traditional topical analysis to offer deeper insights into crisis discourses. We approached this by separately investigating the topics of different notably peaking acts, analyzing how topics across acts differed during the two first significant peaks in acts during Phase 1, for both NDTV and CTV data (first and third peaks for CTV, as the second one included only two topics and 94 comments). The topic modeling results for these are presented in Table 4.

With a closer analysis of Phase 1, in NDTV comments (Table 4), requests are mostly related to requesting politicians and organizations to help (poor) people, proposing how to mitigate the situation e.g. by closing airports and banning travel, using masks and staying home. Challenges conveyed dissatisfaction with measures taken, or politicians and authorities (e.g. BJP i.e. Bharatiya Janata Party), the media, or specific public figures (Kumar Singh) or other people, and claims about 'fake news', CTV topics, then again, included a lot of questions mostly about the virus and its effects on Canada, the situation in Italy, how the situation will affect work, how the virus affects people's lungs, and whether there is need to wear masks. Statements were most common and mostly related to opinions about the actions of politicians like Trudeau, how businesses would be affected by the Pandemic, how wearing masks helps, origin of the virus, how the virus might affect individuals with asthma, as well as opportunities to earn a living. To sum, topics prevalent for different acts and for the two channels differ notably: although there are some similar topics across acts (doctors, the media), their positioning was often different (appreciation of doctors vs. accusing doctors). The analysis of acts may thus provide more insights into the specific functions and purposes of comments – even ones that may have quite similar topics. Such an analysis can also reveal differences in prevalent topics across country contexts during the same period.

8 Discussion

We have shown in this paper that analyzing the acts of news comments can provide meaningful insights into crisis discourses on social media. We showed that there is significant variance in the use of specific acts along a crisis timeline. This can be highly interesting for social media monitoring during crises, indicating how the public is positioned to the crisis at a given time, and what concerns or questions they might have. This could help experts to anticipate what tactics might be needed for mitigating the crisis, or what problems might emerge in the near future. Such insights can also help reveal significant changes in the frequencies of manipulative or possibly problematic content (e.g. attacks against authorities). We showed that acts reveal more context for more fine-grained content analyses. This helps in analyzing what the purpose of comments is, which is crucial for understanding the meaning of comments and what they are about. Thus, the functions and purposes of acts are important in gaining a deeper understanding of how certain topics are talked about (e.g. politicians, doctors or the media) when analyzing large sets of social media data.

Few-shot learning models enable rapid deployment of applied models in novel settings, where often few resources are available and thus no large training datasets can be found for supervised modeling. This is very often the case for social media analyses during emerging events. The opportunity to rapidly develop novel models with few resources is crucial for crisis analytics. We have shown that it is possible to implement an applied model with new classes for empirical analysis, and to reach good performance even with a small manually annotated dataset. Acts such as challenges (and accusations) are interesting to study, especially to investigate discursive struggle related to the crisis, and we feel models including these should be developed further in future research.

One interesting finding in this paper is that although classic crisis research [41] stresses that discursive struggle occurs in the beginning of a crisis, we illustrated that the early public response to crisis is somewhat different as contrasted to the classic crisis theory [41]. Although classic crisis research on discursive struggle [41] has been conducted from the crisis management point-of-view, our results highlight the differences between these two, and how audience reactions might differ from how they are expected to react to crises. Public responses might notably emphasize resolution acts, rather than merely being thrown into a struggle of panic and chaos. Thus, audiences might also be a positive driving force in crisis mitigation. Discursive struggle also occurred during the mid-phase (Phase 2) and prolonged phase (Phase 3) of the crisis. Thus, crisis discourses during a long crisis might involve multiple points of discursive struggle – 'mini crises'.

8.1 Limitations

We acknowledge that perhaps other types of annotation procedures than what we have used might better represent the meanings of the comments in asynchronous conversations, e.g. their ambiguous nature, or multimodal and multilingual content. These could not be accounted for in this paper, but could be investigated in the future. Also, a comment with more than one act is counted more than once in the time series analysis, which might affect further analyses. The concept of discursive struggle, in addition, is a highly abstract concept. We did our best to represent the concept in our computational analysis, but it needs to be noted that our approach may be a limited interpretation of the concept.

As for ethical considerations and data privacy, we anonymized our data to not reveal the identities of participants in the comment section discussions. The data is stored on secure servers provided by Aalto University and handled with care. Our code, and an anonymized and de-identified version of our dataset and annotated data will be made available via application through our GitHub page. This is to ensure researcher use of the data but not to endanger the identification of individuals in the dataset.

9 Conclusions and Future Work

We have shown that our approach allows an analysis of how online discourses evolve during a crisis. Further research could investigate if similar developments in crisis discourses can be found in different cultural contexts, and in relation to different crises, also complementing computational modeling with qualitative methods. How to best annotate and model acts in asynchronous rapidly changing crisis conversations is still a developing area of work, e.g. to better represent multiple meanings of comments, multimodal and multilingual content. This research advances computational analysis methods for social media, elaborating how communicative acts can support analyses of the development of crisis discourses in large social media datasets. Such methods are needed for multifaceted understandings of crises and social media engagement. Further, since social media is used as a venue for influencing public opinion and spreading disinformation, the discursive conflicts taking place in this arena are essential for e.g. crisis communicators, researchers, and journalists to both understand and manage. These insights will be increasingly important in pre-emptively mitigating disorder, confusion, and manipulation on social media during times of crisis.

Acknowledgments. This research has been funded by the Academy of Finland project "Reconstructing Crisis Narratives for Trustworthy Communication and Cooperative Agency", project number 339931. In this paper, we utilized computer resources provided by the Aalto University School of Science "Science-IT" project. We also used the Finnish CSC – IT Center for Science LTD's high-performance computing resources. We would like to thank Nitin Sawhney, Salla-Maaria Laaksonen and Minttu Tikka for feedback and support during our research.

Disclosure of Interests. The authors have no competing interests to declare that are relevant to the content of this article.

References

1. Akiba, T., Sano, S., Yanase, T., Ohta, T., Koyama, M.: Optuna: a next-generation hyperparameter optimization framework. In: Proceedings of the 25th ACM SIGKDD Conference on Knowledge Discovery & Data Mining, pp. 2623–2631. ACM, USA (2019)
2. Artstein, R., Poesio, M.: Intercoder agreement for computational linguistics. Comput. Linguist. **34**(4), 555–596 (2008)
3. Bellutta, D., King, C., Carley, K.M.: Deceptive accusations and concealed identities as misinformation campaign strategies. Comput. Math. Organ. Theory **27**, 302–323 (2021)
4. Bird, S., Klein, E., Loper, E.: Natural language processing with Python: analyzing text with the natural language toolkit. Inc, O'Reilly Media (2009)
5. Bracewell, D.B., Tomlinson, M., Wang, H.: Semi-supervised modeling of social actions in online dialogue. In: IEEE Seventh International Conference on Semantic Computing, pp. 168–175. IEEE (2013)
6. Brown, T., Mann, B., Ryder, N., et al.: Language models are few-shot learners. In: NIPS'20: Proceedings of the 34th International Conference on Neural Information Processing Systems, pp. 1877–1901. ACM, USA (2020)
7. Cheng, S.W.: A corpus-based approach to the study of speech act of thanking. Concentric: Stud. Linguist. **36**(2), 257–274 (2010)

8. Clark, A., Popescu-Belis, A.: Multi-level dialogue act tags. In: Proceedings of the 5th SIGdial Workshop on Discourse and Dialogue at HLT-NAACL, pp. 163–170, ACL, USA (2004)
9. Clark, H., Schaefer, E.F.: Contributing to discourse. Cogn. Sci. **13**(2), 259–294 (1989)
10. Colin, R., Shazeer, N., Roberts, A., et al.: Exploring the limits of transfer learning with a unified text-to-text transformer. J. Mach. Learn. Res. **21**(1), 1–67 (2020)
11. Couper-Kuhlen, E.: What does grammar tell us about action? Pragmatics **24**(3), 623–647 (2014)
12. Couper-Kuhlen, E., Selting, M.: Interactional Linguistics: Studying Language in Social Interaction. Cambridge University Press (2017)
13. Devlin, J., Chang, M., Lee, K., Toutanova, K.: BERT: pre-training of deep bidirectional transformers for language understanding. In: Proceedings of the 2019 Conference of the North American Chapter of the Association for Computational Linguistics, pp. 4171–4186. ACL, Minnesota (2019)
14. Edwards, D., Potter, J.: Language and causation: a discursive action model of description and attribution. Psychol. Rev. **100**(1), 23–41 (1993)
15. Fleiss, J. Levin, B. Cho Paik, M.: The measurement of interrater agreement: statistical methods for rates and proportions, pp. 598–626. John Wiley & Sons (1981)
16. Forsyth, E., Martell, C.: Lexical and discourse analysis of online chat dialog. In: International Conference on Semantic Computing, pp. 19–26. IEEE (2007)
17. Gao, W., Li, P., Darwish, K.: Joint topic modeling for event summarization across news and social media streams. In: Proceedings of the 21st ACM International Conference on Information and Knowledge Management, pp. 1173–1182. ACM, USA (2012)
18. Geng, R., Li, B., Li, Y., et al.: Dynamic memory induction networks for few-shot text classification. In: Proceedings of the 58th Annual Meeting of the Association for Computational Linguistics, pp. 1087–1094. ACL, online (2020)
19. Ghosh, S., Ghosh, S.: Classifying speech acts using multi-channel deep attention network for task-oriented conversational search agents. In: Proceedings of the 2021 Conference on Human Information Interaction and Retrieval, pp. 267–272. ACM, USA (2021)
20. Giglietto, F., Righetti, N., Rossi, L., Marino, G.: Coordinated link sharing behavior as a signal to surface sources of problematic information on Facebook. In: International Conference on Social Media and Society, pp. 85–91. ACM (2020)
21. Giles, D., Stommel, W., Paulus, et al.: Microanalysis of online data: The methodological development of "Digital CA". Discourse, Context & Media **7**, 45–51 (2015)
22. Godfrey, J., Holliman, E., McDaniel, J.: SWITCHBOARD: Telephone speech corpus for research and development. In: Proceedings of the 1992 IEEE International Conference on Acoustics, Speech and Signal Processing, pp. 517–520. IEEE, USA (1992)
23. Grootendorst, M.: BERTopic: neural topic modeling with a class-based TF-IDF procedure. arXiv preprint arXiv:2203.05794 (2022)
24. Halder, K., Akbik, A., Krapac, J., Vollgraf, R.: Task-aware representation of sentences for generic text classification. In: Proceedings of the 28th International Conference on Computational Linguistics, pp. 3202–3213. International Committee of Computational Linguistics, Spain (2020)
25. Herring, S., Das, A., Penumarthy, S.: CMC act taxonomy, 2007 edition. Indiana University. https://homes.luddy.indiana.edu/herring/cmc.acts.html. Accessed 15 Aug 2024

26. Hou, Y., Lai, Y., Wu, Y., et al.: Few-shot learning for multi-label intent detection. In: The Thirty-Fifth AAAI Conference on Artificial Intelligence 35, pp. 13036–13044. AAAI (2021)
27. Hu, E., Shen, Y., Wallis, P., et al.: LoRA: low-rank adaptation of large language models. arXiv preprint arXiv:2106.09685 (2021)
28. Jin, Y., Fisher Liu, B., Austin, L.: Examining the role of social media in effective crisis management: the effects of crisis origin, information form, and source on publics' crisis responses. Commun. Res. **41**(1), 74–94 (2014)
29. Joffe, H.: Risk: from perception to social representation. Br. J. Soc. Psychol. **42**(1), 55–73 (2003)
30. Jørgensen, M., Phillips, L.: Discourse analysis as theory and method. Sage (2002)
31. Joulin, A., Grave, E., Bojanowski, P., et al.: Fasttext.zip: compressing text classification models. arXiv preprint arXiv:1612.03651 (2016)
32. Labov, W., Waletzky, J.: Narrative analysis: oral versions of personal experience. J. Narrative Life Hist. **7**(1–4), 3–38 (1997)
33. Mahabadi, R., Ruder, S., Dehghani, M., Henderson, J.: Parameter-efficient multi-task fine-tuning for transformers via shared hypernetworks. In: Proceedings of the 59th Annual Meeting of the Association for Computational Linguistics and the 11th International Joint Conference on Natural Language Processing, pp. 565–576. ACL, online (2021)
34. Meng, L., Huang, M.: Dialogue intent classification with long short-term memory networks. In: Huang, X., Jiang, J., Zhao, D., Feng, Y., Hong, Yu. (eds.) Natural Language Processing and Chinese Computing, pp. 42–50. Springer International Publishing, Cham (2018). https://doi.org/10.1007/978-3-319-73618-1_4
35. Meredith, J.: Analysing technological affordances of online interactions using conversation analysis. J. Pragmat. **115**, 42–55 (2017)
36. Newman, N., Fletcher, R., Schulz, A., et al.: Reuters institute digital news report 2021. Reuters Institute for the study of Journalism (2021)
37. Nguyen, T., Shirai, K.,: Topic modeling based sentiment analysis on social media for stock market prediction. In: Proceedings of the 53rd Annual Meeting of the Association for Computational Linguistics and the 7th International Joint Conference on Natural Language Processing, pp. 1354–1364. ACL, China (2015)
38. Paakki, H., Vepsäläinen, H., Salovaara, A.: Disruptive online communication: how asymmetric trolling-like response strategies steer conversation off the track. Comput. Support. Coop. Work **30**(3), 1–37 (2021)
39. Reimers, N., Gurevych, I.: Sentence-BERT: sentence embeddings using Siamese BERT-networks. In: Proceedings of the 2019 Conference on Empirical Methods in Natural Language Processing and the 9th International Joint Conference on Natural Language Processing, pp. 3982–3992. ACL, China (2019)
40. Schick, T., Schütze, H.: Exploiting cloze questions for few-shot text classification and natural language inference. arXiv preprint arXiv:2001.07676 (2020)
41. Sellnow, T:, Sellnow, D., Helsel, E., et al.: Risk and crisis communication narratives in response to rapidly emerging diseases. J. Risk Res. **22**(7), 897–908 (2019)
42. Stivers, T.: Sequence organization. In: Handbook of Conversation Analysis, pp. 191–209 (2013)
43. Tunstall, L., Reimers, N., Jo, U., et al.: Efficient few-shot learning without prompts. arXiv preprint arXiv:2209.11055 (2022)
44. Virtanen, M., Vepsäläinen, H., Koivisto, A.: Managing several simultaneous lines of talk in Finnish multi-party mobile messaging. Discourse, Context Media **39**, 100460 (2021)

45. Zelenkauskaite, A., Toivanen, P., Huhtamäki, J., Valaskivi, K.: Shades of hatred online: 4chan duplicate circulation surge during hybrid media events. First Monday **26**(1) (2020)
46. Zhang, A., Culbertson, B., Paritosh, P.: Characterizing online discussion using coarse discourse sequences. In: AAAI on Web and Social Media 11(1), pp. 357–366 (2017)
47. Zhuang, L., Wayne, L., Ya, S., Jun, Z.: A robustly optimized BERT pre-training approach with post-training. In: Proceedings of the 20th Chinese National Conference on Computational Linguistics, pp. 1218–1227. ACL, China (2021)
48. Gil de Zúñiga, H., Diehl, T., Weeks, N.: Political persuasion on social media: tracing direct and indirect effects of news use and social interaction. New Media Soc. **18**(9), 1875–1895 (2016)

A Lightweight Approach for User and Keyword Classification in Controversial Topics

Ahmad Zareie(✉), Kalina Bontcheva, and Carolina Scarton

Department of Computer Science, The University of Sheffield, Sheffield, UK
{a.zareie,k.bontcheva,c.scarton}@sheffield.ac.uk

Abstract. Classifying the stance of individuals on controversial topics and uncovering their concerns is crucial for social scientists and policymakers. Data from Online Social Networks (OSNs), which serve as a proxy to a representative sample of society, offers an opportunity to classify these stances, discover society's concerns regarding controversial topics, and track the evolution of these concerns over time. Consequently, stance classification in OSNs has garnered significant attention from researchers. However, most existing methods for this task often rely on labelled data and utilise the text of users' posts or the interactions between users, necessitating large volumes of data, considerable processing time, and access to information that is not readily available (e.g. users' followers/followees). This paper proposes a lightweight approach for the stance classification of users and keywords in OSNs, aiming at understanding the collective opinion of individuals and their concerns. Our approach employs a tailored random walk model, requiring just one keyword representing each stance, using solely the keywords in social media posts. Experimental results demonstrate the superior performance of our method compared to the baselines, excelling in stance classification of users and keywords, with a running time that, while not the fastest, remains competitive.

Keywords: Users Classification · Keyword Classification · Stance Detection · Random Walk

1 Introduction

In society, individuals may hold varying opinions on controversial topics. Controversial topics refer to subjects that create polarisation in individuals' stances, leading to dichotomous opinions (e.g., Leave vs. Remain in the Brexit context, Pro-vaccine vs. Anti-vaccine in the COVID-19 context, or support for a specific candidate in an election); this resulting position is known as the stance. In controversial topics, classifying users based on their stance, discovering the concerns within each class and tracking the evolution of classes and concerns over time are regarded as crucial for social scientists and policymakers [2]. Data collected

L. M. Aiello et al. (Eds.): ASONAM 2024, LNCS 15212, pp. 243–253, 2025.
https://doi.org/10.1007/978-3-031-78538-2_21

from Online Social Networks (OSNs) provide valuable insights for this crucial task since it can be seen as a proxy for opinions in societies. In these networks, individuals express and discuss their viewpoints to disseminate their opinions and potentially influence others' perspectives. Therefore, the stance of users can be inferred based on their activities in OSNs; additionally, the keywords they use in their posts can express their concerns. In this paper, we propose a **random walk approach** to classify users and keywords in OSNs based on their **stance on a controversial topic**. This approach is lightweight, relying solely on keywords shared by users for the classification task. This approach only requires one keyword to represent each stance as prior knowledge, without the need for pre-determined thresholds. Hereafter, we use 'hashtag' interchangeably with 'keyword', but the method can be used with any token as keywords.

Analysing data in OSNs to discover the stance of users and the generated content (posts) has increasingly become the focus of researchers from various disciplines, and a multitude of methods have been proposed for this task [2,3]. Previous work focuses mainly on supervised models, which rely on a set of data known as labelled (training) data as prior knowledge. These supervised methods could be divided into two groups: (i) content-based [5,11]: requiring a set of texts as labelled data and utilising the text of posts to understand the stance, and (ii) interaction-based [1,13,14]: relying on a set of users representing each stance as labelled data and using the interactions between pairs of users (e.g., likes, retweets, replies, and mentions on Twitter – now X) to understand the stance. The labelled data required in supervised methods is typically generated through a manual process, involving human judgments, which is time-consuming (particularly when tracking the evolution of public opinion over time [2]) and highly affected by annotators' biases. Data labelling may also lead to language- and domain-dependant models, requiring further data annotation for handling new domains and languages. Unsupervised methods [6,10,12] have also been proposed for classifying individuals' stances without the need for labelled data. These methods utilise linguistic features extracted from post content, such as lexicons, grammatical dependencies, and the latent meanings of words in context. Alternatively, they may leverage users' profile features to infer their stance. However, methods relying on linguistic features may encounter challenges in multilingual domains due to their dependency on language-specific characteristics [3]. Meanwhile, those utilising profile features may face difficulties in generalising across diverse social media platforms, given the variability in profile features offered by each platform.

Our method employs a lightweight unsupervised approach, that is language- and profile-agnostic, setting it apart from these previous studies. Moreover, it does not rely on user interactions, which may not always be available on certain social media platforms (e.g., users sharing opinions in a Telegram group). To our knowledge, Darwish et al. [5] and Coletto et al. [4] are the only previous work addressing a classification task similar to our work. Darwish et al. [5] propose an approach that relies on predefined sets of users, each representing a particular stance. The stance of each user is then determined based on their similarity to these predefined sets. Although this method relies on keywords (hashtags) shared by users to detect stances, similar to ours, it differs in that it requires a set of users for each stance as labelled data, whereas our method relies on only one keyword for each stance. Coletto et al. [4] propose an approach that classifies keywords and users iteratively. Similar to our approach, this method relies on the keywords shared by users and requires only one keyword for each stance as prior knowledge. However, this method requires pre-defined thresholds for classification, and determining the optimal thresholds for various domains is not trivial and, in some cases, not feasible. Additionally, our experimental evaluation demonstrates that our lightweight method outperforms both of these methods.

The main contribution of this paper is the lightweight classification method, which utilises a tailored random walk approach to effectively classify users and hashtags with minimal user data and prior knowledge. We apply our method, called the Lightweight Random Walk Method (LRM), to analyse conversations about the UK General Election 2019 and Brexit on Twitter (now X). Our approach shows superior performance compared to baseline methods in accurately classifying users into different stance classes and identifying the most prominent hashtags within each class. Our method also offers competitive running time compared to baseline methods. Additionally, the proposed model is utilised to track the evolution of stance classes over time in both the general election and Brexit datasets. Results obtained by tracking the evolution in the Brexit data are consistent with previous studies applying content and text analysis [8,9]. Although there is a lack of previous studies on analysing stance in the 2019 general election, our work corroborates with reports [7] showing quantitative analysis about this election in social media.

2 Lightweight Random Walk Method (LRM)

Let $U = \{u_1, u_2, \cdots, u_n\}$ be a set of users sharing a set of hashtags $H = \{h_1, h_2, \cdots, h_m\}$, where n and m indicate the number of users and hashtags, respectively. Let $R_{n \times m}$ be the sharing records with R_{ki} denoting the number of times u_k shares h_i; $R_{ki} = 0$ if u_k does not share h_i. The notation $\mathcal{R}_i$ indicates the total number of times that hashtag h_i is shared, calculated as the sum of sharing instances across all users, i.e., $\mathcal{R}_i = \sum_{k=1}^{n} R_{ki}$, and notation $\mathscr{R}_k$ represents the total number of times u_k has shared hashtags, calculated as the sum of sharing instances of u_k across all hashtags, i.e., $\mathscr{R}_k = \sum_{i=1}^{n} R_{ki}$.

Suppose there are t stances (represented as t classes) towards a controversial topic, with each stance c (where $c \in \{1, \cdots, t\}$) associated with a given hashtag s_c, referred to as the seed hashtag or simply the seed. The objective is to classify hashtags and users into these t classes using only one hashtag (seed) associated with each stance.

The classification in the proposed method is carried out as follows: (1) *Hashtag classification:* for each hashtag, we calculate the similarity between the hashtag and each seed. Then, the hashtag is assigned to class c where s_c has the highest similarity to the hashtag. The set of hashtags assigned to class c is denoted by

C_c. *User classification:* For each user, based on the class of the hashtags shared by the user, we determine the user's inclination towards each class. The user is assigned to class c when they exhibit the highest inclination towards the class. The method comprises four steps described in the following sub-sections.

2.1 Generation of a Hashtag-Sharing Graph

To construct the hashtag-sharing graph, sharing records are projected onto a graph where nodes represent hashtags and edges denote relationships between pairs of hashtags. This graph is defined using a matrix $A_{m \times m}$ where A_{ij} denotes the strength of the relationship between hashtags h_i and h_j. The value of A_{ij} is computed using Eq. (1).

$$A_{ij} = \frac{\sum_{u_k \in U} \min(R_{ki}, R_{kj})}{\min(\mathcal{R}_i, \mathcal{R}_j)} \tag{1}$$

The value A_{ij} falls within the range of $[0, 1]$, with a larger value indicating a stronger relationship.

2.2 Determination of Similarity Between Hashtags and Seeds

We apply a modified local random walk algorithm to determine the similarity between each hashtag $h_i \in H$ and each seed s_c (where $c \in \{1, \cdots, t\}$). In a local random walk algorithm, given a transition matrix A (where A_{ij} is regarded as the probability that a random walker will go to h_j from h_i), a walker starts walking from one node and continues traversing the graph until a specified stopping criterion is met. The general equation of the random walk is as Eq.(2):

$$\pi_c(z) = A^T \cdot \pi_c(z-1) \tag{2}$$

where $\pi_c(z)$ is an $m \times 1$ vector in which the i-th element, denoted by $\pi_{ci}(z)$, represents the probability of visiting node h_i in the z-th step of the walk. The notation A^T indicates the transpose of the transition matrix, and $\pi_c(0)$ is a vector with all zero values except the c-th element, which is 1.

In our modified random walk algorithm, a walker starts from a seed hashtag s_c and traverses the graph to calculate the similarity between each hashtag and the seed. However, to avoid assigning undue similarity between the given hashtag s_c and other hashtags strongly related to other seed hashtags, the walker should be prevented from visiting hashtags with strong relationships to other seeds. This visiting may occur in two cases if a walker starting from s_c: (i) visits a different seed s_j and then proceeds to hashtags strongly related to s_j, or (ii) visits a general hashtag, which is used by individuals from different stances (e.g., the hashtag *#ge2019* in the context of the 2019 UK general election), and then proceeds to hashtags strongly related to other seeds. To address case (i), we set all the values in the corresponding row and column of $j \in \{1, \cdots, t\} - \{c\}$ in the transition matrix to zero to prevent the walker from visiting hashtags through seed s_j, ensuring that s_c is the only seed involved in the walk. To address (ii), we utilise the notation of entropy to adjust the transition probabilities through every hashtag h_i, considering the generality of the hashtag across various stances. For this purpose, we first compute an entropy E_i, as shown in Eq. (3). The value of E_i falls within the range of $[0, 1]$, with a larger value indicating that the hashtag h_i is more neutral.

$$E_i = \frac{-\sum_{c=1}^{t} \frac{A_{ic}}{\mathscr{A}_i} \cdot log(\frac{A_{ic}}{\mathscr{A}_i})}{log(t)} \tag{3}$$

In Eq. (3), $\mathscr{A}_i$ indicates the sum of the relationships between h_i and all seed hashtags. Then, the value A_{ic} is decreased as $A_{ic} = (1 - E_i)/d(i) \cdot A_{ic}$, where d_i is the degree of the node h_i in the hashtag-sharing graph (the number of non-zero values in the i-th row of the matrix $A_{m \times m}$).

The random walk approach is iterated for each seed hashtag s_c, to calculate the similarity between hashtags and seeds. In each iteration, the matrix A is row-normalised to ensure that the sum of values in each row is 1. Then, a walker traverses the graph by starting from a seed s_c. In a random walk approach, the similarity of every hashtag $h_i \in H$ to s_c can be calculated using Eq. (4).

$$S_{ci} = \sum_{z=1}^{\rho} \pi_{ci}(z) \tag{4}$$

In this equation, $\pi_{ci}(z)$ represents the i-th element of vector $\pi_c(z)$, which can be calculated using Eq. (2), and ρ denotes the number of steps in the random walk approach. To balance between running time and classification performance, we empirically set this value to 10 in our experiments.

2.3 Classification of Hashtags

After determining the similarity between hashtags and all seeds, each hashtag $h_i \in H$ is assigned to class c where the hashtag has the largest similarity value to the seed s_c. We use the notation $\mathcal{C}_c$ to refer to the set of hashtags assigned to class c. Additionally, we determine the stance intensity of each hashtag h_i as shown in Eq. (5).

$$\mathcal{I}(h_i) = \frac{-\sum_{c=1}^{t} \frac{S_{ci}}{i} \cdot log(\frac{S_{ci}}{i})}{log(t)} \tag{5}$$

where denotes the sum of the similarity values between h_i and all seeds. The stance intensity $\mathcal{I}(h_i)$ falls within the range of $[0, 1]$, where a larger value indicates a greater stance intensity.

2.4 Classification of Users

In this step, the stance and intensity of the hashtags used by each user u_k are utilised to determine the user's inclination towards each class. The user is assigned to class c where they exhibit the greatest inclination towards the hashtags in $\mathcal{C}_c$. The inclination of u_k towards class c is calculated using Eq. (6).

$$\mathcal{L}_{kc} = \sum_{h_i \in \mathcal{C}_c} \mathcal{I}(h_i) \cdot \frac{R_{ki}}{\mathscr{R}_k} \tag{6}$$

In this equation, $\mathscr{R}_k$ represents the total number of times u_k has shared hashtags, and R_{ki} represents the number of times user u_k has shared hashtag h_i. The value of $\mathcal{L}_{kc}$ ranges form 0 to 1, with a larger value indicating a stronger inclination of u_k towards class c.

3 Experimental Evaluation

3.1 Datasets

To assess the effectiveness of the proposed method, we employ two Twitter (X) datasets collected via Twitter Streaming APIs. For each dataset, initially, we compile a list of the most frequently used hashtags in the relevant context and collect tweets containing at least one of these hashtags.

(i) The UK General Election Dataset (*Election* dataset) covers the UK general election held on 12 December 2019. This dataset was collected over four weeks, from 16 November 2019 to 12 December 2019, and includes 234,800 unique hashtags shared by 929,441 unique accounts, totalling 14,233,501 sharing records. In this dataset, the five major political parties are treated as distinct classes: Labour (*LAB*), Conservative (*CON*), Liberal Democrats (*LDM*), Scottish National (*SCN*), and Green (*GRE*). For each party, a hashtag that explicitly represents the party's stance is selected as a seed. The chosen seeds are *#votelabour*, *#voteconservative*, *#votelibdem*, *#votesnp*, and *#votegreen* for the *LAB*, *CON*, *LDM*, *SCN*, and *GRE* classes, respectively.

(ii) Brexit Referendum Dataset (*Brexit* dataset) pertains to the Brexit referendum, held on 23 June 2016, where the UK voted to decide whether to leave or remain in the European Union. This dataset was collected over four weeks, from 27 May 2016 to 23 June 2016. It includes 158,673 unique hashtags shared by 1,242,530 unique accounts, totalling 15,446,338 sharing records. In this dataset, two classes are considered: Remain (*RMN*) and Leave (*LVE*). The seed hashtags chosen for this dataset are *#voteremain* and *#voteleave* for the *RMN* and *LVE* classes, respectively.

For each dataset, we establish golden sets of users and hashtags. The golden set of users comprises accounts of members of parliament, candidates, or government ministers known for supporting each stance present in the collected dataset, acting as representatives of their respective classes. A hashtag qualifies as a golden hashtag for a class if it is shared at least 30 times by more than one account in that class and at least five times more frequently than by accounts in any other class. These criteria ensure the establishment of a reasonable golden hashtag set with a practical size. Note that the golden sets of users and hashtags are used exclusively for evaluation purposes. Table 1 summarises the number of golden users and hashtags identified for evaluation in each stance (class) and dataset. During the experiments on both datasets, hashtags with low engagement are filtered out by excluding those with sharing frequencies lower than the average frequency of all hashtags. Similarly, users with sharing records fewer than the average sharing records of all users are also removed.

3.2 Setting

We compare our proposed LRM method with the following five methods: (i) The Iterative Classification Method (ICM) [4] which uses iterative classification to determine the stance of hashtags and users. (ii) The Hashtag Similarity Method (HSM) [5], originally designed for user classification using seed users, has been adapted in this paper due to the lack of seed users. It now classifies hashtags based on similarity to seed hashtags. (iii) Random Method (RDM) that randomly assigns hashtags and users to classes. (iv) The Label Propagation Method (LPM) which constructs a graph of hashtag co-occurrences and uses label propagation to assess the similarity between each hashtag and seeds. (v) Simple Random Walk (SRM) that generates a graph of hashtag co-occurrences and applies a random walk to measure the similarity to the seeds. ICM uses iterative classification for both hashtags and users, while RDM assigns them randomly. HSM, LPM, and SRM classify hashtags based on similarity to seeds. For user classification, these methods assign users to the class with the most shared hashtags.

Table 1. The golden sets (users and hashtags)

Dataset	Election dataset					Brexit dataset	
Class	*LAB*	*CON*	*LDM*	*SCN*	*GRE*	*RMN*	*LVE*
User	349	446	602	104	367	445	145
Hashtag	80	109	53	29	32	36	28

Three experiments were conducted for evaluation. The first assesses performance in classifying golden users and hashtags using F_1 scores for each class and F_{Macro} score as an indicator of overall performance. The second evaluates method efficiency based on average running time. Lastly, the third uses the proposed method to track the evolution in the weeks leading up to the events (election/referendum).

3.3 Results

In the first experiment, each method is applied to classify the users and hashtags using the seed hashtags for both datasets; and evaluated using the golden users and hashtags sets. Table 2 shows the results in terms of F_1 and F_{Macro} for the Election dataset, whilst Table 3 summarises the results for the Brexit dataset. The best performance in each class (F_1) and across all classes (F_{Macro}) is shown in **bold face** in these tables.

Table 2. Classification performance in the Election dataset

	Hashtag Classification						User Classification					
Score	ICM	HSM	RDM	LPM	SRM	LRM	ICM	HSM	RDM	LPM	SRM	LRM
F_1 (LAB)	**0.919**	0.710	0.225	0.609	0.741	0.909	0.944	0.659	0.248	0.569	0.827	**0.975**
F_1 (CON)	**0.913**	0.870	0.249	0.688	0.850	**0.913**	0.911	0.891	0.237	0.532	0.861	**0.969**
F_1 (LDM)	0.901	0.516	0.191	0.351	0.739	**0.913**	0.831	0.367	0.198	0.140	0.811	**0.938**
F_1 (SCN)	0.897	**0.981**	0.134	0.963	0.732	0.912	0.852	**0.944**	0.069	0.727	0.459	0.897
F_1 (GRE)	0.867	0.595	0.138	0.514	0.639	**1.000**	0.826	0.022	0.173	0.011	0.765	**0.956**
F_{Macro}	0.899	0.734	0.188	0.625	0.740	**0.929**	0.873	0.577	0.185	0.396	0.745	**0.947**

Table 3. Classification performance in the Brexit dataset

	Hashtag Classification						User Classification					
Score	ICM	HSM	RDM	LPM	SRM	LRM	ICM	HSM	RDM	LPM	SRM	LRM
F_1 (RMN)	0.909	0.750	0.541	0.293	0.933	**0.941**	0.907	0.580	0.617	0.071	0.905	**1.000**
F_1 (LVE)	0.880	0.767	0.432	0.613	0.878	**0.917**	**1.000**	0.446	0.280	0.337	0.258	**1.000**
F_{Macro}	0.895	0.758	0.487	0.453	0.906	**0.929**	0.954	0.513	0.448	0.204	0.582	**1.000**

Table 4. Running time of the different methods for both datasets (in minutes)

Dataset	ICM	HSM	RDM	LPM	SRM	LRM
Election dataset	144.03	20.18	02.20	57.23	38.51	26.02
Brexit dataset	60.08	10.51	01.15	44.15	33.01	12.10

Regarding F_{Macro} scores, the proposed method (LRM) outperforms the other methods, with ICM providing the second-best results. Notably, SRM using a simple random walk also exhibits competitive performance, suggesting that employing the random walk approach is effective for calculating hashtag similarities for the task outlined in this paper. However, the modifications implemented in the random walk approach in LRM further enhance classification performance. The difference between the F_{Macro} scores obtained by LRM and those of ICM (the second-best method) indicates that LRM exhibits greater superiority in user classification compared to hashtag classification. This observation underscores the

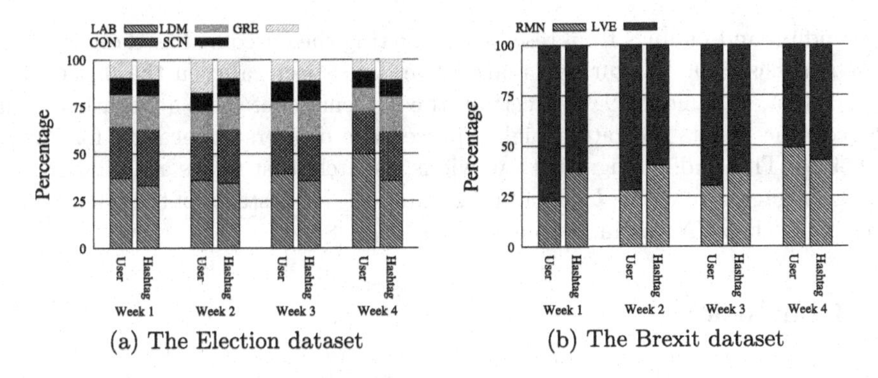

(a) The Election dataset (b) The Brexit dataset

Fig. 1. Percentage of users and hashtags in each class

effectiveness of our heuristic, which leverages hashtag stance intensity for user classification, enhancing performance. Regarding per-class F_1 scores, the LRM method consistently outperforms other methods, with few exceptions, where it shows comparable performance. The worst performance of LRM is in the Election dataset for users and hashtags in the *SCN* class. This may be because *SCN* is a relatively small class when compared to the other classes (see Table 1), and the proposed method may not properly determine the similarity between the hashtags in this class and the associated seed hashtag.

The second experiment evaluates the efficiency of methods by measuring their running time (Table 4). LRM demonstrates superior efficiency compared to LPM, SRM, and ICM. This efficiency stems from LRM's approach of constructing the hashtag-sharing graph based solely on user hashtag-sharing frequencies, unlike LPM and SRM, which process all tweets to find hashtag co-occurrences for graph generation. The iterative processing of posts by ICM renders it the slowest method. While RDM and HSM are faster than LRM, they exhibit significantly lower classification performance, as observed in the first experiment.

In the third experiment, we employ LRM to analyse the evolution of the number of users and hashtags in each class over the four weeks leading up to the event (election/referendum). To conduct this analysis, we extract data collected in each week and utilise LRM to classify users and hashtags. Figure 1 illustrates the percentage of all users and hashtags in each class over the four weeks. In the Election dataset (Fig. 1a) it is observed that *LAB* consistently maintains a higher percentage of users and hashtags compared to other classes, with *CON* following closely. Interestingly, even in the week preceding the election (week 4), the percentage of *LAB* users and hashtags remains higher than that of *CON*. This observation aligns with reports suggesting greater engagement with Labour accounts on social media platforms [7]. In the Brexit dataset (Fig. 1b), the activity of users supporting *LVE* is higher than that of users advocating for *RMN*. However, as the referendum date approaches, the percentage of *RMN* users and hashtags increases. By week 4, the percentage of users advocating *RMN* and *LVE* is 48.80% and 51.20%, respectively, closely reflecting the outcome of the referendum and findings from research predicting the outcome based on Twitter post analysis [8,9]. We further analyse the data specifically on the day of the referendum, 23 June 2019, to ascertain the percentage of users advocating each stance. The results indicate a higher percentage of users advocating for *RMN* (59.59%). This finding aligns with previous research, that shows a similar trend: a large engagement of the *LVE* campaign since the early stages of the referendum process, with *RMN* catching up at a later stage [8].

4 Conclusion

This paper introduced a novel lightweight method for understanding public opinions on controversial topics, discovering the hashtags used to advocate each stance, and tracking the evolution over time. The method requires only one hashtag representing each stance and relies solely on the hashtags shared by the users. Unlike methods that rely on text analysis techniques or user interactions, our approach is language-agnostic and can be applied across various social networks without requiring user interactions. It also aligns with a real-world scenario, where a social scientist or other stakeholders would be interested in monitoring a given event, while they can easily identify seed hashtags. Experimental results demonstrate that our method outperforms similar methods and baselines.

Acknowledgment. This work is supported by the UK's innovation agency (InnovateUK) grant number 10039039 (approved under the Horizon Europe Programme as VIGILANT, EU grant agreement number 101073921) (https://www.vigilantproject.eu).

References

1. Aldayel, A., Magdy, W.: Your stance is exposed! analysing possible factors for stance detection on social media. In: Proceedings of the ACM on Human-Computer Interaction, vol. 3 (2019)
2. ALDayel, A., Magdy, W.: Stance detection on social media: state of the art and trends. Inf. Process. Manage. **58**(4), 102597 (2021)
3. Alturayeif, N., Luqman, H., Ahmed, M.: A systematic review of machine learning techniques for stance detection and its applications. Neural Comput. Appl. **35**(7), 5113–5144 (2023)
4. Coletto, M., Lucchese, C., Orlando, S., Perego, R.: Polarized user and topic tracking in twitter. In: Proceedings of the 39th International Conference on Research and Development in Information Retrieval, pp. 945–948 (2016)
5. Darwish, K., Magdy, W., Zanouda, T.: Improved stance prediction in a user similarity feature space. In: Proceedings of the 2017 IEEE/ACM International Conference on Advances in Social Networks Analysis and Mining, pp. 145–148 (2017)
6. Darwish, K., Stefanov, P., Aupetit, M., Nakov, P.: Unsupervised user stance detection on twitter. In: Proceedings of the International AAAI Conference on Web and Social Media, vol. 14, issue 1, pp. 141–152 (2020)

7. Fletcher, R.: Did the conservatives embrace social media in 2019. UK election analysis, pp. 1–123 (2019)
8. Grčar, M., Cherepnalkoski, D., Mozetič, I., Kralj Novak, P.: Stance and influence of Twitter users regarding the Brexit referendum. Comput. Soc. Netw. **4**, 1–25 (2017)
9. Khatua, A., Khatua, A.: Leave or Remain? Deciphering Brexit deliberations on twitter. In: 2016 IEEE 16th International Conference on Data Mining Workshops, pp. 428–433 (2016)
10. Kobbe, J., Hulpuş, I., Stuckenschmidt, H.: Unsupervised stance detection for arguments from consequences. In: Proceedings of the 2020 Conference on Empirical Methods in Natural Language Processing, pp. 50–60 (2020)
11. Mohammad, S.M., Sobhani, P., Kiritchenko, S.: Stance and sentiment in tweets. ACM Trans. Internet Technol. **17**(3) (2017)
12. Samih, Y., Darwish, K.: A few topical tweets are enough for effective user stance detection. In: Proceedings of the 16th Conference of the European Chapter of the Association for Computational Linguistics: Main Volume, pp. 2637–2646 (2021)
13. Williams, E.M., Carley, K.M.: TSPA: efficient target-stance detection on twitter. In: 2022 IEEE/ACM International Conference on Advances in Social Networks Analysis and Mining, pp. 242–246 (2022)
14. Zhou, L., Zhou, K., Liu, C.: Stance detection of user reviews on social network with integrated structural information. J. Intell. Fuzzy Syst. **44**(2), 1703–1714 (2023)

Centrality in Directed Networks

Gordana Marmulla(✉) and Ulrik Brandes

Social Networks Lab, ETH Zurich, Zürich, Switzerland
{gmarmulla,ubrandes}@ethz.ch

Abstract. The identification of important nodes in a network is a pervasive task in a variety of disciplines from sociology and bibliometry to geography and chemistry, and an ever growing number of centrality indices is proposed for this purpose. While such indices are often adhoc, preservation of the vicinal preorder has been identified as the core axiom shared by centrality rankings on undirected graphs. We extend this idea to directed graphs by defining vertex preorders based on directed neighborhood-inclusion criteria. While, for the undirected case, the vicinal preorder is total on threshold graphs and preserves all standard centrality indices, we show that our generalized preorders are total on certain subclasses of threshold digraphs. We thus provide a consistent formalization of the hitherto rather conceptual notions of radial, medial, and hierarchical centralities.

Keywords: centrality · network analysis · threshold digraph · neighborhood inclusion

1 Introduction

Network science is the study of empirical phenomena involving data on overlapping dyads [14]. Such data can often be represented as graphs and a particularly common task is to identify nodes that are of structural importance, because they are centrally involved in relationships with others [13,21]. For this purpose, centrality indices are used, which assign values to the vertices of a graph. There is a vast literature on the question of what constitutes a centrality [9,10,21]. Formal attempts to settle it are largely focused on the axiomatization of indices [1,3,7,26,29], which serves to characterize indices by virtue of graph modification effects on the values assigned to vertices. The definition of every centrality index, however, incorporates a wealth of assumptions. For instance, they are generally defined to be automorphism invariant, which is therefore also the most common axiom. This makes it difficult, however, to adapt an index to situations in which there are additional attributes or multiple relations.

In a positional approach to network science [12], centrality is therefore not defined by an index, which establishes a complete ranking of all vertices, but in successively extended partial rankings. This allows for the gradual introduction of assumptions until a sufficiently complete ranking is established, or can be

L. M. Aiello et al. (Eds.): ASONAM 2024, LNCS 15212, pp. 254–263, 2025.
https://doi.org/10.1007/978-3-031-78538-2_22

ruled out. Attention then shifts to the minimum criteria sufficient for a vertex to be considered more central than another. For undirected graphs, the vicinal preorder [20] has been identified as the common basis of standard variants of centrality [30,31], leading to the following proposition [31, Proposition 4].

Proposition 1. *A vertex index is a centrality, if and only if it respects the vicinal preorder.*

As threshold graphs [24] are characterized by a complete vicinal preorder, centrality rankings necessarily agree on these graphs [32]. The relevance of the bottom-up approach has been limited by its restriction to undirected graphs, and while ideas for an extension to directed graphs have been floated [13], their implications have not been studied.

We propose three preorders based on directed neighborhood inclusion. They are consistent with the approach for undirected graphs and formalize the conceptual distinction of radial and medial centralities proposed by Borgatti and Everett [10], adding a subclass of radial centralities that we call hierarchical. Radial centralities measure the centrality of a node based on its accessibility, while medial centralities do so based on its bridging function. The claim that they represent "distinct intuitive conceptions" [21, p. 215] of centrality was already corroborated by low correlations found for centrality indices informally categorized as radial and medial [8,33].

We also identify classes of digraphs on which the preorders yield complete rankings and which can therefore be seen as idealized representations for different notions of centrality. These are shown to be subclasses of threshold digraphs [15], and hence establish a correspondence between certain threshold digraphs and types of neighborhood-inclusions akin to the situation for undirected threshold graphs.

The remainder is organized as follows. After some basic definitions in Sect. 2, we introduce a family of preorders based on directed neighborhood inclusion in Sect. 3. Subclasses of threshold digraphs on which these yield complete rankings are determined in Sect. 4 and their relation to each other and their common superclass are discussed in Sect. 5. We conclude with suggestions for future work in Sect. 6.

2 Notation

We consider simple directed graphs (digraphs) $D = (V, E)$ consisting of a finite set V of vertices and a set $E \subseteq (V \times V) \setminus \{(i,i) : i \in V\}$ of directed edges. For a vertex $i \in V$, its in- and out-neighborhood are denoted by $N^-(i) := \{j : (j,i) \in E\}$ and $N^+(i) := \{j : (i,j) \in E\}$. The corresponding closed neighborhoods are defined by $N^-[i] := N^-(i) \cup \{i\}$ and $N^+[i] := N^+(i) \cup \{i\}$. A binary relation $\leqslant$ on V is called a preorder (or partial ranking) if it is reflexive and transitive. A preorder is total (or complete), if $i \leqslant j$ or $j \leqslant i$ for any $i, j \in V$. A total preorder is also called (complete) ranking.

3 Directed Neighborhood-Inclusion Criteria

We propose three extensions of undirected neighborhood inclusion to directed graphs. The first two appeared already in an earlier proposal [13], but without further study. The third, medial, relation defined below is new, because the version proposed previously applies only to pairs of non-adjacent vertices, so that no complete ranking can be obtained unless the graph is empty.

The relations defined in the following are supposed to capture the intuition that having more and better relationships can not make a vertex less central. In the radial case, this means having access to (being accessed by) others, and in the hierarchical case this condition is tightened to also having fewer reverse relationships. This is different in the medial case, where it is advantageous to have both incoming and outgoing edges, and thus be located between others. In the medial case, for comparing an adjacent pair, additional requirements have to be met. The additional requirements ensure that the advantage of one node, which is created through the adjacency, can be compensated by the respective other node. For example, if $(i, j) \in E$, the betweenness of i is enhanced through every walk passing i via a node of $N^-(i)$ and then ends after (i, j) in j. With $N^-(i) \subseteq N^+[j]$, for each of these walks w.r.t. i, there is another walk starting at i and passing j, thus, enhancing the betweenness of j. Visualizations of the neighborhood-inclusion relations are displayed in Fig. 1.

Definition 1. *Let $D = (V, E)$ be a digraph.*

(i) The radial outwards and inwards neighborhood-inclusion relations $\leqslant^+, \leqslant^- \subseteq V \times V$ are defined by

$$i \leqslant^+ j \;:\Longleftrightarrow\; N^+(i) \subseteq N^+(j), \quad \text{and} \quad i \leqslant^- j \;:\Longleftrightarrow\; N^-(i) \subseteq N^-(j)\ .$$

(ii) The hierarchical downwards and upwards neighborhood-inclusion relations $\leqslant^\downarrow, \leqslant^\uparrow \subseteq V \times V$ are defined by

$$\begin{aligned} i \leqslant^\downarrow j \;&:\Longleftrightarrow\; N^+(i) \subseteq N^+(j) \;\wedge\; N^-(i) \supseteq N^-(j) \\ i \leqslant^\uparrow j \;&:\Longleftrightarrow\; N^-(i) \subseteq N^-(j) \;\wedge\; N^+(i) \supseteq N^+(j)\ . \end{aligned}$$

(iii) The medial neighborhood-inclusion relation $\leqslant^\bowtie \subseteq V \times V$ is defined by

$$i \leqslant^\bowtie j \;:\Longleftrightarrow\; \begin{cases} \quad N^+(i) \subseteq N^+[j] & \\ \wedge\, N^-(i) \subseteq N^-[j] & \\ \wedge\, N^-(i) \subseteq N^+[j] & \text{if } (i, j) \in E \\ \wedge\, N^+(i) \subseteq N^-[j] & \text{if } (j, i) \in E\ . \end{cases}$$

Lemma 1. *The binary relations of Definition 1 are preorders.*

Proof. For the radial (outwards as well as inwards) and hierarchical (downwards as well as upwards) relations, it is straightforward to see that all neighborhood-inclusion relations are reflexive and transitive and, therefore, preorders. For

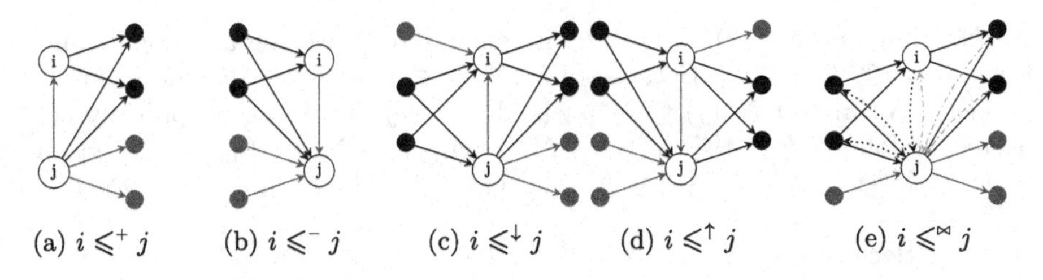

(a) $i \leqslant^{+} j$ (b) $i \leqslant^{-} j$ (c) $i \leqslant^{\downarrow} j$ (d) $i \leqslant^{\uparrow} j$ (e) $i \leqslant^{\bowtie} j$

Fig. 1. Visualization of the proposed neighborhood-inclusion relations for directed networks. Gray edges may or may not be present. Dotted and dashdotted edges represent inclusion requirements given an edge between i and j. (Color figure online)

the medial neighborhood-inclusion relation, reflexivity is also straightforward. It remains to show that the medial relation is transitive. Let $i, j, k \in V$ be distinct vertices with $i \leqslant^{\bowtie} j$ and $j \leqslant^{\bowtie} k$. We consider cases based on the connectivity between i and j; solid edges are present, dashed edges are absent.

$i \mathrel{\substack{\dashrightarrow \\ \dashleftarrow}} j$: Since $i \leqslant^{\bowtie} j$ and because i and j are disconnected, the neighborhood-inclusions hold in particular for the open neighborhoods, such that $N^{+}(i) \subseteq N^{+}(j) \subseteq N^{+}[k]$ and $N^{-}(i) \subseteq N^{-}(j) \subseteq N^{-}[k]$. What remains to show is that the potential additional conditions hold, dependent on the connection between i and k. Nothing is left to be shown if there is no edge between i and k. If $(i,k) \in E$, then it is $(j,k) \in E$ due to $N^{+}(i) \subseteq N^{+}(j)$. Then, because $j \leqslant^{\bowtie} k$, the additional condition $N^{-}(j) \subseteq N^{+}[k]$ holds, i.e., $N^{-}(i) \subseteq N^{-}(j) \subseteq N^{+}[k]$. If $(k,i) \in E$, then it is $(k,j) \in E$ due to $N^{-}(i) \subseteq N^{-}(j)$. Because $j \leqslant^{\bowtie} k$, the additional condition $N^{+}(j) \subseteq N^{-}[k]$ holds, i.e., $N^{+}(i) \subseteq N^{+}(j) \subseteq N^{-}[k]$. In total, $i \leqslant^{\bowtie} k$.

$i \mathrel{\substack{\rightarrow \\ \dashleftarrow}} j$: Because $j \notin N^{-}(i)$, it holds $N^{-}(i) \subseteq N^{-}(j)$. From $j \leqslant^{\bowtie} k$, it follows then $N^{-}(i) \subseteq N^{-}(j) \subseteq N^{-}[k]$. Since $i \in N^{-}(j) \subseteq N^{-}[k]$ it is $i \in N^{-}[k]$, i.e., $k \in N^{+}(i) \subseteq N^{+}[j]$. Then, for $j \leqslant^{\bowtie} k$ to hold, the additional condition $N^{-}(j) \subseteq N^{+}[k]$ must be fulfilled. This condition entails $i \in N^{-}(j) \subseteq N^{+}[k]$ and thus $k \in N^{-}(i) \subseteq N^{-}[j]$, or equivalently $j \in N^{+}(k)$. With the latter, we have $N^{+}(i) \subseteq N^{+}[j] = N^{+}(j) \cup \{j\} \subseteq N^{+}[k] \cup \{j\} = N^{+}[k]$. Since both $(k,i), (i,k) \in E$, it remains to prove the additional conditions for $i \leqslant^{\bowtie} k$ to hold. Because $(j,k) \in E$ and $j \leqslant^{\bowtie} k$, it holds $N^{-}(j) \subseteq N^{+}[k]$, which entails $N^{-}(i) \subseteq N^{-}(j) \subseteq N^{+}[k]$. Because $(k,j) \in E$ and $j \leqslant^{\bowtie} k$ it holds $N^{+}(j) \subseteq N^{-}[k]$, which entails $N^{+}(i) \subseteq N^{+}[j] = N^{+}(j) \cup \{j\} \subseteq N^{-}[k] \cup \{j\} = N^{-}[k]$. In total, $i \leqslant^{\bowtie} k$.

$i \mathrel{\substack{\dashrightarrow \\ \leftarrow}} j$: Because $j \notin N^{+}(i)$, it holds $N^{+}(i) \subseteq N^{+}(j)$. From $j \leqslant^{\bowtie} k$, it follows then $N^{+}(i) \subseteq N^{+}(j) \subseteq N^{+}[k]$. Since $i \in N^{+}(j) \subseteq N^{+}[k]$, it holds in particular $i \in N^{+}[k]$ and, therefore, equivalently $k \in N^{-}(i)$. The latter entails $k \in N^{-}(j)$ as well, i.e., $(k,j) \in E$. Then, for $j \leqslant^{\bowtie} k$ to hold, the additional condition $N^{+}(j) \subseteq N^{-}[k]$ must be fulfilled. Since $i \in N^{+}(j) \subseteq N^{-}[k]$, this corresponds to the presence of the edge $(i,k) \in E$, as well as $(j,k) \in E$ due to $N^{+}(i) \subseteq N^{+}[j]$. With the latter edge, it follows $N^{-}(i) \subseteq N^{-}[j] = N^{-}(j) \cup \{j\} \subseteq N^{-}[k] \cup \{j\} =$

$N^-[k]$. Since both $(i,k),(k,i) \in E$, it remains to prove the additional conditions for $i \leqslant^{\bowtie} k$ to hold. Because $(j,k),(k,j) \in E$ and $j \leqslant^{\bowtie} k$, the inclusions $N^-(j) \subseteq N^+[k]$ and $N^+(j) \subseteq N^-[k]$ hold. Since j and k are reciprocally connected, it follows $N^-(i) \subseteq N^-[j] = N^-(j) \cup \{j\} \subseteq N^+[k] \cup \{j\} \subseteq N^+[k]$ and $N^+(i) \subseteq N^+[j] = N^+(j) \cup \{j\} \subseteq N^-[k] \cup \{j\} = N^-[k]$.

$i \rightleftarrows j$: Because $i \in N^-(j) \subseteq N^-[k]$ and $i \in N^+(j) \subseteq N^+[k]$, vertices i and k are reciprocally connected as well. In particular, $k \in N^-(i) \subseteq N^-[j]$ and $k \in N^+(i) \subseteq N^+[j]$ meaning that j and k must be also reciprocally connected. Therefore, both $N^-(i) \subseteq N^-[j] = N^-(j) \cup \{j\} \subseteq N^-[k] \cup \{j\} = N^-[k]$ and $N^+(i) \subseteq N^+[j] = N^+(j) \cup \{j\} \subseteq N^+[k] \cup \{j\} = N^+[k]$ hold true. The required additional conditions for the pair i,k follow with those from pair j,k and their mutual connection: $N^-(i) \subseteq N^-[j] = N^-(j) \cup \{j\} \subseteq N^+[k] \cup \{j\} = N^+[k]$ and $N^+(i) \subseteq N^+[j] = N^+(j) \cup \{j\} \subseteq N^-[k] \cup \{j\} = N^-[k]$.

4 Uniquely Ranked Directed Graphs

We next determine the classes of digraphs on which the above preorders yield complete rankings. These are important because all centralities preserving the respective preorder will agree. All of them can be defined via forbidden subgraphs depicted in Fig. 2.

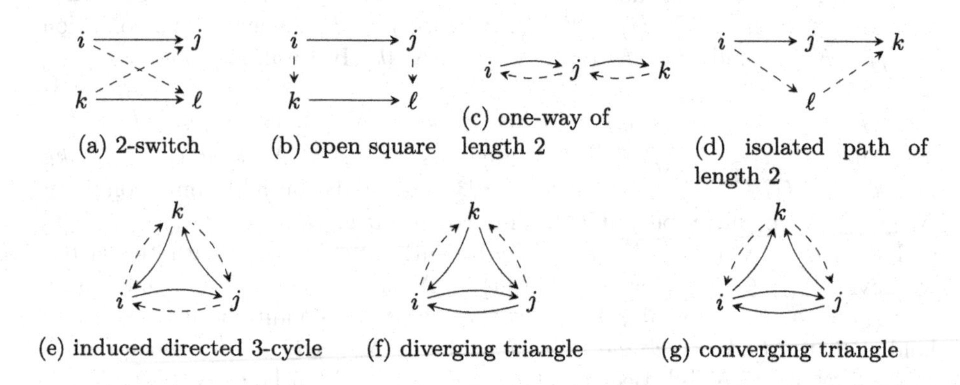

Fig. 2. Forbidden configurations. Solid edges are required, dashed edges are forbidden, and no conditions are placed on absent edges.

Radial. Recall that an interval order [19,34] is an irreflexive binary relation $(V, \prec)$ satisfying the condition $i \prec j \wedge k \prec \ell \implies i \prec \ell \vee k \prec j$. From the graph perspective, interval orders are transitive and acyclic digraphs without a 2-switch [24]. The graphs of interval orders can also be characterized as follows:

for all $i, j \in V$ it holds that (*i*) $N^+(i) \subseteq N^+(j)$ or $N^+(j) \subseteq N^+(i)$, and (*ii*) $N^-(i) \subseteq N^-(j)$ or $N^-(j) \subseteq N^-(i)$ [5,17,18]. In other words, both the out- and in-neighborhoods are linearly ordered by nesting, and therefore interval orders yield exactly those digraphs on which the radial neighborhood-inclusion relations $\leqslant^+$ and $\leqslant^-$ are total.

Interval orders represent extreme cases of acyclicity, essentially reducing centrality to a matter of being positioned at the beginning or the end of a relationship. Due to their transitivity, there is no notion of passing through other vertices via shortest paths, resulting in each vertex conceptually occupying the same medial position. Another implication of transitivity is that the issue of reachability becomes synonymous with determining which vertices belong to which (i.e., out- or in-) neighborhood, rendering distance insignificant while emphasizing the critical role of the chosen direction.

Hierarchical. In general, $\leqslant^+$ need not be the reversed (partial) ranking of $\leqslant^-$, unlike for $\leqslant^\downarrow$ and $\leqslant^\uparrow$ [17]. Digraphs in which $\leqslant^\downarrow$ ($\leqslant^\uparrow$) yields a ranking are the semiorders [23], also called unit interval orders [6]. These are special interval orders satisfying the additional condition $i \prec j \wedge j \prec k \implies i \prec \ell \vee \ell \prec k$ for any other ℓ. From the graph perspective, this means that semiorders do not contain an isolated path of length 2.

Although the choice of direction of edges may influence a node's position in the ranking for interval orders, the hierarchical ranking remains unique up to reversal. Consequently, it only needs to be determined which end of the ranking is deemed more central.

Medial. The medial preorder is total on a class of graphs that, to the best of our knowledge, has not been discussed previously.

Lemma 2. *The medial preorder is total if and only if the digraph contains neither a 2-switch, an open square, a one-way of length* 2*, or a diverging or converging triangle.*

Proof. We prove both directions by contraposition.

$\implies$: Assume D contains one of the forbidden configurations. We refer to the vertex labels as shown in Fig. 2. In a 2-switch, vertices i and k are not comparable with respect to $\leqslant^{\bowtie}$ because $j \in N^+(i)\backslash N^+(k)$ while $\ell \in N^+(k)\backslash N^+(i)$. In an open square, on the one hand $N^-(j) \nsubseteq N^-[k]$ because $i \in N^-(j)\backslash N^-(k)$. On the other hand, $N^+(k) \nsubseteq N^+[j]$ because $\ell \in N^+(k)\backslash N^+(j)$. Thus the only possibility is $N^+(j) \subseteq N^+[k]$ and $N^-(k) \subseteq N^-[j]$ implying that j and k are not comparable with respect to $\leqslant^{\bowtie}$. The same line of arguments apply to the pair i and k in a one-way of length 2. In a diverging triangle, the pair i and j requires additional constraints due to their mutual connection. However, neither $N^-(i) \subseteq N^+[j]$ nor $N^-(i) \subseteq N^+[i]$ because $k \in N^-(i), N^-(j)$ but $k \notin N^+[j], N^+[i]$. The analogous argumentation applies to the converging triangle.

$\Longleftarrow$: Assume that the medial neighborhood-inclusion relation is not total. Then, there exist a pair of vertices $i, j \in V$ with $i \not\leqslant^{\bowtie} j$ and $j \not\leqslant^{\bowtie} i$. There can be five distinct reasons why a pair is not comparable. In fact, there are more non-distinct cases, but they can be matched to one of the five following unique cases by changing the roles of i and j accordingly.

$N^-(i) \not\subseteq N^-[j]$ and $N^-(j) \not\subseteq N^-[i]$: Then, there exists a $k \in N^-(i) \backslash N^-[j]$ for which it follows $k \notin \{i, j\}$. At the same time, there is an $\ell \in N^-(j) \backslash N^-[i]$ for which it holds $\ell \notin \{i, j, k\}$. This situation corresponds to a 2-switch.

$N^+(i) \not\subseteq N^+[j]$ and $N^+(j) \not\subseteq N^+[i]$: With the same line of arguments as in the previous case it follows that D contains a 2-switch.

$N^-(i) \not\subseteq N^-[j]$ and $N^+(j) \not\subseteq N^+[i]$: The former implies the existence of $k \notin \{i, j\}$ with $k \in N^-(i) \backslash N^-[j]$. The latter implies the existence of $\ell \notin \{i, j\}$ with $\ell \in N^+(j) \backslash N^+[i]$. If $\ell = k$, this configurations forms an one-way of length 2. If $\ell \neq k$, this configurations forms an open square.

$N^-(i) \subseteq N^-[j]$ and $N^+(i) \subseteq N^+[j]$, but $N^-(i) \not\subseteq N^+[j]$ when $(i, j) \in E$: Then, there is a $k \in N^-(i) \backslash N^+[j]$. In particular, $k \notin \{i, j\}$ and $k \in N^-(i) \subseteq N^-[j]$. Since $k \notin N^+[j]$ and $N^+(i) \subseteq N^+[j]$, it holds $k \notin N^+(i)$. If $(j, i) \in E$, the configuration corresponds to a converging triangle. If $(j, i) \notin E$, the configuration corresponds to an one-way of length 2.

$N^-(i) \subseteq N^-[j]$ and $N^+(i) \subseteq N^+[j]$, but$N^+(i) \not\subseteq N^-[j]$ when $(j, i) \in E$: With the same line of arguments as in the third case it follows that D contains either a diverging triangle or an one-way of length 2.

Medial relations correspond to a notion of centrality, in which central nodes form cores connecting peripheral ones. The medial centrality criterion is related to the level of reciprocity, i.e., the proportion of dyads with edges in both directions. Reciprocity as a network property was found to be significantly associated with the correlations between all different centrality measures (between symmetric, between asymmetric, and between the combination of both), i.e., the more bi-directional connections, the less distinct measures become [33]. This association supports the idea that medial and radial notions of centrality are indiscriminate on undirected graphs. In contrast, directed graphs allow for formal criteria to define these concepts.

5 Threshold Digraphs

All above classes of digraphs on which the our preorders yield rankings are subclasses of threshold digraphs.

Definition 2 (Threshold Digraph *[15]***).** *A digraph is called threshold, if it contains neither a 2-switch nor an induced directed 3-cycle.*

Interval orders, being acyclic and lacking a 2-switch, are clearly threshold, and so is their subclass of semiorders. Lemma 2 asserts that digraphs on which $\leqslant^{\bowtie}$ is total contain neither a 2-switch nor a one-way path of length 2. This

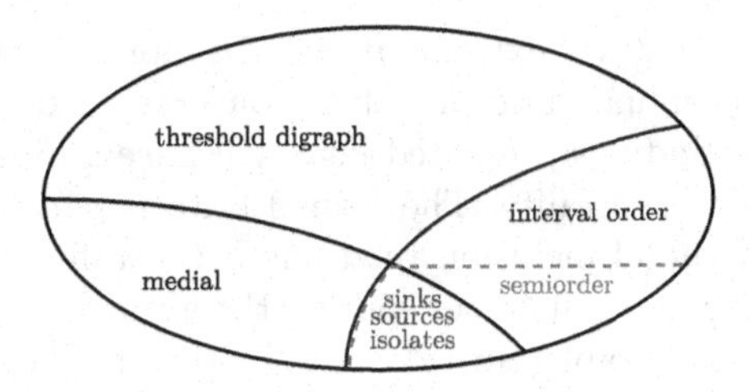

Fig. 3. Unique ranked graphs as subclasses of threshold digraphs.

condition is stronger than the absence of induced directed 3-cycle, and assures that members of our new class of digraphs are threshold, too.

As the name suggests, threshold digraphs were introduced as a generalization of threshold graphs. There have been various other approaches for extending threshold graphs for the directed case: oriented threshold graphs [4], which are generalized by directed threshold graphs [22], which in turn are a subclass of threshold digraphs [15].

For interval orders and semiorders, recognizing whether a digraph belongs to one of these types was resolved decades ago and can both be done in linear time $\mathcal{O}(|V| + |E|)$ [25,27]. Whether this can be done similarly efficiently for the class of digraphs on which $\leqslant^{\bowtie}$ yields a ranking remains to be investigated.

For undirected graphs, threshold graphs rank vertices uniquely with respect to the vicinal preorder and thus the distinctness of a given graph to a threshold graph is of interest [32]. Analogously, the determined subsets of threshold digraphs, on which the preorders are complete, might help to estimate how consistent centrality measures will perform on the given digraph. The work of Cloteaux et al. [15], linking degree sequences and threshold digraphs, could serve as a starting point. For semiorders and interval orders, in particular, already established results such as [2,11,16] may be relevant.

There are more equivalent characterizations for threshold digraphs [28] which reveal that these digraphs are Ferrers digraphs [16] with loops removed. A characterization for Ferrers digraphs is that both their out- and in-neighborhoods (but now potentially with loops) are linearly nested. An alternative opportunity to examine the extent and kind of centrality of a threshold digraph could be then the (minimum) number of loops required to transform it into a Ferrers digraph and the type(s) of neighborhood-inclusion that are possible through transformations.

6 Conclusion

In order to establish a common basis for centrality in directed graphs, we proposed three neighborhood-inclusion relations on directed graphs. They provide formalizations of different notions of centrality, namely radial, hierarchical, and medial. As the relations are reflexive and transitive, the resulting rankings are partial in general, and the classes of digraphs on which the rankings are total are non-trivial. These distinct extreme directed graphs suggest that notions of centrality can be discriminated despite their convergence on undirected graphs.

We hope that the findings presented here will have an impact on the definition and application of centrality. The formal criteria can be used to examine which of the neighborhood-inclusion relations a centrality index does or does not preserve. There are also, however, purely theoretical collateral insights. All uniquely ranked directed graphs turned out to be threshold digraphs. Since there are threshold digraphs not covered by any of the neighborhood-inclusion criteria (cf. Fig. 3), there may be additional notions of centrality or some other type of ranking hiding. While digraphs with complete rankings obtained from radial and hierarchical preorders are the well-studied interval orders and semiorders, the medial preorders warrant further exploration.

Among the many interesting directions for future work we mention the completeness of our partial rankings on specific graphs, graph-modification distances to uniquely ranked digraphs, and the degree to which select graph characteristics separate different centrality indices.

References

1. Altman, A., Tennenholtz, M.: Axiomatic foundations for ranking systems. J. Artifi. Intell. Res. **31**, 473–495 (2008). https://doi.org/10.1613/jair.2306
2. Avery, P.: An algorithmic proof that semiorders are representable. J. Algorithms **13**(1), 144–147 (1992). https://doi.org/10.1016/0196-6774(92)90010-A
3. Bloch, F., Jackson, M.O., Tebaldi, P.: Centrality measures in networks. Soc. Choice Welfare **61**(2), 413–453 (2023). https://doi.org/10.1007/s00355-023-01456-4
4. Boeckner, D.: Oriented threshold graphs. Australasian J. Combinat. **71**(1), 43–53 (2018)
5. Bogart, K.P.: An obvious proof of Fishburn's interval order theorem. Discret. Math. **118**(1–3), 239–242 (1993). https://doi.org/10.1016/0012-365X(93)90065-2
6. Bogart, K.P., West, D.B.: A short proof that 'proper = unit'. Discret. Math. **201**(1–3), 21–23 (1999). https://doi.org/10.1016/S0012-365X(98)00310-0
7. Boldi, P., Vigna, S.: Axioms for centrality. Internet Math. **10**(3-4), 222–262 (2014). https://doi.org/10.1080/15427951.2013.865686
8. Bolland, J.M.: Sorting out centrality: an analysis of the performance of four centrality models in real and simulated networks. Soc. Netw. **10**(3), 233–253 (1988). https://doi.org/10.1016/0378-8733(88)90014-7
9. Borgatti, S.P.: Centrality and network flow. Soc. Netw. **27**(1), 55–71 (2005). https://doi.org/10.1016/j.socnet.2004.11.008
10. Borgatti, S.P., Everett, M.G.: A Graph-theoretic perspective on centrality. Soc. Netw. **28**(4), 466–484 (2006). https://doi.org/10.1016/j.socnet.2005.11.005
11. Bousquet-Mélou, M., Claesson, A., Dukes, M., Kitaev, S.: (2 + 2) -free posets, ascent sequences and pattern avoiding permutations. J. Combinatorial Theory, Ser. A **117**(7), 884–909 (2010). https://doi.org/10.1016/j.jcta.2009.12.007
12. Brandes, U.: Network positions. Methodol. Innovat. **9**, 1–19 (2016). https://doi.org/10.1177/2059799116630650
13. Brandes, U.: Central positions in social networks. In: Fernau, H. (ed.) Proceedings of the 15th International Computer Science Symposium in Russia (CSR 2020), vol. 12159, pp. 30–45 (2020).https://doi.org/10.1007/978-3-030-50026-9_3

14. Brandes, U., Robins, G., McCranie, A., Wasserman, S.: What is network science? Netw. Sci. **1**(1), 1–15 (2013) https://doi.org/10.1017/nws.2013.2
15. Cloteaux, B., LaMar, M.D., Moseman, E., Shook, J.: Threshold digraphs. J. Res. Nat. Instit. Standards Technol. **119**, 227–234 (2014) https://doi.org/10.6028/jres.119.007
16. Cogis, O.: Ferrers digraphs and threshold graphs. Discret. Math. **38**(1), 33–46 (1982). https://doi.org/10.1016/0012-365X(82)90166-2
17. Felsner, S.: Interval Orders: Combinatorial Structure and Algorithms. Dissertation, Technische Universität Berlin (1992)
18. Fishburn, P.C.: Intransitive indifference with unequal indifference intervals. J. Math. Psychol. **7**(1), 144–149 (1970). https://doi.org/10.1016/0022-2496(70)90062-3
19. Fishburn, P.C.: Interval orders and interval graphs: a study of partially ordered sets. Wiley, New York (1985)
20. Foldes, S., Hammer, P.L.: The dilworth number of a graph. Annals Dis. Mathem. **2**, 211–219 (1978). https://doi.org/10.1016/S0167-5060(08)70334-0
21. Freeman, L.C.: Centrality in social networks conceptual clarification. Soc. Netw. **1**(3), 215–239 (1978). https://doi.org/10.1016/0378-8733(78)90021-7
22. Gurski, F., Rehs, C.: Comparing linear width parameters for directed graphs. Theory Comput. Syst. **63**(6), 1358–1387 (2019). https://doi.org/10.1007/s00224-019-09919-x
23. Luce, R.D.: Semiorders and a theory of utility discrimination. Econometrica **24**(2), 178–191 (1956). https://doi.org/10.2307/1905751
24. Mahadev, N.V.R., Peled, U.N.: Threshold graphs and related topics. No. 56 in Annals of discrete mathematics. Elsevier, Amsterdam; New York (1995)
25. Mitas, J.: Minimal representation of semiorders with intervals of same length. In: Bouchitté, V., Morvan, M. (eds.) ORDAL 1994. LNCS, vol. 831, pp. 162–175. Springer, Heidelberg (1994). https://doi.org/10.1007/BFb0019433
26. Nieminen, U.: On the centrality in a directed graph. Soc. Sci. Res. **2**(4), 371–378 (1973). https://doi.org/10.1016/0049-089X(73)90010-0
27. Papadimitriou, C.H., Yannakakis, M.: Scheduling interval-ordered tasks. SIAM J. Comput. **8**(3), 405–409 (1979). https://doi.org/10.1137/0208031
28. Reilly, E., Scheinerman, E., Zhang, Y.: Random threshold digraphs. The Electr. J. Combinat. **21**(2), P2.48 (2014). https://doi.org/10.37236/4050
29. Sabidussi, G.: The centrality index of a graph. Psychometrika **31**(4), 581–603 (1966). https://doi.org/10.1007/BF02289527
30. Sadler, E.: Ordinal Centrality. J. Polit. Econ. **130**(4), 926–955 (2022). https://doi.org/10.1086/718191
31. Schoch, D., Brandes, U.: Re-conceptualizing centrality in social networks. Eur. J. Appl. Math. **27**(6), 971–985 (2016). https://doi.org/10.1017/S0956792516000401
32. Schoch, D., Valente, T.W., Brandes, U.: Correlations among centrality indices and a class of uniquely ranked graphs. Soc. Netw. **50**, 46–54 (2017). https://doi.org/10.1016/j.socnet.2017.03.010
33. Valente, T.W., Coronges, K., Lakon, C., Costenbader, E.: How correlated are network centrality measures? Connections **28**(1), 16–26 (2008)
34. Wiener, N.: A contribution to the theory of relative position. Proc. Cambridge Philos. Soc. **17**, 441–449 (1914)

Mitigating the Spread of COVID-19 Misinformation Using Agent-Based Modeling and Delays in Information Diffusion

Mustafa Alassad and Nitin Agarwal(✉)

COSMOS Research Center, University of Arkansas – Little Rock, Little Rock, AR 72204, USA
{Mmalassad,nxagarwal}@ualr.edu

Abstract. The rapid spread of COVID-19 misinformation on social media poses challenges in detection and analysis. There has been extensive discussion about the roles of online and offline campaigns in spreading misinformation. Recognizing the analytical gap between online and offline behaviors during the COVID-19 pandemic, we propose a systematic and multidisciplinary approach. This approach utilizes agent-based modeling to interpret the spread of misinformation and the actions of users/communities on social media networks. Our model was tested on a Twitter network concerning a demonstration against COVID-19 lockdowns in Michigan in May 2020. We implemented the one-median problem to categorize and simplify the Twitter network, measured the response time to the spread of misinformation, employed a cybernetic organizational method to manage the process of mitigating misinformation spread in the network, and optimized the allocation of agents to reduce the response time to misinformation spread. The study demonstrates the effectiveness of our proposed approach in delaying information diffusion, thereby mitigating the spread of COVID-19 misinformation on social media.

Keywords: Systems Thinking · Organizational Cybernetics · Stochastic One-median Problem · Misinformation · COVID-19 · Information Diffusion Delay

1 First Section

During the COVID-19 pandemic, social platforms faced an "infodemic" of misinformation, increasing global risks. Millions shared COVID-19-related content online, but WHO officials warned that fake news was spreading faster than the virus [1]. President Biden and US officials criticized social media for spreading misinformation about COVID-19 and vaccines, threatening efforts to control the pandemic [2, 3].

The spread of COVID-19 misinformation on social media can harm individuals and society [4, 5]. Efforts to combat this include collaborations and online tools for verifying news [6]. However, advanced methods and systematic approaches are needed to understand the context and intentions behind true or false information, conspiracy theories, and coordinating groups [7–9].

L. M. Aiello et al. (Eds.): ASONAM 2024, LNCS 15212, pp. 264–273, 2025.
https://doi.org/10.1007/978-3-031-78538-2_23

Many tools, like fact-checkers and automatic conspiracy theory detection tools [10], are inadequate for tracking malicious activities and influencing attempts in dynamic social networks. These tools often fail to consider information flow among communities, conspiracy diffusion over time, and the evolving network. Fact-checker applications need enhanced methodologies and systematic approaches to limit conspiracy theories, investigate influential spreaders, respond to incidents, and monitor information flow in real time. Advanced methods like game theories [11], graph theories [12], and information theory [13] are necessary to improve analysis and limit conspiracy theories' spread in real time.

In this paper, we used systematic modeling to enhance analysis in dynamic social networks. Our approach describes organizational behavior and operations in response to information spread, optimizing time and resources. We employed the Organizational Cybernetic Approach (OCA) [14] to control communications between communities and the stochastic one-median problem [15] to minimize response time to malicious information. This approach improves the operation level's response to abnormal information spread and aids better decision-making. Next, we describe the problem statement.

2 Problem Definition

Consider a social network $G = (N, A)$ consisting of the distinct node sets $N = \{1, 2, \ldots, n\}$ and the set of edge (links)$A = \{(i, j), (k, l), \ldots, (s, t)\}$ represented by directed node pair combinations going from community i to community j. Communities i, and j are associated with numerical values representing the number of intra edges $d_{j,i}$ between communities i and j. These numbers represent the actual number of users from the community i linked to users in community j at every time window. Also, h_j means the communities' rate of the misinformation spread in the worst-case scenario, and $\left(h_j\right)$ represents the communities' malicious information spread rate of $\left(h_j = \frac{|N_j|}{N}\right)$, where (λ_i) is the proportion rate of operation level that can be monitored in the network. This research aims to develop an agent-based model that interacts with online social networks and offline environments to mitigate COVID-19 misinformation spread and tackle the complexity of social media analysis. The following research questions guide the development of the model.

- RQ1. What strategies can an agent employ to confront and overcome the challenges of curbing the spread of COVID-19 misinformation?
- RQ2. How does an agent navigate the complexities of overseeing communities and of identifying and reporting the rapid spread of unusual information in real-time?
- RQ3. What is the agent's primary objective in addressing the spread of misinformation related to COVID-19?

Next, we discuss the research methodology.

3 Methodology

This research aims to integrate advanced systems thinking and modeling into dynamic social network analysis. By using systems design modeling, we can visualize interactions among communities and assess social network conditions. The challenge is to develop solutions that meet agents' real-time needs and constraints. Integrating social network analysis techniques allows for a detailed examination of network dynamics, identifying key influencers and mapping information flow. This helps pinpoint intervention areas to counter misinformation effectively. Systems design modeling optimizes solutions, enhances agent performance, and analyzes causes and effects within the network, ensuring precise and impactful actions [16].

Assumptions required for the solution procedure, which are necessary for the operation level strategies, are: the operation level response time to malicious information spread is deterministic, (V = 55) messages per hour; the time required to respond to a stochastic malicious information spread case between communities is normally distributed, (β = 2) case per hours; the time of the operation level to prepare for a new stochastic malicious information spread is deterministic; communities are assumed to be active over time, including users able to link with users in other communities; and communities are assumed to be either spreading malicious information or not.

Organizational Cybernetic Approach (OCA). The Systems Thinking method manages communication between a system and its environments, flags feedback, and accounts for unexpected behaviors from users and communities. It also simplifies analyzing the system's growing complexity and referenced in [14]. The research emphasizes the operational level (system one in OCA), stressing its need for flexibility to facilitate efficient interaction with diverse environments [14]. This level ensures equitable monitoring of large communities across networks, addressing abnormal behaviors, radical posts, cyber actions, and collective behaviors [17]. Operational activities require regular reporting to management and control levels following responses to malicious posts or before initiating new actions. Processes are continually monitored to promptly address abnormal information dissemination, managing each instance in a first-in-first-out (FIFO) sequence. OCA aims to monitor, record feedback, and develop strategies to improve agents' communication with the social environment by developing and managing a plan to limit the misinformation spread; assigning agents to interact with the environment and monitoring the sources of the spread of misinformation; evaluating the performance of the agents; and developing a new strategy based on the environment's reported negative or positive feedback.

The management level in OCA ensures smooth misinformation mitigation processes, reporting these and agents' performance to the control level. The control level within OCA enhances dynamic communication through policy adjustments during urgent scenarios, rectifying feedback and adapting mitigation plans to minimize unforeseen issues in the social network. The development level integrates critical data from operational and control levels, monitoring changes in online and offline environments. Acting as a command center, it oversees the execution of misinformation mitigation strategies, ensuring coordinated efforts among agents and the broader environment. The policy level adapts the system to misinformation and environmental changes, maintaining the effectiveness of mitigation efforts and articulating the network's strategy.

Stochastic One-Median Problem. This operational method effectively tackles stochastic information spread in dynamic networks, especially in scenarios where responses to malicious information might be overlooked [18]. Using the stochastic one-median problem improves performance, optimizes operational efforts, and aids in selecting optimal community combinations for enhanced monitoring processes. This stochastic monitoring approach integrates expected response times for abnormal behaviors as detailed in Eqs. (1) to (4).

$$MinTR_j(C)\ \forall j \in I \tag{1}$$

$$TR(C_j) = \overline{Q}_{C_j} + \overline{t}_{C_j}\ \forall j \in I \tag{2}$$

$$\overline{Q}_{C_j} = \frac{\lambda_i \overline{S_2}(C_j)}{2(1-\lambda_i \overline{S}(C_j))} \begin{matrix} \forall j \in I \\ \forall i \in M \end{matrix} \tag{3}$$

$$\overline{t}_{C_j} = \sum_{j=1}^{I} h_j d(C_j, I)\ \forall j \in I \tag{4}$$

TR is the sum of the mean-queuing-delay $\overline{Q}$ and the mean response time $\overline{t}$ as shown in (2). Equation (3) is to define $\overline{Q}$, where C_j is the community j in the network. λ_i is the proportion rate for an agent to handle misinformation transferred between communities; for this purpose, the network will be divided into sub-networks, as discussed in the next section. $\overline{S}(C_j)$ is the mean total response time (starting from the first moment the abnormal behavior is detected), and $\overline{S_2}(C_j)$ is the second stochastic moment of the total response time to any new misinformation spread in the network. Equation (4) defines $\overline{t}$, where I is the number of communities in the network, h_j is the the rate of the misinformation spread from community C_i in the worse scenario, and $d(C_j, I)$ is the shortest path to transfer information between community C_j and community C_i.

The four outlines in the stochastic one-median problem and the operation level that can be translated into the OCA structure and the social network analysis are – (1) The operation level is the sole system level responsible for responding to abnormal behaviors in the social network. (2) This level would record and report the interactions of the users/communities in online/offline social environments. (3) The operation level will report any changes in the behaviors of the networks to the higher levels in the OCA. (4) This level must implement the off-scene setup time to respond to any stochastic and new abnormal behaviors in the network.

Agent Level Performance. This research presents a linear multi-objective problem [19] to assess agents' effectiveness in combating COVID-19 misinformation on social media. The model considers information flow between communities, environmental feedback, and integrates an information diffusion delay to enhance agent performance in reducing false information spread. It comprises three objectives aimed at mitigating COVID-19 misinformation in the network: minimizing misinformation spread between communities (Z1), reducing the distance or nodes through which information travels (Z2), and minimizing information diffusion delay for network reliability (Z3). These objectives are supported by a set of associated constraints.

$$MinZ1 = \sum_{p \in O} \sum_{q \in D} w^{pq} Y^{pq} \tag{5}$$

$$MinZ2 = \sum_{u \in M} \sum_{p \in O} \sum_{q \in D} l_u X_u^{pq} \tag{6}$$

$$MinZ3 = \sum_{u \in M} \left[\frac{\sum_{p \in O} \sum_{q \in D} X_u^{pq}}{C_u - \sum_{p \in O} \sum_{q \in D} X_u^{pq}} \right] \tag{7}$$

Subject To:

$$\sum_{p \in O} \sum_{q \in D} X_u^{pq} \leq \frac{Delay \cdot \gamma \cdot \mu \cdot C_u}{|u|} \; \forall u \in M \tag{8}$$

$$\sum_{u \in \Gamma(i)} X_u^{pq} - \sum_{u \in \Gamma-(i)} X_u^{pq} = d^{pq} - Y^{pq} \; if \, i = p \tag{9}$$

$$\sum_{u \in \Gamma(i)} X_u^{pq} - \sum_{u \in \Gamma-(i)} X_u^{pq} = -d^{pq} + Y^{pq} if \, i = q \tag{10}$$

$$\sum_{u \in \Gamma(i)} X_u^{pq} - \sum_{u \in \Gamma-(i)} X_u^{pq} = 0 \; otherwise \tag{11}$$

$$X_u^{pq}, Y^{pq} \geq 0, \; \forall u \in M, \; p \in O, \; q \in D$$

Equation (8) ensures that X_u, the actual number of connections between two communities connected by arc u, is less than or equal to C_u (the maximum possible connections between communities connected by arc u when no delay is applied). γ is the network throughput, similar to the total number of misinformation units in the network. $\mu = 1$ is a unit of information sent from a user in community i to a user in community j, and *Delay* is the information diffusion delay factor. Equations (9) through (11) measure the information unit conservation flow between communities in the network, where d^{pq} is the number of monitored users between the source community and destination community. Y^{pq} is the number of unmonitored users between the source community and destination community. $\Gamma(i)$ is the set of arcs whose source community is i. $\Gamma - (i)$ is the set of arcs whose destination community is i. w^{pq} is the amount of weight given to the communities p and q. l_u is the length of the path for information to transfer between communities. $p \in O$ is the source communities. $q \in D$ is the destination community setting. N is the communities set and M is the arc (link) set for the network.

4 Related Work

Several empirical studies underscore Systems Thinking and modeling as advanced methods for tackling complex social network issues [20]. The Casual Feedback method [21] identifies key variables for complex systems modeling. Muchnik et al. [22] highlight power laws in social networks and large-scale communities. Du et al. [23] discuss boundary conditions in systems thinking and modeling. Control theory [11] and information theory [13] enhance decision-making strategies based on game theory in social networks [19], providing optimized solutions for diverse agent-based complex problems. The complexities of unpredictable user behaviors, community dynamics, and rapid social network growth challenge effective analysis, limiting traditional methods. Integrating various fields is crucial for developing strategies to examine local user interactions and inter-community communication over time. This integrated approach helps categorize online and offline user feedback. We propose a systematic agent-based approach to enhance communication between online environments and organizational operations, mitigate malicious information spread, monitor information flow among communities and users, and enable real-time responses to abnormal information.

To optimize real-time decision-making, various operational methods model stochastic incidents and interactions between teams and environments across different domains. Examples include enhancing fire department operations during emergencies, managing rush-hour traffic congestion, and quickly restoring telecommunication services after network failures [24]. Hakimi [24] introduced the ρ-median location model to optimize depot placement in infrastructure networks during failures, minimizing node distances and hub numbers. Patterson et al. [25] proposed a relaxed ρ-median model with overlapping service regions to reduce call losses in telecommunication networks. Love et al. [26] illustrated operational interactions using a bipartite graph representation of the ρ-median algorithm. On the contrary, Odoni [27] highlighted queuing challenges with the ρ-median model due to extended arc operations, while Chan et al. [17] noted its analytical complexity increases with multiple operation centers. Ahituv et al. [15] proposed partitioning the network into smaller, independently operating subnetworks. Our study uses the stochastic one-median problem to improve operational performance in addressing abnormal information spread within social networks.

5 Results and Findings

This section presents results from a real-world Twitter dataset used to assess the accuracy and feasibility of our proposed approach. Assumptions regarding the policy level in OCA include employing two operation levels ($M = 2$), each tasked with monitoring less than 55% of the network. According to [4], the policy level functions akin to stakeholders, deciding on strategy implementation and adjusting operation levels based on environmental reports and feedback. The development level acts as an operations hub, executing organizational strategies, liaising with other levels, and monitoring network conditions to ensure effective strategy implementation or necessary updates.

Dataset. We collected a co-hashtag network from Twitter during an armed protest demonstration against COVID-19 lockdown in Michigan state in May 2020. Data was gathered using Twitter API for hashtags #MichiganProtest, #MiLeg, #Endthelockdown, and #LetMiPeopleGo from April 1 to May 20, 2020, resulting in 16,383 tweets and 9,985 unique user IDs. The network analysis depicted 3,632 nodes (Fig. 1), consolidating minor communities into one node [28]. The modularity analysis identified 382 communities, focusing here on the top 5 largest (C_1 through C_5) due to their user count, while smaller communities were grouped into a single node (C_6) to account for their interactions.

Experiment Results. To evaluate the performance of the agent levels in mitigating the spread of COVID-19-related misinformation in social media networks, we developed a simulation model that accounts for the misinformation spread between communities, with a maximum of 250 instances of false information being propagated across the network. In other words, the network's throughput is the number of misinformation units spread between communities. Each misinformation unit includes sets of messages transferred from the community C_i (source) to community C_j (target). For example, the misinformation units spread from community C_1 (source) to users in C_2 and C_4 (targets), where C_1 sent 100 misinformation units, fifty per target community. Likewise, 50 misinformation units spread from C_2 to C_6. Another misinformation spread was logged from C_5 to C_3. Finally, 50 misinformation units spread from communities in C_6 to target users in C_2.

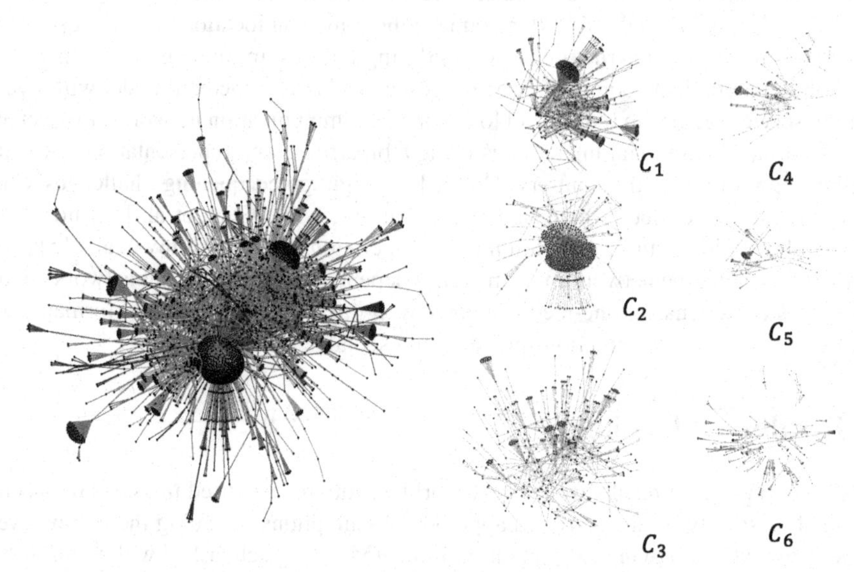

Fig. 1. Twitter network for COVID-19 anti-lockdown protest in Michigan and the six communities identified using modularity.

The model assesses how agents respond to the spread of information over time, incorporating an information diffusion delay factor into Eqs. (8) from Sect. 3, with data delay thresholds set at 0.01–0.1 per hour. This approach gauges agents' effectiveness in curbing misinformation propagation across the network. Increasing the delay threshold enables agents to manage more users and misinformation units exchanged between communities. Optimizing this delay factor hinges on network throughput, user connections between communities, and message transfers among them.

Table 1 compares the handling of misinformation units in the network with and without an applied delay factor. Various thresholds for the delay factor were tested to assess agents' effectiveness in mitigating misinformation spread. Figure 2 illustrates the efficient frontier resulting from these tests. Our findings show that implementing a delay factor did not effectively reduce misinformation spread when the threshold was below 0.05 h. However, we observed a notable improvement in agents' performance when the delay threshold was set to 0.05 h or higher. This strategy can enhance the performance of agents in handling the maximum number of misinformation units spread between different communities on social networks.

Table 1. Information Diffusion Delay Factor.

C_i	C_j	Information Diffusion Delay Thresholds						
		None	0.01	0.02	0.03	0.04	0.05	0.1
1	5	8	1	2	4	5	7	13
1	4	8	1	2	4	5	7	13
1	6	25	4	8	12	17	21	42
1	2	59	10	19	29	38	50	98
1	3	74	12	24	37	48	62	123
3	2	40	7	13	20	26	33	67
3	5	13	2	4	6	8	11	22
3	4	18	3	5	9	11	15	30
3	6	42	7	13	21	27	35	70
2	6	125	21	41	62	82	104	208
4	6	8	1	2	4	5	7	13
5	6	25	4	8	12	17	21	42
4	5	4	0	1	2	2	3	7
2	5	13	2	4	6	8	11	22
2	4	29	9	9	14	19	24	48

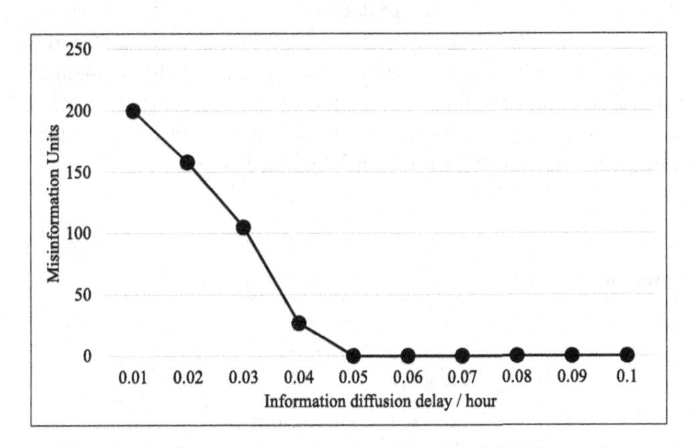

Fig. 2. Agents' performance in mitigating misinformation.

6 Conclusion

This research tackles COVID-19 misinformation on social networks using multidisciplinary methods like Systems Thinking and operational techniques. It employs cybernetic strategies to counter misinformation with practical front-end and back-end solutions. A systematic agent-based model is introduced to simplify social network complexities, reduce misinformation, and optimize performance using real-world feedback. By applying an Organization Cybernetic Approach (OCA) and the one-median problem, the model enhances resource efficiency, comprehends network structures, and boosts agent effectiveness. Evaluations demonstrate effective mitigation of COVID-19 misinformation on Twitter, highlighting the value of agent-based modeling, delay factor methodology, and Systems Thinking. This approach allows for real-time communication control, community behavior monitoring, agent performance measurement, and dynamic strategy adjustment. It handles complex social network analysis efficiently, regardless of network size or user count, using minimal resources (two agent levels). Managers can optimize resource allocation based on network size, community metrics, and real-time environmental feedback. In the future, we intend to examine whether the spread of disinformation is increasing or decreasing in dynamic social networks and evaluate the OCA's strategies to mitigate misinformation spread after considering different approaches, such as reinforcement learning and training maze-agent.

Acknowledgments. This research is funded in part by the U.S. National Science Foundation (OIA-1946391, OIA-1920920), U.S. Office of the Under Secretary of Defense for Research and Engineering (FA9550–22-1–0332), U.S. Army Research Office (W911NF-23–1-0011, W911NF-24–1-0078), U.S. Office of Naval Research (N00014–21-1–2121, N00014–21-1–2765, N00014–22-1–2318), U.S. Air Force Research Laboratory, U.S. Defense Advanced Research Projects Agency, Arkansas Research Alliance, the Jerry L. Maulden/Entergy Endowment at the University of Arkansas at Little Rock. Any opinions, findings, and conclusions or recommendations expressed in this material are those of the authors and do not necessarily reflect the views of the funding organizations. The researchers gratefully acknowledge the support.

Disclosure of Interests. The authors have no competing interests.

References

1. Munich Security Conference. https://www.who.int/director-general/speeches/detail/munich-security-conference. Accessed 19 Jun 2024
2. Egan, L.: They're killing people: Biden blames Facebook, other social media for allowing Covid misinformation. https://www.nbcnews.com/politics/white-house/they-re-killing-people-biden-blames-facebook-other-social-media-n1274232. Accessed 19 Jun 2024
3. Stolberg, S.G., Alba, D.: Surgeon General Assails Tech Companies Over Misinformation on Covid-19 - The New York Times. https://www.nytimes.com/2021/07/15/us/politics/surgeon-general-vaccine-misinformation.html. Accessed 19 Jun 2024
4. Alassad, M., Hussain, M., Agarwal, N.: How to control coronavirus conspiracy theories in twitter? A systems thinking and social networks modeling approach. In Proceedings of 2020 IEEE International Conference on Big Data (Big Data), pp. 4293–4299 (2020)

5. Alassad, M., Hussain, M., Agarwal, N.: Systems Thinking and Modeling in Social Networks Systems Thinking and Modeling in Social Networks: A Case Study of Controlling COVID-19 Conspiracy Theories. IDEAS Social CyberSecurity Symposium (2020)
6. Spann, B., Maleki, M., Mead, E., Buchholz, E., Agarwal, N., Williams, T.: Using diffusion of innovations theory to study connective action campaigns. In: Proceedings of the SBP-BRiMS, pp. 131–140 (2021)
7. Fact Check Tools. https://toolbox.google.com/factcheck/explorer. Accessed 19 Jun 2024
8. Zhou, C., Xiu, H., Wang, Y., Yu, X.: Characterizing the dissemination of misinformation on social media in health emergencies: an empirical study based on COVID-19. Inf. Process. Manag. **58**(4), 102554 (2021)
9. Shajari, S., Agarwal, N., Alassad, M.: Commenter behavior characterization on YouTube channels. In: Proceedings of the Fifteenth International Conference on Information, Process, and Knowledge Management (eKNOW 2023), Venice, Italy (2023)
10. Søe, S.O.: Algorithmic detection of misinformation and disinformation: Gricean perspectives. J. Documentation **74**(2), 309–332 (2018)
11. Weng, L., et al.: The role of information diffusion in the evolution of social networks. In: Proceedings of the 19th ACM SIGKDD International Conference on Knowledge Discovery and Data Mining, pp. 356–364 (2013)
12. Chan, Y., McCarthy, J.: Game-theoretic paradigms in collaborative research: part 1 - theoretical background. Int. J. Soc. Syst. Sci. **6**(4), 331–347 (2014)
13. Peng, S., Yang, A., Cao, L., Yu, S., Xie, D.: Social influence modeling using information theory in mobile social networks. Inf. Sci. **379**, 146–159 (2017)
14. Mann, C.J.H.: Systems Thinking – Creative Holism for Managers. Kybernetes **33**(8) (2004)
15. Ahituv, N., Berman, O.: Operations Management of Distributed Service Networks (1988)
16. Wasson. C.: System analysis, design, and development: Concepts, principles, and practices (2005)
17. Chan, Y.: Measuring Spatial Separation: Distance, Time, Routing, and Accessibility. In Location, Transport and Land-Use, pp. 120–209, Springer-Verlag (2005)
18. Even, S., Itai, A., Shamir, A.: On the complexity of timetable and multicommodity flow problems. SIAM J. Comput. **5**(4), 691–703 (1976)
19. Chan, Y., McCarthy, J.: Game-theoretic paradigms in collaborative research: part 2-experimental design. Int. J. Soc. Syst. Sci. **6**(4), 348–364 (2014)
20. Losty, P.A., Weinberg, G.M.: An introduction to general systems thinking. J. R. Stat. Soc. Ser. A **139**(4), 544 (1976)
21. VanderWeele, T.J., An, W.: Social Networks and Causal Inference. Handbook of causal analysis for social research, pp. 353–374, Springer, Dordrecht (2013)
22. Muchnik, L., et al.: Origins of power-law degree distribution in the heterogeneity of human activity in social networks. Sci. Rep. **3**(1), 1–8 (2013)
23. Du, B., Lian, X., Cheng, X.: Partial differential equation modeling with Dirichlet boundary conditions on social networks. Bound. Value Probl. **2018**(1), 1–11 (2018)
24. Hakimi, S.L.: Optimum locations of switching centers and the absolute centers and medians of a graph. Oper. Res. **12**(3), 450–459 (1964)
25. Patterson, T.S.: Dynamic Maintenance Scheduling for a Stochastic Telecommunication Network: Determination of Performance Factors. Air Force Institute of Technology Air University (1995)
26. Love, R. F., Morris, J. G., Wesolowsky, G. O.: Facilities location (1988)
27. Whitehead, A. N: Facility-Location Models. In: Location, Transport and Land-Use, pp. 21–119, Springer-Verlag (2005)
28. Spann, B., Johnson, O., Agarwal, N.: A computational framework for analyzing social behavior in online connective action: a COVID-19 lockdown protest case study. In: Americas Conference on Information Systems (AMCIS) (2022)

Provenance for Longitudinal Analysis in Large Scale Networks

Andrei Stoica[1] and Mirela Riveni[2(✉)]

[1] University of Groningen, Groningen, The Netherlands
[2] Information Systems Group, University of Groningen, Groningen, The Netherlands
m.riveni@rug.nl

Abstract. Concerns related to the veracity and originality of the content on social networks are at an ongoing rise. Considerable work has been done on information spreading, and tools have been built, while approaches with provenance-based analysis are rare. We are of the opinion that provenance-based analysis and visualization tools can make (mis-)information spreading analysis more efficient. Thus, we study provenance, and present a provenance pipeline for data analytics, where users are able to interact with multiple network analysis modules through a graphical user interface, and describe a proof-of-concept system. Although provenance visualization can suffice in capturing all the necessary metadata, integration with other network visualization modules suited to the same data enhanced our results analysis and conclusions. Having designed distinct provenance models, we captured and analysed lineage of information on community dynamics. We tested our proposed prototype with a real-world dataset comprising of more than 10 million filtered tweets, focused on COVID-19 vaccinations, and conducted an analysis on community dynamics with network science metrics and NLP.

Keywords: Large Scale Networks · Network Science · Data Analysis · Misinformation · NLP · Provenance

1 Introduction

The volume and complexity of the data that we are dealing with in every area of our social lives is only increasing. Thus, ways of structuring, analysing and visualizing data is of crucial importance for valuable insights when we deal with Big Data in research. And, a critical challenge emerges: how do we guarantee the reliability and traceability of data, and how can we improve the ways of analysing and visualising data? This is where the notion of provenance comes into play. The authors in [14] have classified provenance systems into database-oriented, service-oriented and miscellaneous categories, depending on their application area. [3] introduced the concepts of "why-provenance", i.e. the process that generates the data, and [4] defined the "where-provenance" as the origin of data. [2] present an

L. M. Aiello et al. (Eds.): ASONAM 2024, LNCS 15212, pp. 274–285, 2025.
https://doi.org/10.1007/978-3-031-78538-2_24

in-depth study on workflow provenance, in relation with workflows in scientific computing. Provenance is not studied enough in data science and its benefits in Big Data analytics research, and more specifically in social network analysis. We focus on case studies of social network analysis data, and specifically in analysing misinformation, from influential accounts, group structures around influential accounts, to network structure as parameters of influence in (mis-)information spreading. Furthermore, we analyse how provenance-based analysis can help in these types of analysis. Considerable work has been done on misinformation detection and spreading. However, work on provenance-based analysis lack, and we believe that this is an important approach for analysis and visualization of network data, as proven by the results of this work. This work focuses on the research questions posed in the following. A) How can we best model provenance data for efficient studies of large-scale social network data, such as misinformation spread analysis? B) What are the benefits that provenance-based analysis can bring to social-network data analytics? C) What are the components and how best to integrate them for a holistic social network analysis pipeline with provenance-based analysis and visualisation? We introduce a proof-of-concept prototype, which includes modules that receive network data as input in multiple formats, provide functionality of running various network-science metrics and clustering algorithms, and get the results in a file, in graph-based visual forms, in network view and in provenance-based model graphs. This paper is organized as follows. In Sect. 2 we discuss related work on provenance and its application areas. Section 3 describes the implementation of a module integrated in the data analytics pipeline, i.e., for opinion changes with NLP. In Sect. 4 we discuss our framework for provenance-based data analysis and visualisation. Section 5 discusses our provenance models and experiments with our proof-of-concept prototype. In Sect. 6 we conclude the paper including a discussion on future work.

2 Related Work

2.1 Provenance of Data in Different Application Areas

Provenance in social computing is discussed by the authors in [13], where they present provenance models for crowdsourcing for teams. The authors have conducted experiments with synthetic and real world data of software engineering teams that collaborate online, and have modeled provenance for task assignment and task execution. They also provide visualisation of their provenance models based on logs files. Our work is similar to this work in terms of granularities, as we also present models considering individual and network level, but for a different application area. Our approach is specifically tailored to social networks data, and we also bring the novelty with providing a proof-of-concept prototype. Provenance is also explored for crowdsourcing, and is discussed in [8,16]. Provenance for process mining is described in [20]. In the area of Web Data, the author in [7] proposes a provenance model suitable for systems that consume linked data.

2.2 Provenance in Social Networks

A methodology for misinformation detection on Twitter[1] data is presented in [1]. Provenance data is used to determine the original source of a piece of information. The study introduces both user- and content-related metrics based on social provenance attributes captured in a social provenance database and the fuzzy analytic hierarchy process algorithm determines the weights of these metrics in order to assess the credibility of information which circulates on the network. However, the authors in this paper work with synthetic data, while we use a real dataset, and in addition provide a framework. Tracking provenance data with retweets is presented in [10]. In this work we consider all types of reactions and also properties of accounts and tweets, details which give us more effective analysis results about the important indicators in information spread. The networks field of study with regard to provenance is not studied enough, especially with respect to possible provenance models specifically tailored to social networks interactions. This is where our work is important. We provide provenance models for modeling social network data from different perspectives and different granularity as can be seen from Sect. 4 and describe an implemented proof of concept pipeline for social network analysis, where data is mapped to the PROV-O ontology, analysed, and visualised with different views.

2.3 Provenance Visualization

The authors in [19] have presented a visualization approach that takes graphs as input and outputs provenance-based results from various types of analysis such as graph comparison, summarising, and stream data. Another graph-based, provenance data visualisation tool named, Prov Viewer is presented by authors in [9]. Our proposed framework differs from these two tools in that it is a more holistic architecture proposal, on which users can also run different network-science based metrics and algorithms, and specify if they want to have them shown as network-based results or as provenance graphs.

3 Misinformation Spreading Analysis: Opinion Changes with NLP

3.1 Module Overview

One module which is integrated into the data analytics pipeline described throughout the paper is the detection of opinion changes with the help of natural language processing techniques. The dataset [5] we apply our algorithms on is structured in a similar fashion to how the official Twitter API generates the data. More precisely, Twitter generates a JSON tweet object. Some of the keys in

[1] We use the term Twitter because the dataset we work with has been collected before its acquisition, it should be noted that when we refer to Twitter we refer to the platform before its acquisition and renaming to X.

the tweet dictionary had been refactored (either renamed, recomputed, removed or changed the type of) prior to us working with the dataset. After performing the changes, all recordings had been merged into 20 .csv files, counting for more than 10 million tweets, most of them corresponding to 20 days, starting around the beginning of March 2021, roughly one year after the COVID-19 pandemic gained momentum. To accomplish the desired results, we keep track of the situations where one user posted multiple reactions, i.e. replies, quotes, retweets, or any combination of them, to a source tweet. Opinion change is obviously a comprehensive task. In the context of social media, resources are even more limited. We only have access to a user's activity, which in most cases turns out to be reactive rather than proactive and not necessarily well argumented. Users can either react to certain source tweets by adding textual comments or they can like or retweet the original post. On top of that, users can interact by following or unfollowing each other, which may also signal approval or disapproval of certain users' views. Therefore, due to the subjective nature of opinion changes, especially within social media, detecting them is a delicate task. There is a plethora of NLP libraries which compute text sentiments. Our choice was SentiStrength [15]. The choice was made based on the thorough dictionary used to analyse the texts, as well as the possibility to compute a scale score, not only a binary score (negative or positive), meaning sentiment scores range from −4 (most negative) to 4 (most positive). This allowed us to analyse the intensity of the opinion changes further down the line. The SentiStrength algorithm is lexicon-based, meaning it makes use of a pre-defined sentiment lexicon. This is a collection of words that were assigned sentiment scores. Its documentation mentions the algorithm is designed to estimate the strength of positive and negative sentiment in short texts in English, even for informal language. The authors point out that it has human-level accuracy for short social web texts in English, except political texts. Nevertheless, there are limitations to the algorithm, such as a limited context sensitivity. Lexicon-based algorithms do not spot that words can have distinct meanings in different contexts. Also, sarcasm or irony pose serious problems. For example, the text *"Side effects: sore arm, and an overwhelming dread of having to go back to work tomorrow."* is an ironic text which was assigned the minimum score of −4, where in fact it reflects a positive reaction to a source tweet. The lexicon should be updated regularly in order to include new words, as well. One way to overcome these issues is to use machine learning algorithms. They use learning techniques, such as (deep) neural networks, support vector machines or logistic regressions to identify the relationship between text features and sentiment labels from labeled training data. However, the disadvantage is the lack of a large, up-to-date labeled training dataset, relevant to our topic, to train such models on. Moreover, machine learning algorithms can be much more computationally demanding that lexicon-based, albeit the execution of the latter was challenging in itself given our large dataset.

3.2 Text Cleaning

Although the SentiStrength library is built to interpret relatively informal social media texts, there is a specific jargon that needs to be treated manually beforehand, so that the algorithm yields the best possible outcome. Therefore, the tweets suffered the following alterations: all words were converted to lowercase, new line characters were replaced by white spaces, tags were removed, URLs were removed, punctuation was removed, contractions were converted to full forms, emojis (emoticons, symbols and pictographs, flags, etc.) were removed, stopwords were removed. Lastly, we opted to create a custom, smaller list of stopwords that ought to be removed from tweet texts before being passed on to the sentiment analysis algorithm for processing. We studied an already built-in list of stopwords provided by another popular NLP module, namely **NLTK**, and it served general purposes, thus results for our particular case were worse than expected. This is due to the list being too extensive and a lot of words having an impact in computing the sentiment of a tweet, such as "not" or "all".

3.3 Detecting Opinion Changes

Before describing the process of detecting opinion changes, we need to define the design choices and assumptions we considered. First, a group of recordings may contain a considerable number of tweets and within the respective group, there may be more than one opinion change. Hence, for the sake of capturing the most relevant information, we opted to only track the largest opinion change within the group, which we refer to as the opinion change of the group, i.e. the difference in sentiment score between the largest score and the lowest score. Furthermore, even though more tweets' texts may generate the same maximum or minimum score, we only take the earliest occurrence of each (minimum and maximum scores) into account. Lastly, we base our analysis on the assumption that retweets represent a full and undisputed agreement with the original post and we artificially assigned the sentiment score of 4 (the maximum value of the SentiStrength range). Having established the ground rules we take into consideration, we describe the process of observing changes in opinions. For each group of tweets, we compute the tweets' sentiment scores. If there is a change within a group, i.e. there is at least one negative score and at least one positive score, then we identified an opinion change and we add it to a dictionary with the shape: keys representing a tuple of **(author_id, reference_id)**, and values representing a list of sentiment scores. Note that the chronological order of the tweets in each group was retained, meaning we can spot if we have a negative or positive opinion change within a group. Our analysis revolves around the sentiments of reactions as we follow the dynamics of interactions and whether these have an impact on people's interpretation of the respective topics.

4 Social Networks Data Analysis with Provenance

4.1 Background, Problem Statement and Methodology

This study defines and implements social network provenance models, specifically tailored to large-scale social network analysis, i.e., Twitter interactions. On top of that, it also provides a graphical user interface to further ease interactions with the created provenance models, as well as other types of network visualizations. Different provenance representations of lineage data in social networks can highlight trends, they can uncover important features of the entities and agents participating in such interactions. Both textual and visual provenance information can further be used to detect misinformation spread, predict user behaviour or classify tweet content, e.g., for opinion change and sentiment analysis. Provenance can be defined as the documented history of an entity's origin, transformations and interactions within a given context. A formalized way to create provenance representations is the PROV Data Model (PROV-DM) [17], and the provenance ontology (PROV-O) [18] standardised by W3C. The main elements of the data model are the entities, activities and agents [17]. They are also encoded in the PROV ontology (PROV-O) [18], which is meant to represent, exchange and integrate provenance information retrieved from different systems and from within distinct environments.

Technology Stack: The dataset we use is based on the ID's provided by authors in [6], which we hydrated to get the tweet contents that we were interested in. As far as the set of technologies is concerned, throughout the implementation of this project, various ones are used. The provenance models are implemented using the ProvToolbox Java library [11,12]. With respect to the actual information contained in the provenance representations, it is retrieved from the original dataset, using Python, due to its build-in capabilities to perform fast, vectorized operations on large datasets. Another implementation of this study presents an interactive software solution which integrates a number of social network analysis modules, namely provenance visualizations, network graphs and opinion changes analysis. They comprise the logic behind a GUI with the following architecture: the frontend component is a React application. There are two backend applications, a Spring Boot one for the provenance Java module and a Flask one for the Python-based network generation algorithm.

4.2 Provenance Modeling

Model 1. Individual Tweets: The first model we design aims to cover information about individual tweets. Therefore, at the center of the representation there is a source/original tweet. One can choose any number of tweets from the subnetwork of reactions associated with the original tweet and display them. Relevant characteristics are chosen for every type involved in the model, i.e.

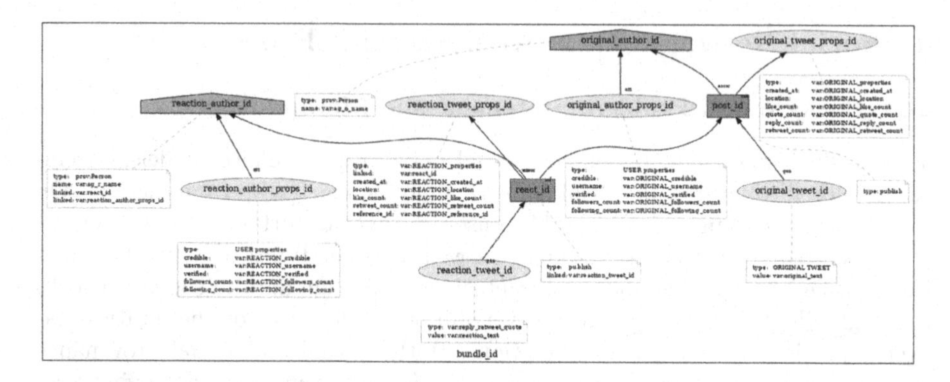

Fig. 1. Model 1. Original tweet and a reaction

original/reaction tweet and properties entities, post or react activities, author agents and properties. It aims to present a detailed overview of atomic participants within a social network. The complete model, displaying a reaction, along with the original tweet, is shown in Fig. 1. The figure is a SVG representation of provenance information created with the ProvToolbox library.

Model 2. Aggregate Statistics About Reactions over Time: Model 2 aims to capture a broader picture than Model 1, it places a source tweet at the heart of the representation and displays aggregate statistics about reactions to the source tweet occurring after certain amounts of time. Thus, it is designed to track time dynamics as well. The main particularity of Model 2 is the switch of focus towards groups of reactions, instead of individual ones. This enables us to get a better understanding of the general trends within social networks, and not only limiting the study to the behaviour and properties of the most popular tweets. The graphical representation of Model 2 is given in Fig. 2.

Model 3. All Tweets at a Given Time-Point: Unlike Models 1 and 2, Model 3 takes into consideration all tweets at a given time-point, and it captures time dynamics as well. The representation's aim is to outline essential information on as many textual tweets as possible, regardless of the type (i.e., original tweets, replies and quotes).

4.3 Experiments, Results and Visualisation

All provenance experiments are conducted on the dataset described in Section III. A few notable results were seen when using the provenance pipeline to analyse our dataset. An instance of Model 2 was captured, in a way that the tweet with the most reactions was selected, and its subnetwork studied. The chosen subnetwork includes all reactions, i.e. all replies, quotes and retweets to the original tweet, posted at a time difference indicated by the time interval. In order

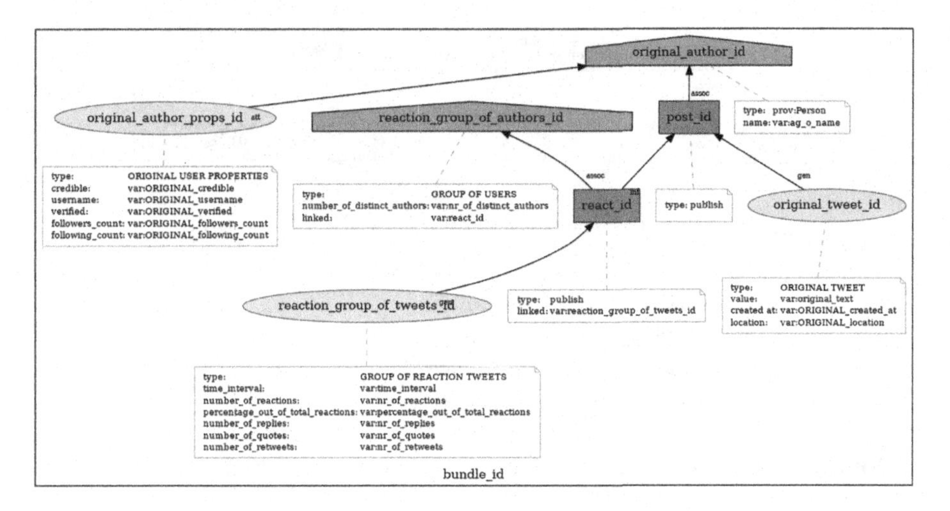

Fig. 2. Model 2. Original tweet, and aggregate reactions

to draw a conclusion on how fast and how many users respond to a tweet, the interval between the first and last reactions was considered, and split into a few shorter ones. For each interval, the reactions were filtered. Having a subset of reactions, the required aggregate statistics could be computed. For example, the number of reactions per type, the percentage of reactions posted in each respective interval out of all reactions to that specific original tweet and the number of distinct users who posted the reactions of the selected time interval need the whole subset to be computed. Around 5% of the reactions were posted in that time, namely 2380. During the first three hours, roughly half of the reactions occur. The number is impressive, as one can consider this interval to be relatively short. After three hours, but within the first day after the moment the source tweet is posted, the percentage of reactions is equally impressive, almost half of the reactions are released. After the first day, reactions are sparse, only around 6–7%. It is clear to see that the vast majority of reactions occur right after a source tweet is posted. Interest on news or opinions in social networks plummets in a very short time.

Figure 3 displays an instance of Model 3. Tweets posted on March 1, 2021, which is a regular non-holiday Monday, were selected, during the interval 9AM - 5PM, denoted as the working hours. All timestamps have been converted to their local timezone.

5 Proof-of-Concept Prototype for Provenance-Based Data Analytics

Although provenance visualizations can suffice in capturing all the necessary metadata to prove a point, integration with other visualization modules suited to the same data available can only enhance the conclusions. Therefore, a complete

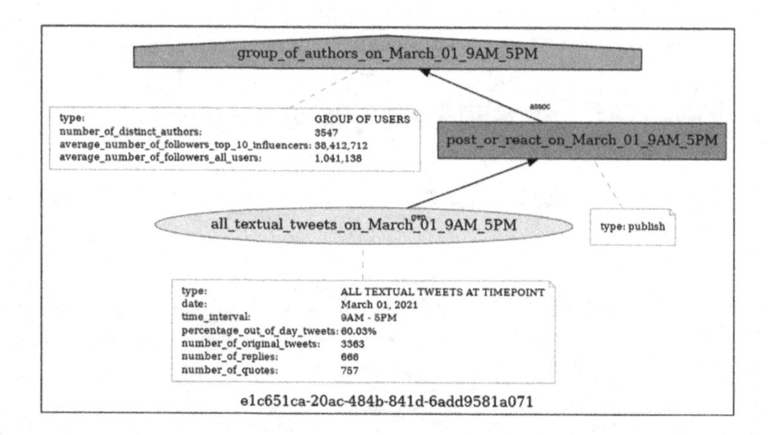

Fig. 3. Provenance graph with results for Model 3: March 1 - business hours

pipeline was created, where users are able to interact with multiple social network analysis modules through a graphical user interface (GUI). In an attempt to offer an alternative perspective view on the data, the proposed implementation includes the option to construct network graphs, using the same filters as for the provenance visualizations. This can be observed in another snapshot of the front-end application in Fig. 4.

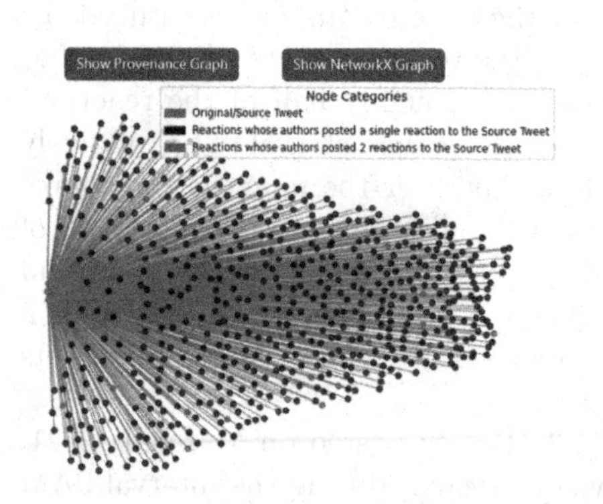

Fig. 4. GUI - reaction network graph, after applying filters

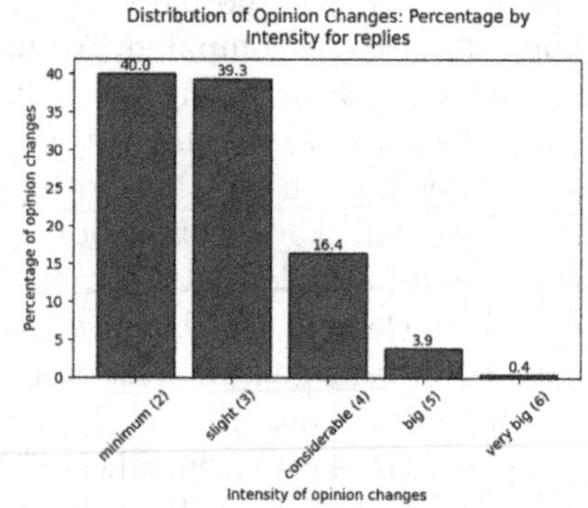

Fig. 5. GUI - intensity distribution of opinion changes for replies subnetwork

Within our GUI, users can depict distributions of opinion changes within Twitter interactions, obtained using the Matplotlib Python module. The user input is the textual reaction types(s) to be included in the analysis, i.e. any combination of replies and/or quotes. The operation is performed by the algorithm described in Section III. An example representation of the GUI, where a user selects a subnetwork of replies, is given in Fig. 5. Our prototype implementation and the experiments we conducted show that provenance data, if modeled properly, can give valuable insights in (mis-)information spread analysis, and the important properties of accounts and tweet types influential in spreading misinformation. Stakeholders can get different detailed views on multiple properties of networks, and provenance can help in better network property analysis when applied together with the well known network-science metrics and algorithms. In general, from our experiments, and related work, we can conclude that provenance is of benefit to network analysis as it provides detailed and varied views of data, we thus take this opportunity to make a call for provenance to be studied wider.

6 Conclusion and Future Work

Provenance modeling and analysis in this study aim to offer a data analysis mechanism and framework, tailored to large scale network data, with an example from Twitter interactions. With the integration of provenance, along with network graphs and opinion change distributions, described in Chapter IV, trends or prominent features of the entities or agents, i.e. tweets or authors, participating in such interactions, can be effectively analysed. More specifically, they refer to the lineage of data, community detection and dynamics, optimal posting times for maximum exposure and common reaction time intervals. The pipeline can be of use to both researchers who appreciate alternative views on data or practitioners who would like to develop machine-learning algorithms to predict user behaviour within social networks or classify posts based on their documented lineage. Albeit not the sole focus of this research, as part of the application of the provenance module, we found from the real-life dataset a tendency for an almost immediate response to new information released through the network. Our future work includes testing our pipeline with experiments on eco-chambers on Mastodon, we will analyse group dynamics on this platform and see what insights we can get with provenance visualization with the purpose of further proving the generality of the prototype for graph-based large network datasets.

Acknowledgment. We thank the Center for Information Technology of the University of Groningen for their support and for providing access to the Hábrók high performance computing cluster.

References

1. Baeth, M.J., Aktas, M.S.: Detecting misinformation in social networks using provenance data. Concurrency Comput. Pract. Exper. **31**(3), e4793 (2019). https://doi.org/10.1002/cpe.4793
2. Barga, R.S., Digiampietri, L.A.: Automatic generation of workflow provenance. In: Proceedings of the International Provenance and Annotation Workshop (IPAW), pp. 1–9 (2006)

3. Buneman, P., Khanna, S., Tan, W.C.: Data provenance: Some basic issues. In: FST TCS 2000: Proceedings of the 20th Conference on Foundations of Software Technology and Theoretical Computer Science, pp. 87–93. Springer-Verlag, London (2000). https://doi.org/10.1007/3-540-44450-5_6
4. Buneman, P., Khanna, S., Wang-Chiew, T.: Why and where: a characterization of data provenance. In: Van den Bussche, J., Vianu, V. (eds.) ICDT 2001. LNCS, vol. 1973, pp. 316–330. Springer, Heidelberg (2001). https://doi.org/10.1007/3-540-44503-X_20
5. DeVerna, M.R., et al.: Covaxxy: A collection of english-language twitter posts about covid-19 vaccines. In: Proceedings of the International AAAI Conference on Web and Social Media, vol. 15(1), pp. 992–999 (2021). https://doi.org/10.1609/icwsm.v15i1.18122
6. DeVerna, M.R., et al.: Covaxxy: a collection of english-language twitter posts about covid-19 vaccines. In: Proceedings of the International AAAI Conference on Web and Social Media, vol. 15, pp. 992–999 (2021)
7. Hartig, O.: Provenance information in the web of data (2009)
8. Huynh, T.D., Ebden, M., Venanzi, M., Ramchurn, S.D., Roberts, S.J., Moreau, L.: Interpretation of crowdsourced activities using provenance network analysis. In: Hartman, B., Horvitz, E. (eds.) Proceedings of the First AAAI Conference on Human Computation and Crowdsourcing, HCOMP 2013, 7-9 November 2013, Palm Springs, CA, USA, pp. 78–85. AAAI (2013). https://doi.org/10.1609/hcomp.v1i1.13067
9. Kohwalter, T., Oliveira, T., Freire, J., Clua, E., Murta, L.: Prov viewer: a graph-based visualization tool for interactive exploration of provenance data. In: Mattoso, M., Glavic, B. (eds.) IPAW 2016. LNCS, vol. 9672, pp. 71–82. Springer, Cham (2016). https://doi.org/10.1007/978-3-319-40593-3_6
10. Migliorini, S., Gambini, M., Quintarelli, E., Belussi, A.: Tracking social provenance in chains of retweets. Knowl. Inform. Syst., 1–28 (2023)
11. Moreau, L.: ProvToolbox. https://lucmoreau.github.io/ProvToolbox/, Accessed 10 June 2023
12. Moreau, L.: ProvToolbox GitHub Repository. https://github.com/lucmoreau/ProvToolbox, Accessed 10 June 2023
13. Riveni, M., Nguyen, T., Aktas, M.S., Dustdar, S.: Application of provenance in social computing: a case study. Concurr. Comput. Pract. Exp. **31**(3) (2019). https://doi.org/10.1002/cpe.4894
14. Simmhan, Y.L., Plale, B., Gannon, D.: A survey of data provenance in e-science. SIGMOD Record **34**(3), 31–36 (2005). https://doi.org/10.1145/1084805.1084812
15. Thelwall, M., Buckley, K., Paltoglou, G.: Sentiment strength detection for the social web. J. Assoc. Inf. Sci. Technol. **63**(1), 163–173 (2012). https://doi.org/10.1002/ASI.21662
16. Willett, W., Ginosar, S., Steinitz, A., Hartmann, B., Agrawala, M.: Identifying redundancy and exposing provenance in crowdsourced data analysis. IEEE Trans. Visual Comput. Graph. **19**(12), 2198–2206 (2013). https://doi.org/10.1109/TVCG.2013.164
17. World Wide Web Consortium: PROV-DM Specification. https://www.w3.org/TR/2013/REC-prov-dm-20130430/ (2013), Accessed 10 June 2023
18. World Wide Web Consortium: PROV-O Specification. https://www.w3.org/TR/2013/REC-prov-o-20130430/ (2013), Accessed 26 May 2023
19. Yazici, I.M., Aktas, M.S.: A novel visualization approach for data provenance. Concurr. Comput. Pract. Exp. **34**(9) (2022) https://doi.org/10.1002/CPE.6523, https://doi.org/10.1002/cpe.6523

20. Zerbato, F., Burattin, A., Völzer, H., Becker, P.N., Boscaini, E., Weber, B.: Supporting provenance and data awareness in exploratory process mining. In: International Conference on Advanced Information Systems Engineering. pp. 454–470. Springer (2023). https://doi.org/10.1007/978-3-031-34560-9_27

Culture Fingerprint: Identification of Culturally Similar Urban Areas Using Google Places Data

Fernanda R. Gubert[1(✉)], Gustavo H. Santos[1], Myriam Delgado[1], Daniel Silver[2], and Thiago H. Silva[1]

[1] Universidade Tecnológica Federal do Paraná, Curitiba, Brazil
{fernandagubert,gustavohenriquesantos}@alunos.utfpr.edu.br,
{myriamdelg,thiagoh}@utfpr.edu.br,
[2] University of Toronto, Toronto, Canada
dan.silver@utoronto.ca

Abstract. This study investigates methods using a global data source, Google Places, to identify culturally similar urban areas without relying on difficult-to-access data like user preferences shown through check-ins. We propose and assess a simple method requiring only information about place types and their frequency in the studied areas, and a more advanced method that enhances venue categories using Scenes Theory - it helps us understand the cultural significance of everyday urban life. We tested our methods in 14 cities worldwide and all US states. The results suggest that a straightforward approach based on category frequencies can highlight major cultural differences. However, the Scenes Theory-based method provides a better understanding of cultural nuances, as the ones supported by survey data.

Keywords: Cultural signature · large scale assessment · Google Places

1 Introduction

Traditional methods like surveys and interviews are important data sources for studying culture in its complexity. However, these methods have drawbacks (e.g. high costs and time-consuming), which limit their scalability. To remedy this situation, some works evaluate alternative geolocalized data sources from the web to study culture. These sources exist on a global scale and are faster to obtain. Studies have shown the usefulness of these data sources in several domains [6, 16,19,21], including the cultural ones [2,3,8,15,17,18].

Bancilhon et al. [2] explore an approach to quantifying a society's culture through city street names, revealing that these names reflect cultural values. Using Foursquare data, Senefonte et al. [15] examine how regional and cultural characteristics affect the mobility patterns of both tourists and residents. The results indicate that the tourist's origin significantly influences their behavior,

L. M. Aiello et al. (Eds.): ASONAM 2024, LNCS 15212, pp. 286–297, 2025.
https://doi.org/10.1007/978-3-031-78538-2_25

especially in large cultural differences between the origin and destination. Silva and Silver [18] introduce a graph neural network method for predicting local culture. They evaluate their approach on Yelp data, showing that it could help predict local culture even when traditional local information is unavailable.

When aiming to provide methods based on geolocalized web data to describe local culture, some research indicates that eating and drinking habits can be a valuable option [3,8,17]. These studies illustrate promising approaches to identifying cultural boundaries and similarities between different societies at different scales. However, they rely on user preferences, typically manifested through check-in data, which is challenging to obtain in practice for many users or with global coverage. Another perspective follows the argument presented in [11], which suggests that the availability of resources and services that meet the population's needs contributes to forming a local identity. What is notable about this approach is the opportunity to consider multiple aspects of culture, as the resources of a region can be associated with various categories like religion, cuisine, and arts, providing a format that is still little explored. Our approach aligns with this direction by exploring Scenes Theory [20], which captures local public cultural dimensions embodied in venues such as cafes, churches, restaurants, and nightclubs. This enables the creation of a cultural description of local areas, allowing comparison with other areas—a step we perform in this study to identify cultural similarities. This differs from previous studies [1,4,9,14], which tend to disregard the cultural component in their analyses.

Extending previous studies, the approaches proposed here to describing local culture rely on simple data from the Google Places API. One can provide an expressive cultural abstraction of any covered urban area, thanks to the mapping to Scenes Theory - see Sect. 2. Unlike studies that explored the cultural characteristics of regions using eating habits and user mobility, this study aims to derive such characteristics from the categories of venues present in a city. This allows us to evaluate whether our proposed approaches can adequately express key cultural aspects without relying on user actions, such as check-ins and evaluations.

We evaluate the approaches using data from 14 cities on different continents and all states of the United States. The results indicate that a simple approach, Approach-Frequency, can capture significant cultural differences satisfactorily. However, a more sophisticated approach, Approach-Scenes, can add extra semantic expressiveness in capturing cultural characteristics. This added expressiveness is evident in our outcomes and survey data comparison, indicating that Approach-Scenes better captures cultural nuances.

2 Cultural Signatures Obtained from Google Places (GP)

2.1 Data From GP

GP is a location-based social network that allows users to discover and share information about local venues, geographic locations or points of interest, such as universities, cafes, bus stations, and parks. No type of location was disregarded. GP API provides geolocated venue data, resulting in one of the world's most accurate, up-to-date, and comprehensive venue models. In addition to latitude and longitude coordinates, venues are associated with at least one category designed to describe the venue type. In this study, we consider two datasets from GP, States and Cities, as described next.

The *Dataset States*, presented in [10,22], includes business metadata (geographic info, category information, etc.) from GP up to September 2021 for all U.S. states. The dataset is composed of 4,963,111 unique venues and has 4,501 unique categories. The District of Columbia has the lowest number of distinct venues, totaling 11,003, while California has the highest count at 513,134 unique venues. We explore this dataset to study states focusing on geographic and category information.

For *Dataset Cities* we have collected data from a set of cities. GP API provides, by default, 141 unique categories. However, these categories do not provide the level of specificity necessary in this study. For example, the API assigns the category "restaurant" to all venues of this type, but it does not offer more specific categories related to cuisine, such as Italian or Japanese, which is necessary for this work. The optional "keyword" parameter is used in requests to the GP API aiming to overcome the limitation. The API documentation[1] guarantees valid results when inputs to this parameter are categories of venues, making it a convenient option for the desired purposes. The categories chosen to use in this parameter are those from the Yelp database due to the higher specificity, e.g., Yelp offers specific types of restaurants, such as Italian Restaurants. Yelp categories have a four-level hierarchical structure, making it suitable for our work to adopt only those at the last level. Some of them were excluded because they were not relevant to the purpose of the study, such as Provencal and Northeastern Brazilian, resulting in a total of 888 categories.

Using the proposed strategy, we have collected data from 14 cities, namely: Curitiba and Rio de Janeiro in Brazil; Toronto and Vancouver in Canada; Chicago and Los Angeles in the USA; Berlin and Frankfurt in Germany; Paris and Lyon in France; Seoul and Busan in South Korea; and Nairobi and Mombasa in Kenya. These cities are important in their respective countries and cover regions with different cultural characteristics. A publicly available tool[2] details the data acquisition process and clarifies the need for a balance between costs and data volume, which leads us to have a summarized set of venues. This tool aids in reproducing our study [5].

2.2 Urban Areas' Cultural Dimensions

Following research on local "scenescapes," we measure local scenes for the urban areas by aggregating the set of available venue categories in terms of qualitative meanings they express. To translate these concepts into measurements, for each

[1] https://developers.google.com/maps/documentation/places/web-service/overview.
[2] https://github.com/FerGubert/google_places_enricher.

venue category (e.g., restaurant, university, or bar), a team of trained coders has assigned a score of 1–5 on a set of 15 cultural dimensions $s_i \in S = \{s_1, s_2, ..., s_{15}\}$, such as transgression, tradition, local, authenticity, or glamour. Each area then receives a score for each of the 15 dimensions, calculated as a weighted average. Detailed descriptions of the theoretical meaning of each dimension can be found in [20].

2.3 Transfer Knowledge Procedure

The categories retrieved from GP need to be mapped to the appropriate set of 15 dimensions scores of the Scenes Theory. Without trained coders for our particular areas (States and Cities) we examine the Scenes' dimension scores of the Yelp categories presented in [19]. This knowledge is then adapted for use with GP/ categories, a transferring knowledge outlined in Fig. 1. It illustrates an example for two different venues, each provided by a different dataset, venue A from *Dataset States* and venue B from *Dataset Cities* .

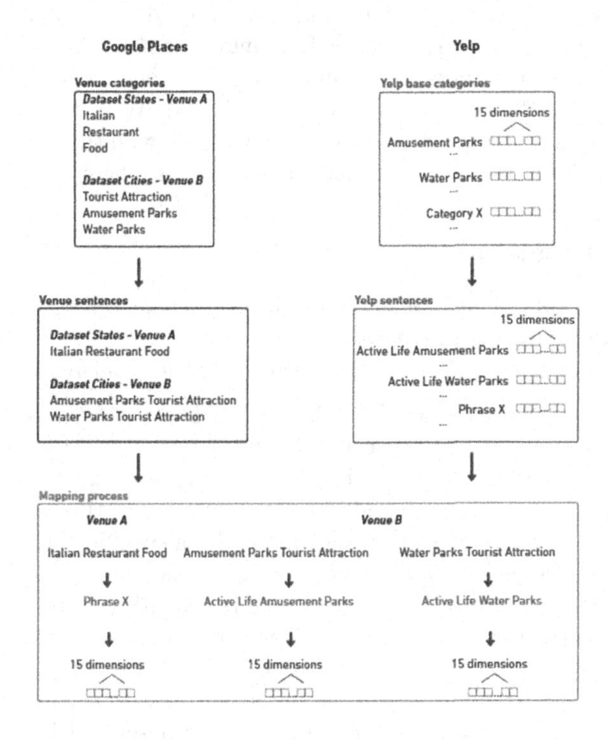

Fig. 1. Overview of mapping GP categories to the local cultural dimensions.

As depicted in Fig. 1, for a better description of the venues, both the selected Yelp categories used in the requests and the broader categories made available by GP are used. To increase semantic capacity and mapping accuracy, sentences are created for each venue, following the procedures for each dataset.

In *Dataset States*, one sentence is created per venue, combining all associated categories. For example, if the venue has the categories "Italian", "Restaurant" and "Food", the sentence is: "Italian Restaurant Food". *Dataset Cities* on the other hand, lacks specific categories by default. Therefore, sentences include a requested Yelp category and all GP categories associated with that venue. For example, if a venue has "Amusement Parks" and "Water Parks" due to Yelp requests and "Tourist Attraction" as a default GP category, the sentences are: "Amusement Parks Tourist Attraction" and "Water Parks Tourist Attraction".

Yelp categories are organized in a 4-level hierarchical structure. To expand semantic capacity, Yelp sentences are created using all hierarchical levels. In other words, for each category at the last level, the associated sentence returns to the first level. This is why "Active Life" was added to the Yelp sentences in Fig. 1; these Yelp categories are immediately below its root category.

In possession of sentences describing the venue, the mapping process is carried out with SBERT, using the Sentence Transformers framework, in which several pre-trained models with a large and diverse dataset of more than 1 billion training pairs are made available and can be used to calculate embeddings from sentences and texts to more than 100 languages [12]. The cosine similarity compares the generated embeddings, and for each sentence related to the venues, the Yelp sentence with the highest score is retrieved. With this mapping, each venue is associated with one or more vectors (depending on the number of related sentences) containing the 15 dimensions of the Scenes Theory.

2.4 Cultural Signatures

We propose two approaches for creating cultural signatures, *Scenes-based approach*and *Frequency-based approach.*

For a particular urban area, the *Scenes-based approach*considers a vector $S_{area} = \{s_1^{area}, s_2^{area}, ..., s_{15}^{area}\}$, where $s_i^{area} = \frac{1}{\omega}\sum_{v=1}^{\omega}\left(\frac{1}{m}\sum_{\phi=1}^{m} S_i^{v,\phi}\right)$, with ω representing the number of unique venues in an urban area, m is the number of categories a venue has, and $S_i^{v,\phi}$ is the i-th element of the vector of cultural dimensions for a certain venue v and one of its category ϕ; thus, s_i^{area} represents the average score of all venues in the urban area for a specific cultural dimension, considering the average scores of all categories for each venue.

We also present an alternative approach, Approach-Frequency, aimed at creating cultural signatures that disregard Scenes information, using only location categories. This approach considers the frequency of the category in the area, i.e., for a particular urban area, a vector describes it by all the unique categories found in that area. For example, an area could be described by the categories [University, Restaurant, Coffee Shop, American Restaurant] and another by [Italian Restaurant, Wine Shop]. The frequency values are normalized per category.

Approach-Frequency helps answer the question: Are the existence and the number of certain types of venues in two different urban areas enough to explain their cultural differences?

3 Cultural Signatures Identify Culturally Similar Areas

3.1 Cities Worldwide

***Scenes* for *Dataset Cities*.** First, we evaluate the results of the cultural signatures generated by the *Scenes-based approach*. We perform hierarchical clustering using Ward's linkage method and Euclidean distance, with the 15 dimensions of Scenes Theory as features. The results are represented in the dendrogram depicted at the top of Fig. 2, where a division into six clusters is identified.

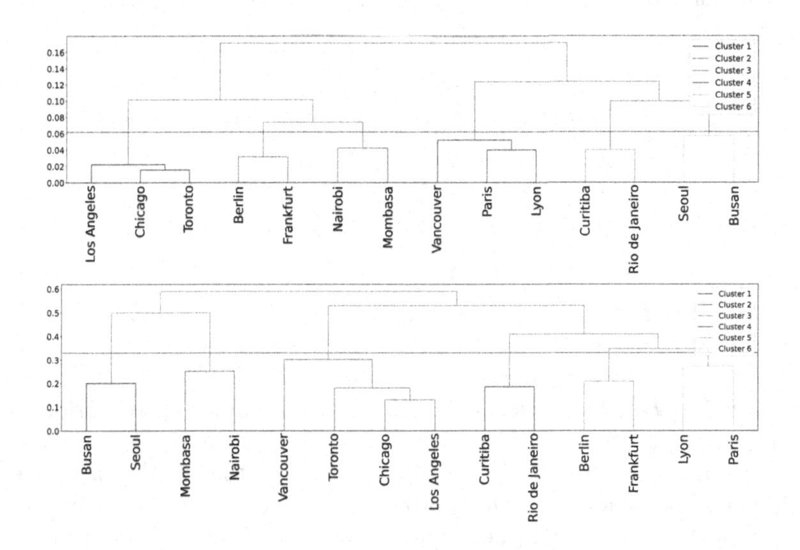

Fig. 2. Hierarchical clustering dendrogram of cities resulting from Approach-Scenes (top) and Approach-Frequency (bottom).

The result aligns with what is expected concerning the cultural characteristics of the areas studied. Most of the clusters coherently grouped cities from the same country - in general, countries have distinct cultural characteristics; the exceptions in this sense are clusters 1 and 4. In cluster 1, Toronto was grouped with Chicago and Los Angeles; note also that Los Angeles is the most dissimilar city in the grouping. The result of Chicago and Toronto being together and more similar makes sense, in that they are often considered to be culturally similar to one another, even compared to Los Angeles. Regarding cluster 4, Vancouver was grouped with Paris and Lyon. We found significant similarities between the most recurrent categories of French cities and Vancouver, such as "Art galleries," which could help explain this result. Although German cities (Berlin and Frankfurt) and French cities (Paris and Lyon) are on the same continent, they are quite distinct culturally, and so their location in separate clusters seems reasonable.

To facilitate a comparative analysis by contrasting the values of each cluster dimension with its corresponding overall average, we calculate the Z-Score, as shown in Fig. 3. The Z-Score is the number of standard deviations concerning the average of what is being observed. This facilitates comparing clusters by extracting the characteristics that stand out in each, compared with a general overview, i.e., the centroid of clusters' centroids. For example, cluster 3, representing Kenya, has one of the lowest values for Tradition. In contrast, for cluster 4 with the cities Vancouver, Paris, and Lyon, this dimension represents one of the most important characteristics. Looking at cluster 1, composed of Chicago, Los Angeles and Toronto, we see that Tradition is not as predominant as in cluster 4. This highlights the potential to identify cultural signatures and provide an overview of geographic areas by extracting their key dimensions.

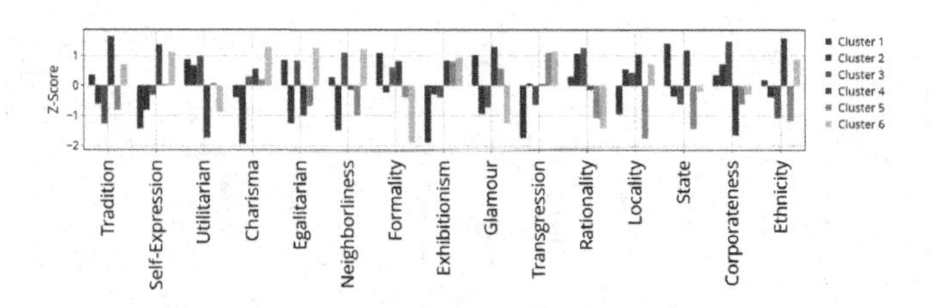

Fig. 3. Z-Score values of Scenes dimensions per cluster.

Frequency* for *Dataset Cities. For Approach-Frequency, we perform hierarchical clustering using the Complete linkage criteria and Cosine distance – the best combination tested. As depicted at the bottom of Fig. 2, the results for Approach-Frequency, as with Approach-Scenes, align with what is expected when grouping cities of the same country. However, using Approach-Frequency differently, Chicago is more similar to Los Angeles, and Vancouver is more related to Toronto than to the French cities.

The results obtained demand reflection because although Toronto and Vancouver are in the same country, they are not necessarily similar in terms of immigration patterns, governance, geography, ecology, and cultural style. Toronto and Chicago, on the other hand, have much in common: they are both Great Lakes cities, with strong industrial heritages and are now in the midst of a post-industrial transformation. Hence, they are often compared as similar cases [7,13].

We can reveal specific characteristics of each cluster by extracting the five most distinct categories for each of them – we do that by calculating the distance of the category from its cluster centroid. After that, we calculate the Z-Score for the selected categories against the overall average. The result of this process is illustrated in Fig. 4. Certain categories in some clusters stand out so notably that they not only significantly deviate from their overall average, but also emerge as the sole positive value compared to others. For example, in French cities, "municipality", in Brazilian cities, "hang gliding", and in Korean cities, "face painting" exhibits this distinct characteristic. Making a comparison with the Z-Score values illustrated in Fig. 3, we can relate these specific findings depicted in Fig. 4 to the aspects highlighted in Tradition for cluster 4 (predominantly French), Transgression for cluster 5 (Brazil) and Self-Expression and Charisma for cluster 6 (South Korea).

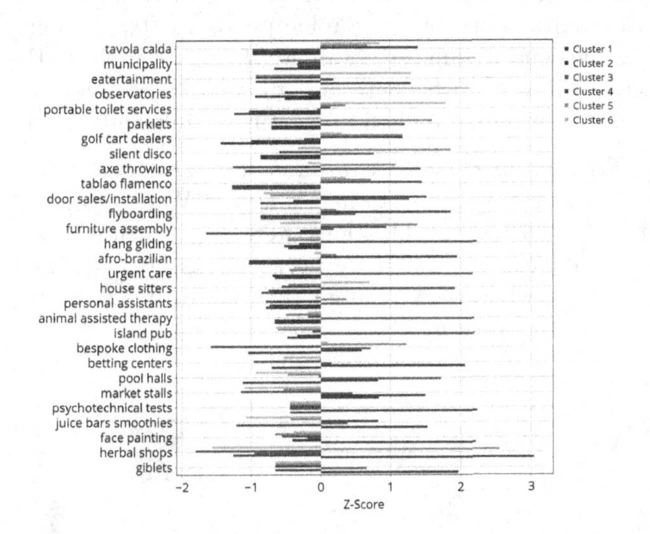

Fig. 4. Z-Score values for the most distinct categories per cluster (Approach-Frequency).

To analyze the clusters that differ between the Approach-Scenes and Approach-Frequency approaches, we examine the most evident characteristics in each. For Approach-Scenes, we focus on clusters 1 and 4, selecting the three most prominent dimensions in each and retrieving the most important sentences for those dimensions. For Approach-Frequency, we look at cluster 3 and identify the 50 most frequent categories. For example, Los Angeles, Chicago, and Toronto have "Business Consulting", "Libraries" and "Gastropubs" in common, whereas Vancouver, Paris, and Lyon are marked by "Antiques Book Store", "Art Gallery", "Comedy and Night Club" and gastronomic diversity, such as "Portuguese Bakery", "Spanish Meal Delivery", "Sushi Bars" and "Tapas Bars". In Approach-Frequency, many categories can be found that summarize these characteristics, such as "Gastropubs", "Art Installation", "Imported Food", "Meal Takeaway" and "Souvenir Shops". The result indicates that, unlike Approach-Frequency, through human knowledge in its dimensions, Approach-Scenes can detect subtle differences among categories with similar meanings.

3.2 All States in the USA

Using *Dataset States*, we apply the transfer knowledge methodology (Sect. 2.3) and create cultural signatures for all states in the country.

Evaluating *Scenes-Based Approach* for *Dataset States*. To analyze cultural signatures in this dataset using Approach-Scenes, we also perform hierarchical clustering with 15 dimensions of the Scenes Theory as features, Ward linkage criteria, and Euclidean distance. By inspecting the dendrogram, we observe a tendency to group regions by geographic proximity. By mapping one of the clearest cuts in the dendrogram, we obtain Fig. 5 (right). It shows that culturally similar regions, such as the US South, are grouped. These results reinforce the effectiveness of the proposed method in identifying culturally similar regions.

Evaluating *Frequency-Based Approach* for *Dataset States*. For this case, we perform hierarchical clustering using the Ward linkage criterion and Euclidean distance. Other combinations were experimented with, but none proved superior. We observe difference between this approach and the results obtained with Approach-Scenes. Figure 5 (left) illustrates the mapped clusters provided by *Frequency-based approach*.

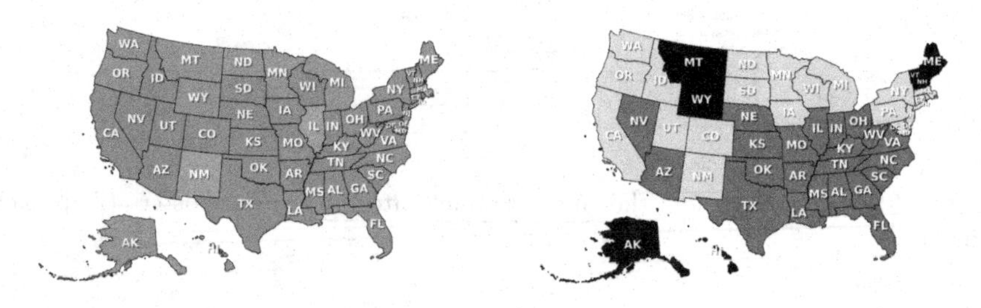

Fig. 5. Results of hierarchical clustering considering all states in the USA represented by Approach-Frequency (left) and Approach-Scenes (right).

It is not possible to detect clear patterns in the Approach-Frequency results, at least as clear as identified by Approach-Scenes, regardless of the number of clusters adopted. Surprisingly, Alaska and Maine are positioned within clusters larger than with Approach-Scenes. Alaska is situated among states such as Washington, Oregon, North Dakota, Minnesota, and Michigan. Maine is part of the largest cluster, which includes most of the remaining states. Thus, Approach-Scenes provides extra semantic expressiveness in smaller dimensions.

4 Comparing with Survey Data

There is no clear way to access the ground truth of our results. However, we explore in this work a source where we expect some correlation: the American Value Survey (AVS, access https://www.prri.org). The survey was conducted among a representative sample of 5,031 adults (age 18 and up) living in all 50 states in the United States, having a statistically valid representation of the USA population, including many minorities or hard-to-reach populations. Interviews were conducted online between September 16-29, 2021 and September 1-11, 2022.

Additional details about the methodology can be found on the Ipsos website[3]. The survey questions include political aspects and basic beliefs. We represent these questions as features to describe states, where the values are the mean answers of all participants for each state. We exclude political questions and focus solely on basic beliefs[4].

To assess the relationship between the results of the AVS and our proposals (Approach-Scenes and Approach-Frequency), we use the Pearson correlation for the Euclidean distance between all pairs of states when describing them by AVS and our approaches. By doing that, we got a moderate correlation of 0.51 ($p < 10^{-4}$) for Approach-Scenes. Using Approach-Frequency, on the other hand, resulted in a Pearson correlation of -0.06 ($p < 10^{-1}$) for the Euclidean distance between all pairs of states.

To better understand the correlation results individually we calculated the Euclidean distance of each state in comparison to all others, considering its descriptions using AVS and each of our proposals. Then, we calculate the Pearson correlation (ρ) of these values. For Approach-Scenes, $\rho \in [-0.221, 0.709]$ and approximately 75% of all states exhibit either a moderate or high correlation. Alaska is the only state with a negative correlation. By looking at the results for Approach-Frequency, with $\rho \in [-0.257, 0.149]$, it is clear that it shows a worse association with another source (AVS) regarding cultural beliefs.

5 Conclusion

In the present work, we examined data from Google Places (GP) and developed two methods to establish cultural signatures of urban areas. The proposals (Approach-Frequency and Approach-Scenes) were then assessed for their effectiveness in cities worldwide and all states in the United States. We obtained evidence that the proposed approaches, even a simple one based on frequency, could capture the cultural character of geographic areas. We gathered evidence based on a comparison with survey data that one of the approaches, based on the Scenes Theory, could capture better cultural nuances. Unlike other approaches that demand proxy data for users' preferences, e.g., user check-ins, our approach only demands simple data, i.e., categories of venues, which are easily obtainable in GP for almost any urban area. Hence, there is significant potential to utilize the proposed methodology for identifying cultural similarities between different locations. This could facilitate the development of numerous new services and applications, such as innovative location recommendation systems based on cultural criteria.

There are several ways to expand this work, such as expanding the dissimilarity analysis to both approaches, *Frequency* and *Scenes*, or testing the proposed methodology with other data sources. Since GP data is not free and acquiring a considerable amount can be costly, this could also allow for expanding the set

[3] https://www.ipsos.com/en-us/solutions/public-affairs/knowledgepanel.

[4] The complete list of questions used can be found at: https://sites.google.com/view/neighbourhood-change.

of venues. Another possibility is to evaluate different levels of granularity, such as neighborhoods and countries.

Acknowledgment. SocialNet project (process 2023/00148-0 of FAPESP) and CNPq (processes 313122/2023-7, 314603/2023-9 and 441444/2023-7).

References

1. Arribas-Bel, D., Fleischmann, M.: Spatial signatures-understanding (urban) spaces through form and function. Habitat Int. **128**, 102641 (2022)
2. Bancilhon, M., Constantinides, M., Bogucka, E.P., Aiello, L.M., Quercia, D.: Streetonomics: quantifying culture using street names. PLoS ONE **16**(6), e0252869 (2021)
3. de Brito, S.A., Baldykowski, A.L., Miczevski, S.A., Silva, T.H.: Cheers to untappd! preferences for beer reflect cultural differences around the world. In: Proceedings of AMCIS 2018, New Orleans, USA (2018)
4. Çelikten, E., Le Falher, G., Mathioudakis, M.: Modeling urban behavior by mining geotagged social data. IEEE Trans. Big Data **3**(2), 220–233 (2016)
5. Gubert, F., Silva, T.: Google places enricher: A tool that makes it easy to get and enrich google places api data. In: Proceedings of WebMedia 2022, Extended Proceedings, pp. 91–94. SBC, Curitiba, PR, Brasil (2022)
6. Hu, L., Li, Z., Ye, X.: Delineating and modeling activity space using geotagged social media data. Cartogr. Geogr. Inf. Sci. **47**(3), 277–288 (2020)
7. Kolpak, P., Wang, L.: Exploring the social and neighbourhood predictors of diabetes: a comparison between Toronto and Chicago. Prim. Health Care Resear. Devel. **18**(3), 291–299 (2017)
8. Laufer, P., Wagner, C., Flöck, F., Strohmaier, M.: Mining cross-cultural relations from wikipedia: a study of 31 european food cultures. In: Proceedings of the ACM WebSci 2015, pp. 1–10, Oxford, UK (2015)
9. Le Falher, G., Gionis, A., Mathioudakis, M.: Where is the soho of rome? measures and algorithms for finding similar neighborhoods in cities. In: Proc. of the ICWSM 2015, Oxford, UK (2015)
10. Li, J., Shang, J., McAuley, J.: UCTopic: Unsupervised Contrastive Learning for Phrase Representations and Topic Mining. In: Proc. of the ACL'22. pp. 6159–6169. ACL, Dublin, Ireland (2022)
11. Mehta, V., Mahato, B.: Measuring the robustness of neighbourhood business districts. J. Urban Design **24**(1), 99–118 (2019)
12. Reimers, N., Gurevych, I.: Sentence-bert: Sentence embeddings using siamese bert-networks. In: Proceedings of the EMNLP 2019. ACL, Hong Kong, China (2019)
13. Robson, K., Anisef, P., Brown, R.S., Nagaoka, J.: A comparison of factors determining the transition to postsecondary education in Toronto and Chicago. Res. Comp. Inter. Educ. **14**, 338–356 (2019)
14. Sen, R., Quercia, D.: World wide spatial capital. PLoS ONE **13**(2), e0190346 (2018)
15. Senefonte, H., Frizzo, G., Delgado, M., Luders, R., Silver, D., Silva, T.: Regional influences on tourists mobility through the lens of social sensing. In: Proceedings of SocInfo 2020, Pisa, Italy (2020)
16. Senefonte, H.C.M., Delgado, M.R., Lüders, R., Silva, T.H.: Predictour: predicting mobility patterns of tourists based on social media user's profiles. IEEE Access **10**, 9257–9270 (2022)

17. Silva, T.H., de Melo, P.O.V., Almeida, J.M., Musolesi, M., Loureiro, A.A.: A large-scale study of cultural differences using urban data about eating and drinking preferences. Inf. Syst. **72**, 95–116 (2017)
18. Silva, T.H., Silver, D.: Using graph neural networks to predict local culture. Environ. Planning B: Urban Analy. City Sci., 12
19. Silver, D., Silva, T.H.: Complex causal structures of neighbourhood change: evidence from a functionalist model and yelp data. Cities **133**, 104130 (2023)
20. Silver, D.A., Clark, T.N.: Scenescapes: how qualities of place shape social life. The University of Chicago (2016)
21. Skora, L.E., Senefonte, H.C., Delgado, M.R., Lüders, R., Silva, T.H.: Comparing global tourism flows measured by official census and social sensing. Online Soc. Netw. Media **29**, 100204 (2022)
22. Yan, A., He, Z., Li, J., Zhang, T., McAuley, J.: Personalized showcases: generating multi-modal explanations for recommendations. In: Proceedings of the SIGIR 2023, pp. 2251–2255. ACM, Taipei, Taiwan (2023)

Utilizing Fractional Order Epidemiological Model to Understand High and Moderate Toxicity Spread on Social Media Platforms

Emmanuel Addai, Niloofar Yousefi, and Nitin Agarwal(✉)

COSMOS Research Center, University of Arkansas – Little Rock, Arkansas, USA
{eaddai1,nyousefi,nxagarwal}@ualr.edu

Abstract. The COVID-19 pandemic has increased social media usage significantly, highlighting its critical role in public statements, information dissemination, news propagation. In this study, we construct and evaluate a fractional-order toxicity contagion model with quarantine intervention in Twitter. The model incorporates different infected groups that account for toxicity intensity and its development, that is, moderate and high infected users, and it is used to investigate the influence of each user in the overall spread of toxic content. We have evaluated the post-free toxic equilibrium point, the reproduction number ($\mathcal{R}_0$), the existence-uniqueness solution, and the stability point. The model, which fits well to (#F*covid) hashtags data, is qualitatively analyzed to evaluate the impacts of different schemes for control strategies. Our findings reveal that the conventional model, which does not differentiate between infected groups, overestimates or underestimates the rate of change in the number of infectious users, resulting in a greater error rate. From analysis, implementing quarantine measures on social media platforms can bring long-term benefits with low risk, affirming platform safety. By quarantining moderate and high toxic active users, the resulting error rates were impressively low, measured at 0.0011 and 0.0012 for the respective groups of infected users. This study will assist network providers in identifying such users, thereby reducing toxic conversations.

Keywords: toxicity spread · SEIQR epidemiological model · social media · fractional derivative

This research is funded in part by the U.S. National Science Foundation (OIA-1946391, OIA-1920920), U.S. Office of the Under Secretary of Defense for Research and Engineering (FA9550-22-1-0332), U.S. Army Research Office (W911NF-23-1-0011, W911NF-24-1-0078), U.S. Office of Naval Research (N00014-21-1-2121, N00014-21-1-2765, N00014-22-1-2318), U.S. Air Force Research Laboratory, U.S. Defense Advanced Research Projects Agency, Arkansas Research Alliance, the Jerry L. Maulden/Entergy Endowment at the University of Arkansas at Little Rock, and the Australian Department of Defense Strategic Policy Grants Program (SPGP). Any opinions, findings, and conclusions or recommendations expressed in this material are those of the authors and do not necessarily reflect the views of the funding organizations. The researchers gratefully acknowledge the support.

L. M. Aiello et al. (Eds.): ASONAM 2024, LNCS 15212, pp. 298–308, 2025.
https://doi.org/10.1007/978-3-031-78538-2_26

1 Introduction

In recent years, online social media have increasingly spread news, election results, and global outbreaks across vast networks, fostering information diffusion and social connections [1]. The spread of negative content on social media platforms has become a significant issue in today's digital age. With the increasing trend of disseminating information through online social networks, understanding the propagation of toxic content is crucial to effectively mitigate its harmful impacts. Because social media platforms have a significant impact on society [2], it has become increasingly important to gain a deeper understanding of their dynamics. For instance, to investigate the toxicity propagation online, to analyze personal attacks on social media, the authors in [3] employed machine learning methods. Alongside these techniques, several epidemiological models have been developed to analyze online toxicity propagation [4].

Recently, mathematical modeling has become crucial in understanding toxic content and information diffusion on social media, predicting how these processes unfold in a network by learning from past diffusion patterns [5]. Modeling how information or toxic content spreads is critical in stopping the spread of toxic content. Fractional-order epidemiological models have garnered considerable attention in studying disease spread. These models present a unique viewpoint by integrating fractional derivatives to capture memory effects and hereditary characteristics of systems, aspects often disregarded in the traditional models. The utilization of fractional calculus enables a more precise alignment with data and affords additional adaptability in modeling intricate systems [1].

In this work, we construct and evaluate a fractional-order toxicity model with quarantine intervention. The model incorporates different infected groups, that is, moderate and high infected users in the SEI_mI_hQR (Susceptible-Exposed-Moderate Infected-High Infected-Quarantined-Recovered) model, and it is used to investigate the influence of/on each user in the overall spread of toxic content. By dividing the Infected state into moderate and high toxicity levels within the model, we can capture the varying degrees of influence users may apply in propagating toxicity on social media platforms. High-toxic individuals might exhibit behaviors that amplify the spread of toxic content at a faster rate compared to those with moderate toxicity levels. This differentiation allows for a more accurate representation of the dynamics of toxicity propagation. We expect that the fractional-order model memory function will be able to more accurately predict both the dynamics of online behavior in the future and the dynamics of toxicity in the past. This work answers the following research questions: **RQ1:**How can a fractional-order toxicity model with quarantine intervention be constructed and evaluated to understand the dynamics of toxic content spread on social media? **RQ2:**How can the index of memory and the quarantine rate be used as preventive measures against the spread of toxic content on social media networks? **RQ3:**What are the effects of different levels of infection (moderate and high) on the overall spread of toxic content on social media platforms?

2 Related Work

The surge of toxicity on social media has become a major challenge, prompting research into its dynamics, impacts, and solutions. Studies have used machine learning to detect toxic content like hate speech [6], examined the impact of toxicity on Reddit discussions [7,8]. Additionally, research has explored the effects of toxicity on user engagement and retention in online communities [9].

2.1 Epidemiological Modeling with Quarantine State

Epidemiological models for online toxicity liken the spread of toxic content to infectious diseases, assuming toxic behavior can 'infect' users. Researchers use these models to forecast toxic users' behavior [10]. For instance, authors in [4] compare different epidemiological models to study the propagation of toxicity.

Incorporating a quarantine state into epidemiological models significantly enhances our understanding and management of infectious diseases, especially in light of recent global health crises. The inclusion of quarantine—both voluntary and enforced—reflects a critical aspect of disease control strategies. This can significantly alter the disease dynamics, potentially leading to a slower spread of the disease, lower peak prevalence, and ultimately fewer cases. This control strategy has been used to study many diseases such as EBOLA [11] and COVID-19 [12]. It has also been adopted in online social networks; for instance, the authors in [13] investigated effective isolation-based strategies for controlling information spread in social networks. In [14], the authors reviewed models, methods and applications where quarantine control strategy were utilized.

2.2 Fractional Calculus

Fractional-order derivatives are a mathematical tool used to analyze phenomena exhibiting non-linear and complex dynamics, such as the spread of toxicity on social media. Unlike classical integer-order derivatives that describe rates of change in constant time intervals, fractional derivatives provide a more nuanced representation by accounting for memory and hereditary properties of processes [15]. This means they can model the way past interactions and behaviors influence the current spread of toxic content.

Fractional-order derivatives enable models that more accurately capture the intricacies of how toxic behaviors evolve over time on social media. These models can consider the long-range dependencies and varying intensities of interactions among users, which are characteristic of social media dynamics. Fractional derivatives enable the creation of sophisticated models to better capture the complexities of toxicity spread, enhancing strategies for monitoring, predicting, and mitigating harmful content on social platforms.

3 Methodology

This section provides an overview of data collection and the methodology used in this paper.

3.1 Fractional Model Formulation (Caputo Derivative)

Previously, $SEIQR$ (Susceptible-Exposed-Infected-Quarantined-Recovered) was used to study an epidemiological model to contain the spread of toxicity using memory-index [16]. In this work, we propose the SEI_mI_hQR model in the Caputo fractional operator sense to study toxicity spread on social media, which is the extension of the SEIQR model. The proposed SEI_mI_hQR model to study toxicity provides powerful tools for understanding and predicting the spread of toxic contents, especially in scenarios where quarantine is the key strategy for network providers' control. We make several assumptions for the SEI_mI_hQR model: the total number of users is not constant, considering recruitment rate (Π) and autonomous exit rate (μ); the users who spread toxic contents are moderate (I_m) and high infected users (I_h); moderate infections (I_m) can become high infections (I_h) and vice versa at rates κ_m and κ_h, respectively; both moderate and high infections can autonomously recover at rates ϕ_m and ϕ_h, respectively; and recovered users become susceptible again at the rate η.

Therefore, we propose Caputo fractional-order differential equations. For the entire population, we define the quantity N:

$$N(t) = S(t) + E(t) + I_m(t) + I_h(t) + Q(t) + R(t). \tag{1}$$

Hence, we have

Our propose Caputo fractional-order derivative is given as;

$$\begin{cases} {}^CD_t^\alpha S(t) = \Pi - (\lambda_{mh} + \mu)S + \eta R, \\ {}^CD_t^\alpha E(t) = \lambda_{mh} S - (\mu + \psi_m + \psi_h)E, \\ {}^CD_t^\alpha I_m(t) = \psi_m E + \kappa_h I_h - (\mu + \phi_m + \theta_m + \kappa_m)I_m, \\ {}^CD_t^\alpha I_h(t) = \psi_h E + \kappa_m I_m - (\mu + \phi_h + \theta_h + \kappa_h)I_h, \\ {}^CD_t^\alpha Q(t) = \theta_m I_m + \theta_h I_h - (\gamma + \mu)Q, \\ {}^CD_t^\alpha R(t) = \gamma Q + \phi_m I_m + \phi_h I_h - (\mu + \eta)R, \end{cases}$$

where $(\lambda_{mh}) = \frac{\beta(I_m+I_h)}{N}$, $t \in [0, \mathcal{T}]$, $\mathcal{T} \in \mathbb{R}$, and ${}_0^CD_t^\alpha$ denotes the Caputo fractional derivative of order, where the memory index is denoted as α, $0 < \alpha \leq 1$, given the initial conditions:

$$S(0) = S\_0 \geq 0, \quad E(0) = E\_0 \geq 0, \quad I_m(0) = I\_m0 \geq 0,$$
$$I_h(0) = I\_h0 \geq 0, \quad Q(0) = Q\_0 \geq 0, \quad R(0) = R\_0 \geq 0.$$

The flow diagram of the model is presented in Fig. 1, while the description of the parameters is presented in Table 1. This table also includes the values that will be used later in our numerical simulations.

It is commonly known that Caputo's derivative offers greater dependability and adaptability in analytical applications. Due to initial condition properties, which are more physically interpretable for most problems, many researchers consider the Caputo operator over other derivatives.

Table 1. Explanation of the model parameters.

Parameter	value	Explanation
Π	100	recruitment rate of human
β	0.0006	effective contact rate
ψ_m	0.009	the rate at which exposed become moderate infected
ψ_h	0.006	the rate at which exposed become high infected
θ_m	0.0067	the rate at which I_m transfer to quarantine class
θ_h	0.017	the rate at which I_h transfer to quarantine class
η	0.04	the rate at which R transfer S
γ	0.002	the rate at which Q transfer to recovery
ϕ	0.1	the rate at which I transfer to recovery
μ	0.1	the rate at which people exit autonomously
κ_m	0.09	the rate at which I_m transfer to I_h
κ_h	0.01	the rate at which I_h transfer to I_m

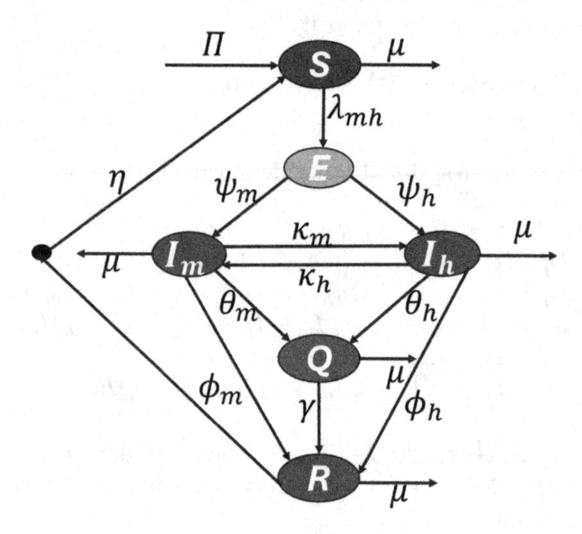

Fig. 1. Transfer diagram for the toxicity spread.

3.2 Data Collection

The Twitter Academic API and the hashtag #Fcovid were used to collect 6,766 COVID-19-related tweets from February 1st, 2020, to December 31st, 2020. A weighted average toxicity of 0.91 was used as a threshold: posts scoring below 0.91 (2,082 posts) were considered moderate, while those scoring above 0.91 (4,684 posts) were deemed highly toxic. To identify the most active users among the infected, we focus on the retweets of posts shared by individuals within the subset of high and moderate toxic users. A higher number of retweets indicates active engagement with the post.

4 Model Analysis

In this section, we conduct the qualitative analysis of the proposed model.

4.1 Basic Reproduction Number

The basic reproduction number represented as $\mathcal{R}_0$, is described as the average number of secondary infections generated by one infected individual within a completely susceptible population. The primary application of $\mathcal{R}_0$ lies in assessing whether a new infectious disease has the potential to proliferate among a population. For the SEI_mI_hQR model, the basic reproduction number $\mathcal{R}_0$ is

$$\mathcal{R}_0 = \frac{\psi_m \beta \mathcal{A}}{\mu(\mu+\psi_m)(\mu+\phi_m+\theta_m)} + \frac{\psi_h \beta \mathcal{A}}{\mu(\mu+\psi_h)(\mu+\phi_h+\theta_h)}.$$

4.2 Existence-Uniqueness and Stability Analysis

This section examined the existence-uniqueness and stability concept of our proposed model. We study the following Theorems to achieve this goal.

Theorem 1. *The Caputo fractional toxicity spread model has a unique solution under the condition that*

$$\frac{\mathcal{T}^{\alpha}}{\Gamma(\alpha)} \mathcal{L}_i < 1, \;\; i = 1, 2, ..., 5$$

when $t \in [0, \mathcal{T}]$ *and* $\mathcal{L}_i$ *satisfy the Lipschitz condition.*

Proof. From Lemma 0.1 and applying the proof from Theorem 7 of [17], our proposed toxicity spread model exists and has a unique solution.

Theorem 2. *The Caputo fractional toxicity spread model is Hyer-Ulam stable, if there exists*

$$\frac{\mathcal{T}^{\alpha}}{\Gamma(\alpha+1)} \mathcal{L}_i < 1, \;\; i = 1, 2, ..., 5$$

Proof. From Lemma 0.1 and applying the proof from Theorem 6.2 of [15], our proposed toxicity spread model is Hyers-Ullam stable.

5 Model Parameterization and Data Fitting Analysis

The least-squares technique is the most commonly used method to estimate parameter values. It is defined as:

$$E_{-\mathrm{rel}} = \frac{\|I_{est}(t_i) - I_{data}(t_i)\|_2}{\|I_{data}(t_i)\|_2}. \tag{2}$$

The relative error 2 is applied to compute the model's error. We rank users according to the average toxicity score of posts and the number of retweets they received. Tables 2 and 3 show users with higher retweet counts and moderate to high average toxicity scores in their posts. These tables indicate the model's error by systematically removing each user one by one. We removed five users based on their retweet counts and average toxicity scores for both moderate and high toxicity groups. The error remains consistent after the removal of 5th user. We aim to minimize user removal while achieving optimal results.

Table 2. Examining the error impact of removing posts from moderate most active and most toxic users.

User	$Error_r$	Average toxicity score	Number of Retweet Received
1st	0.0031	0.883	134
2nd	0.0027	0.862	75
3rd	0.0021	0.881	27
4th	0.0011	0.873	10
5th	0.0011	0.858	8

Table 3. Examining the error impact of removing posts from high most active and most toxic users.

User	$Error_r$	Average toxicity score	Number of Retweet Received
1st	0.011	0.998	1489
2nd	0.009	0.939	227
3rd	0.0073	0.956	67
4th	0.0012	0.997	67
5th	0.0012	0.997	44

6 Numerical Simulation and Discussion

This section presents numerical results showing system trajectories with varying input parameters. We examine different scenarios to understand the system's behavior under diverse conditions. Our analysis of these numerical data is focused on highlighting the system's adaptability and vulnerability to various input changes. For our numerical analysis, it is important to acknowledge that certain assumptions have been made regarding the initial conditions of the state variables ($S = 12000; E = 11000;\ I_m = 3000; I_h = 3000; Q = 7000; R = 2000$) and system parameters (refer to Table 1).

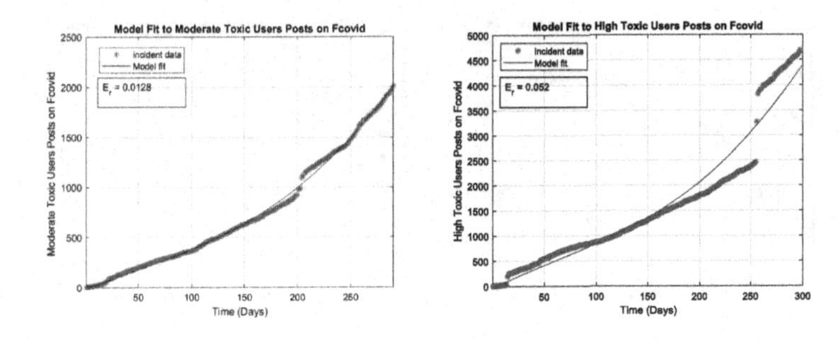

Fig. 2. Fitted SEI_mI_hQR model for moderate/high toxic users posts.

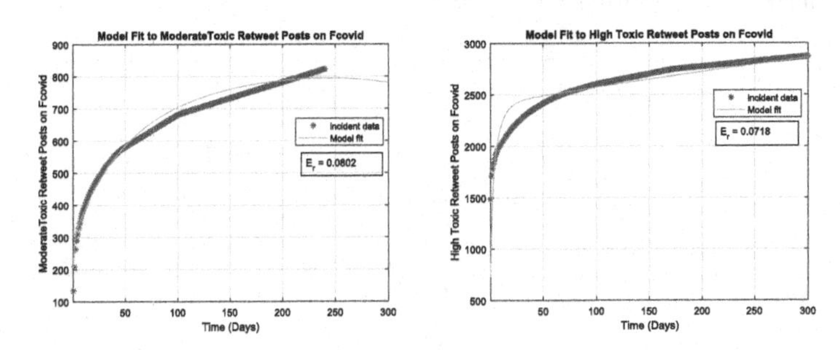

Fig. 3. Fitted SEI_mI_hQR model for moderate/high toxic users with high retweets received.

Simulations were conducted to visualize the SEI_mI_hQR model under varying parameters. By plotting the trajectories of toxicity spread across all the compartments over time, we can understand the progression of toxic posts comprehensively. First, from Figs. 2, 3 and 4, the data is observed to depict the toxicity spread. The data is compared with the proposed model to check the accuracy of the proposed model. Figure 2 shows cumulative moderate toxic posts against the proposed model's moderate infected (I_m) with an observed error of 0.0128. similarly, for high toxic posts with an observed error of 0.052. Figure 3, provides insights into how the model fits the retweet activity of moderate and high toxic users, the lower error of 0.0802 and 0.0718, respectively.

In Fig. 4, our proposed model, which involves transferring users with high and moderate average toxicity, the top five users were removed from moderate and high infected and transferred to quarantine. The resulting errors were 0.0011 and 0.0012, respectively. This result provides evidence that our proposed model, involving grouping the infected into moderate and high and then transferring some of the users to quarantine, yields superior outcomes in error reduction.

The plots in Fig. 5 present simulations of the memory effect using the Adams–Bashforth–Moulton Predictor–Corrector method. Figure 5a and 5b shows the impact of fractional parameters on the Susceptible and Exposed classes. This behavior indicates a large influx of new users and high exposure due to the platform's fractional operator and toxic nature. In Fig. 5c and Fig. 5d, we illustrate

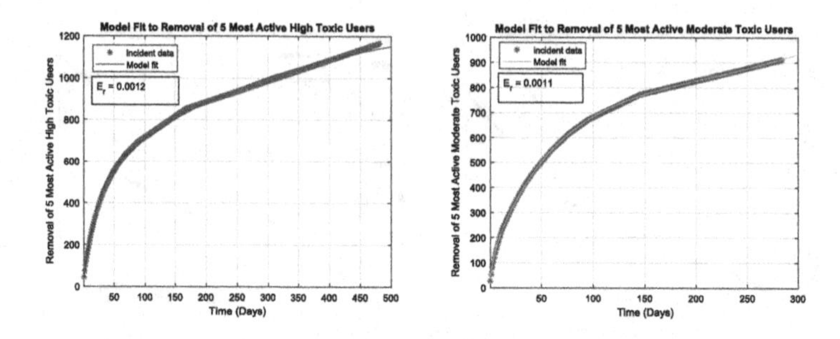

Fig. 4. Fitted SEI_mI_hQR model for the top 5 high/moderate most active and toxic users transferred to quarantine.

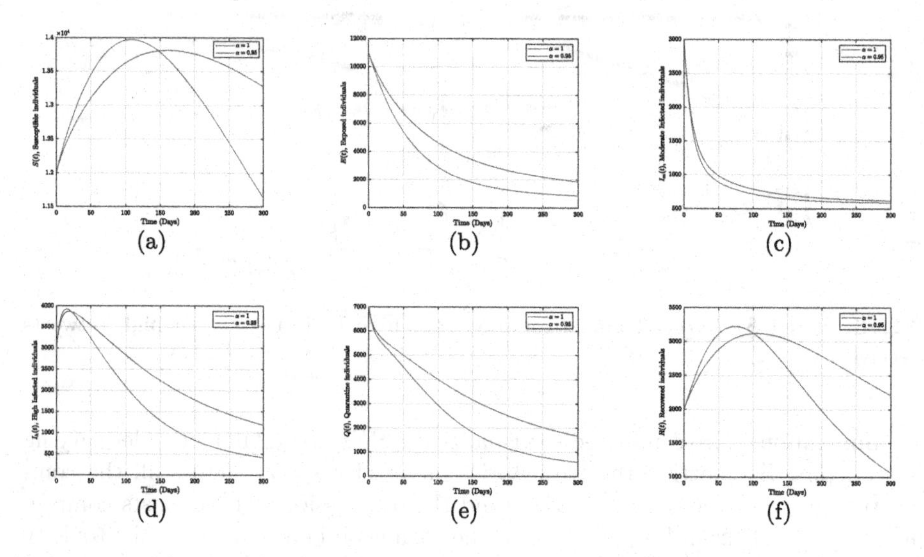

Fig. 5. Numerical dynamics on SEI_mI_hQR model at different fractional order α.

the moderate and high infected classes, respectively, and begin to increase them by lowering the fractional values. Figure 5e shows increased dynamics due to the number of infected individuals increasing, as depicted in Fig. 5c and 5d, and hence a high number of users to quarantine. In Fig. 5f, we observed that the model captured high recovery due to interactions among the network's toxic users and the introduction of quarantine as a control intervention.

7 Conclusion

In this study, we investigated toxicity spread on social media platform using the SEI_mI_hQR epidemiological model, taking social network service provider intervention into account and quarantining users, meanwhile, considering the index of memory nature of parameters of the model, which makes it more appropriate to the real situation. The post-free toxic equilibrium point exists for the model.

The reproduction number ($\mathcal{R}_0$) is determined using the next-generation matrix method, and the existence-uniqueness, as well as stability, are also analyzed using fractional operator techniques. Our findings from this study reveal that the conventional model, which does not differentiate between infected groups, either overestimates or underestimates the rate at which individuals move from the Exposed compartment to the Infected and then to the Quarantine during the rising or falling phases of the number of infectious users, respectively.

Implementing quarantine measures by network service providers can be highly beneficial, showing a low level of risk and affirming platform safety. Quarantining toxic and active users resulted in impressively low error rates of 0.0011 and 0.0012 for the moderate and high toxic users, respectively. Furthermore, the index of memory and the quarantine rate can be used as preventive measures for the toxicity spread on social network. Future research will use different data and methods to improve this model and explore additional control measures.

References

1. Dai, L., Liu, X., Chen, Y.: Global dynamics of a fractional-order SIS epidemic model with media coverage. Nonlinear Dyn. **11**, 19513–19526 (2023)
2. Shajari, S., Alassad, M., Agarwal, N.: Characterizing suspicious commenter behaviors. In: International Conference on Advances in Social Networks Analysis and Mining (2023)
3. Wulczyn, E., Nithum, T., Dixon, L.: Ex machina: personal attacks seen at scale. In: The 26th International Conference on World Wide Web (WWW 2017), pp. 1391–1399 (2017). https://doi.org/10.1145/3038912.3052591
4. N. Yousefi, N., Agarwal, N.: Study the influence of toxicity intensity on its propagation using epidemiological models. In: The Proceeding of 30th Americas Conference on Information Systems (2024)
5. Bettencourt, L.M., Cintrn-Arias, A., Kaiser, D.I., Castillo-Chvez, C.: The power of a good idea: quatitative modeling of the spread of ideas from epidemiological models. Phys. A **364**, 513–536 (2006)
6. Yousefi, N., Noor, N. B., Spann, B., Agarwal, N.: Towards developing a measure to assess contagiousness of toxic tweets. In: Workshop Proceedings of the 17th International AAAI Conference on Web and Social Media, TrueHealth 2023: Workshop on Combating 13 Health Misinformation for Social Wellbeing, 5-8 June, 2023, Limassol, Cyprus, p. 43. (https://doi.org/10.36190/2023.43)
7. Yousefi, N., Noor, N. B., Spann, B., Agarwal, N.: examining toxicity's impact on reddit conversations. In: Cherifi, H., Rocha, L.M., Cherifi, C., Donduran, M. (eds.) Complex Networks & Their Applications XII, pp. 401–411. Springer Nature Switzerland, Cham (2023). (https://doi.org/10.1007/978-3-031-53503-1_33)
8. Falade, T. C. C., Yousefi, N., Agarwal, N.: Toxicity prediction in reddit. In: The Proceeding of 30th Americas Conference on Information Systems (2024)
9. Duggan, M.: Online Harassment (2023) Accessed 24 Dec 2023
10. Yousefi, N., Agarwal, N., Addai, E.: developing epidemiological models with differentiated infected intensity. In: 17th International Conference on Social Computing, Behavioral- Cultural Modeling & Prediction and Behavior Representation in Modeling and Simulation (2024) (In press)

11. Ngwa, G. A., Teboh-Ewungkem, M. I.: A mathematical model with quarantine states for the dynamics of ebola virus disease in human populations. Comput. Maths Methods Med., 9352725 (2016)
12. Alsheri, A.S., Alraeza, A.A., Afia, M.R.: Mathematical modeling of the effect of quarantine rate on controlling the infection of COVID-19 in the population of Saudi Arabia. Alexandria Eng. J. **61**(9), 6843–6850 (2022)
13. Li, T., Wang, S., Li, B.: Research on suppression strategy of social network information based on effective isolation. Proc. Comput. Sci. **131**, 131–138 (2018)
14. Chen, B.L., et al.: Influence blocking maximization on networks: models, methods and applications. Phys. Rep. **976**, 1–54 (2022)
15. Zhang, N., Addai, E., Zhang, L., et al.: Fractional modeling and numerical simulation for unfolding Marburg-monkeypox virus co-infection transmission. Fractals **31** (2350086), (2023). https://doi.org/10.1142/S0218348X2350086X
16. Addai, E., Yousefi, N., Agarwal, N.: SEIQR: an epidemiological model to contain the spread of toxicity using memory-index. In: Fifth International Workshop on Cyber Social Threats, International Conference on Web and Social Media (2024) (In press)
17. Addai, E., Ngungu, M., Omoloye, M.A., Marinda, E.: Modelling the impact of vaccination and environ-mental transmission on the dynamics of monkeypox virus under Caputo operator. In: AIMS MBE (2023)

From Retweets to Follows: Facilitating Graph Construction in Online Social Networks Through Machine Learning

Anahit Sargsyan(✉) and Jürgen Pfeffer

School of Social Sciences and Technology, Technical University of Munich, Munich, Germany
{anahit.sargsyan,juergen.pfeffer}@tum.de

Abstract. Online social networks (OSNs), such as Twitter and Facebook, enable users to create, share, and interact with diverse content, thereby producing intricate pathways for information propagation. This flow, which can be modeled through graphs that capture Follower/Following relationships and various interactions such as retweets and mentions, can offer valuable insights into the dynamics of online social behavior and information sharing. While the Follower/Following networks are important for modeling user characteristics and behaviors, their construction can prove expensive in terms of both time and resources. More importantly, in some OSNs, partial or full restrictions have been posed on the access to users' Follower/Following information, effectively rendering the regular construction process of Following graphs intractable. In this paper, we explore the viability of extracting users' Following connections from their Retweet/Mention networks through predictive models. Taking Twitter as a case study, we train and contrast the performance of five different models, including classical Machine Learning (ML) methods as well as a recently developed Deep Learning (DL) approach, on two different datasets. The difference in prediction results across the models and datasets is traced and analyzed. Lastly, we round up the contributions by providing a carefully curated Twitter dataset compiled from over 9,000 individuals' timelines, encapsulating their retweets, followers, and following networks. Taken together, the results and findings featured herein can aid in paving the way for improved understanding and modeling of online social networks.

Keywords: Online Social Networks · Follow Graph · Link Prediction · Machine Learning · Egocentric Networks

1 Introduction

Online social platforms, such as Twitter (currently X) and Facebook, allow users to form Follower/Following connections, generate content on their timelines, and interact with content generated by other users, e.g., by liking, mentioning,

L. M. Aiello et al. (Eds.): ASONAM 2024, LNCS 15212, pp. 309–320, 2025.
https://doi.org/10.1007/978-3-031-78538-2_27

sharing, commenting, and retweeting. This enables the content to spread from one user to another via different pathways. Each of these interactions can then be used to build graphs that model relationships between users (e.g., the graphs portrayed in Fig. 1) or between a user and content. Studying these graphs could help model and predict the evolution of OSNs from both social and information point of view. The established models, in turn, can be leveraged to improve the current systems and develop new applications in OSNs, assist advertisers and marketers in designing more effective campaigns, and enrich user behavior analysis.

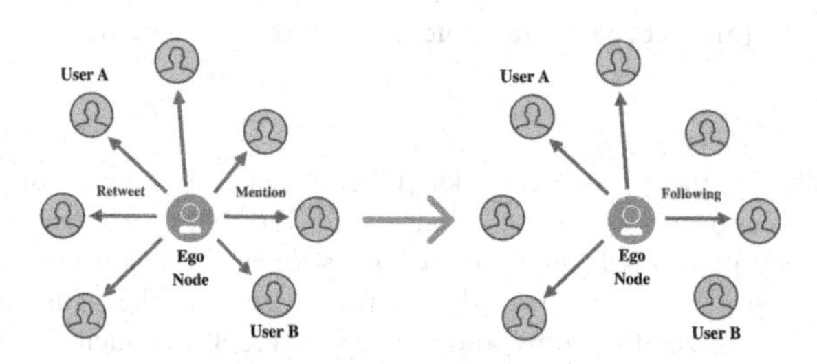

Fig. 1. An abstract illustration of the studied problem of inferring the ego user's Following connections (on the right) from their Retweet/Mention Network (on the left).

Among various OSNs, Twitter stands out for its fast-paced nature, massive user base, and unique format of communication through short, public messages. With millions of active contributors ($\approx$ 40 M.) and hundreds of millions of tweets ($\approx$ 375 M.) posted daily [15], Twitter has become a valuable asset for researchers and analysts seeking to understand social dynamics, information dissemination, and user behavior in online spheres. To this end, prior works have extensively analyzed the structural and topological properties of Follow, Retweet, and Mention graphs on Twitter, revealing interesting insights into connections/correlations between them. In [13], Myers et al. examined the Twitter Follow graph's topological properties, concluding that while the Follower/Following relationship is primarily associated with *information consumption*, it often indicates connections rooted in *social ties*. Further investigations into the Twitter ecosystem have revealed that, as compared to the Follow graph, the Retweet graph can effectively capture *genuine interest and trust connections* among users [4]. Complementing the above two studies, Amati et al. [2] explored the Twitter Mention graph and provided a qualitative and quantitative comparison between the three graphs. As transpired from the analysis, the Mention graph tends to better capture the information spreading on the network and, perhaps more interestingly, can provide a quantitative assessment of the *actual strength* of a Follow relationship.

Being the most natural and intuitive, Follower/Following networks have attracted considerable interest in the literature. Besides influencing the information flow within OSNs, these networks can reveal valuable insights into community structures, influence dynamics, and user preferences. In fact, the graphs constructed from Follower/Following connections have been utilized for enhancing the prediction of users' political preferences [6], fake news identification [18], early prediction of virality of tweets during incidents [21], to name a few.

Despite their utility and potential, however, Follow networks demand *significantly more time and effort* to be constructed in comparison to Retweet or Mention networks. Particularly, unlike the latter two, which can be built without relying on API access, the compilation of Follow networks involves recursive calls, and the number of users returned in each response is typically restricted. Furthermore, in some OSNs, access to users' Follower/Following connections has been artificially *limited or disabled completely.* Case in point, currently in Twitter, in addition to restricted and paywalled access to APIs, limitations have also been imposed on viewing the Follower/Following connections.

In response to these obstacles, and in an effort to facilitate the construction of large-scale Follow networks in OSNs, this paper seeks to answer the question whether information from users' retweets and mentions can be leveraged to accurately predict their Follow connections. Taking Twitter as a case study, we train and evaluate the performance of five different predictive models on two different datasets. The difference in prediction results across the models and datasets is traced and analyzed. In summary, the present work complements and advances the existing research as elaborated below.

First, we propose to extract (infer) users' Following network from their Retweet/Mention network (i.e., a combination of Retweet and Mention graphs) through predictive models. We cast this problem both as a binary classification task and as an edge classification problem within a graph. Five different predictive models are evaluated, and their performance comparison is provided. The results demonstrate that an ensemble tree-based classifier can achieve an average accuracy and an F1-score of nearly 90% for certain network types, yet its performance may significantly vary depending on the degree of the input and output networks. Specifically, the findings suggest a positive correlation between the prediction accuracy and the degree of Retweet/Mention network, whereas with respect to the degree of the Following network the opposite trend is observed.

Second, we provide a Twitter dataset compiled from around 9,000 individuals' timelines, including their retweets, followers, and following networks. The *dataset*, *trained models*, and the complete source code to reproduce the conducted analysis can be accessed online at http://bit.ly/3y5zaRg.

The roadmap of this paper is as follows. In Sect. 2, we briefly survey the related literature on link prediction methods in social networks. Section 3 provides a formal description of the problem and lays out the details of the two considered datasets. In Sect. 4, we overview the employed candidate predictive models, and in Sect. 5 report the results of their performance comparison. Lastly, Sect. 6 concludes the paper with a discussion and an outlook on future work.

2 Related Work

In what follows, we briefly survey the existing studies on link prediction methods, whereas Table 1 provides an overall comparison between the present work and the literature reviewed. For a more exhaustive review on link prediction in social networks, we refer the reader to [10,12].

Prediction of connections in OSNs has been widely studied within the framework of the classical link prediction problem (LPP) [11]. In LPP, given the current snapshot of a social network, the objective is to predict the missing or future links. On the other hand, here, the goal is to infer a different type of link from the current graph (i.e., extracting the Following relationships from a Retweet/Mention graph). This problem can be modeled as an LPP on a heterogeneous graph provided one has partial information regarding Follow connections, whereas in the current setting, we assume a complete absence of the latter information.

Table 1. A comparative summary of present work versus prior studies on link prediction in social networks.

	Objective	Network Type	Graph Structure	Approach	Application Scope
Liben-Nowel et al. [11]	Link prediction	Evolving	General	Structural similarity based	Social networks
Ahmad et al. [1]	Link prediction	Evolving	General	Structural similarity based	Social networks
Perozzi et al. [14]	NRL	Static	General	Random walks	Social networks
Grover et al. [7]	NRL	Static	General	Second-order random walks	General networks
Jin et al. [9]	Link prediction	Static	General	Generative Adversarial Network based	General networks
Hours et al. [8]	Link prediction	Evolving	General	Structural and contextual similarity based	OSN
Zhang et al. [22]	Link prediction	Evolving	General	GNN with network representation embeddings	General networks
Behera et al. [3]	Follower link prediction	Evolving	General	XGBoost	Social networks
Toprak et al. [20]	Link recommendation	Evolving	Egocentric	Social circle-aware and structural similarity based	OSN
Present work	Following link prediction	Static	Egocentric	XGBoost	OSN

The domain of link prediction in social networks has garnered a rich arsenal of techniques and approaches, ranging from topological metric-based analyses to Machine Learning models. Early research, such as the work by Liben-Nowell et al. [11], explored classical topological metrics, including common neighbors, Jaccard's coefficient, and Adamic-Adar for link prediction in co-authorship networks. Following this direction, Ahmad et al. [1] proposed a link prediction algorithm named Common Neighbor and Centrality-based Parameterized algorithm, which can suggest the formation of new links in complex networks. Hours et al. [8] analyzed a large Twitter dataset aiming to predict mention links among users by combining contextual and local structural features.

While topological metrics can offer a simplistic yet insightful way to predict links, there has been a shift towards machine learning-based methods. One subdirection in this line of research employs network representation learning (NRL) based algorithms for link prediction, such as DeepWalk [14], Node2vec [7], and embedGAN [9]. Zhang et al. [22] proposed a Graph Neural Network (GNN) based link prediction framework, referred to as SEAL, that can jointly learn from three types of features (subgraphs, embeddings, and attributes). Praznik et al. [17] applied the SEAL framework to three sets of Twitter data to predict future hashtag co-occurrences. Focusing on follower connections, Behera et al. [3] proposed a link prediction model based on XGBoost classifier. The authors included features such as Katz centrality, Page Ranking for the nodes, and profile-related features. Leveraging social cognitive theories, Toprak et al. [20] proposed supervised and unsupervised social circle-aware link prediction methods for egocentric graphs. The method has been validated on two Twitter datasets comprising a community of video gamers and generic users against several benchmarks, including SEAL and Node2vec.

3 Problem Statement and Datasets

Recall that given a user's Retweet/Mention network, the problem at hand seeks to extract the corresponding Following connections. More formally, let $G_{\mathrm{RT}} = (V, E)$ be a directed parameterized graph denoting the given Retweet/Mention network of an ego user, with V denoting the set of all users (i.e., the ego user and those whom the ego user mentioned or retweeted) and E denoting the connecting directed edges. Note that in our case $|V| = |E|+1$ as the graph G_{RT} is egocentric. Every vertex $v \in V$ is associated with a set f^v of n *user-level features* $f^v \triangleq \{f_1^v, f_2^v, ..., f_n^v\}$ and every edge $e \in E$ is characterized by a set k^e of m interaction related features $k^e \triangleq \{k_1^e, k_2^e, ..., k_m^e\}$ which are detailed in the paragraphs to follow. The objective is to infer a direct mapping $G_{\mathrm{RT}} \rightarrow G_{\mathrm{F}} = (V, \tilde{E})$, where G_{F} defines the corresponding network of Following relationships of the ego user and $\tilde{E} \subseteq E$.

Alternatively, as previously noted, one can formulate the above problem as a binary classification task where the input is represented by a real-valued vector formed by concatenating the user-level, edge-level, and graph-level features. To formalize, let $X \in \mathbb{R}^{2n+m+3}$ denote the input vector constructed by concatenating f^v of the ego-user with that of the alter, along with the corresponding edge-level features k^e as well as the following three additional graph-level features: (1) *ego_degree*: Degree of G_{RT}; (2) *avg_tweet_impression_count*: Average impression count of ego-user's tweets within G_{RT}; (3) *avg_alter_tweet_count*: Average number of tweets of alters within G_{RT}.

With the above notation, the problem then translates into learning a mapping of the form $X \rightarrow Y \in \{0, 1\}$, where the prediction target Y denotes the absence or presence of a Following connection. With this framing, the problem can be tackled by standard ML classification methods, which are discussed in Sect. 4.

3.1 Datasets

As previously noted, to provide a more comprehensive comparison of the selected predictive models (explained in Sect. 4), we test their performance on two different datasets, TwiBot–22 [5] and our manually compiled data corpus, termed CodeSwitchNet. The key difference between these two datasets lies in the approaches to data collection and the intended application, as elaborated below.

TwiBot–22: TwiBot-22 dataset [5] was collected primarily for evaluating and enhancing bot detection algorithms on Twitter. The data collection was conducted in two stages, from January 2022 to March 2022. A rich collection of 1,000,000 user profiles was collected, including bots and genuine users, along with their associated tweets and various relationships, such as followers, following, likes, and retweets. Though these relationships are present, they are not exhaustive, offering a representative rather than a comprehensive view.

CodeSwitchNet: Arabizi (transliterated Arabic) seed words were used to collect tweets from 2020 to 2023 using Twitter's Academic API v2 [16]. From over 1,000,000 users that were in the dataset, we randomly selected 10,000 users who used code–switching and, more precisely, used a mixture of Arabic, English, French, and Arabizi to collect their timelines, Follower, and Following networks. Out of the selected users, 9,155 were public and accessible at the time of data collection which took place from May 2023 to June 2023. Due to the constraints imposed by Twitter API, up to 3,000 latest tweets were collected from each user's timeline, based on which the corresponding Retweet/Mention networks were constructed.

3.2 Data Preprocessing

For data consistency, the users flagged as bots were removed from Twibot-22, whereas for CodeSwitchNet, no user filtering steps were taken in this regard. Table 2 provides further details regarding the two datasets after preprocessing.

Table 2. Statistics of the two datasets.

	Number of egos	Number of nodes	Number of edges	Median G_{RT} degree	Median G_F degree
CodeSwitchNet	9,155	573,075	2,169,388	205	23
TwiBot–22	7,849	350,500	1,337,517	142	40

As seen from Table 2, the average size of ego networks varies considerably among the two datasets, which stems from the nature of data collection for each. Specifically, CodeSwitchNet encapsulates the complete network of the users at the time of data collection, as opposed to TwiBot–22, which captures only a

Table 3. Node and edge level features.

User-level	Edge-level
f_1^v: Account age in years	k_1^e: Avg. no. of retweets per interaction
f_2^v: Followers to Following ratio	k_2^e: Avg. no. of replies per interaction
f_3^v: Total no. of tweets	k_3^e: Avg. no. of quotes per interaction
f_4^v: Account is verified	k_4^e: Avg. no. of likes per interaction
f_5^v: G_{RT} degree	k_5^e: Avg. no. of impressions per interaction
f_6^v: No. of lists the ego has been added to	k_6^e: No. of tweets with the alter mentioned
	k_7^e: No. of times ego retweeted alter
	k_8^e: Avg. no. of tweets by ego in G_{RT}

subset of the connections. Another major difference between these two datasets is related to the median degree of ego users' Following graphs (i.e., G_{F}), which in TwiBot–22 is nearly double the value in CodeSwitchNet. In contrast, the median degree of Retweet/Mention graphs (i.e., G_{RT}) is higher in CodeSwitchNet than in TwiBot–22. These potentially indicate the presence of different Following patterns in these two datasets.

To create the networks for the users, we used their timelines to extract each Twitter handle appearing in a tweet. Each occurrence of a handle can be either a retweet or a mention, hence the graph naming. We further extracted features describing both the profile of a user and the interaction between ego and alter. Specifically, six user level features ($f_1^v, f_2^v, ..., f_6^v$) and eight edge-level features ($k_1^e, k_2^e, ..., k_8^e$) were considered as listed in Table 3. The features k_1^e, k_2^e, k_3^e, k_4^e and k_5^e were not available in TwiBot–22, therefore the predictive models were trained on each dataset separately.

4 Predictive Models

The pool of candidate approaches selected for evaluation involves two categories of models, which are discussed below. For each model, the same split of data, 80%-20%, was used for training and testing, ensuring that all ego networks were present in both sets. The results reported in Sect. 5 are based on the test set.

Statistical ML Models: From this category, we consider both standard classification models, such as Logistic Regression, distance-based method, such as k-Nearest Neighbors (kNN), and two ensemble tree-based models, namely Random Forest (RF) and XGBoost. (kNN) was trained with $k = 11$ with the distance metric set to Euclidean. For RF, the number of estimators was kept at the default value of 100, while the maximum depth was adjusted to 50. For XGBoost, the number of estimators was increased to 500, maximum depth and learning rate were set to 75 and 0.01, respectively.

Graph-Based DL Approach: Directed Graph Convolutional Network (DGCN) [19] was selected, which is tailored to leverage both first-order and second-order proximity information between nodes. Given our focus on direct connections, a single layer of convolution was used, since it was sufficient to capture the relevant information efficiently. Furthermore, to adapt the DGCN for edge classification, we concatenated the embeddings of adjacent nodes with edge-specific features and then applied fully connected layers followed by Softmax activation to derive predictions for edges. The model was trained with a learning rate of 0.001 and weight decay of 0.005. The number of epochs was set to 500 with an early stopping condition.

Baseline Model: As an additional reference, we employ a naive classifier, referred to as Baseline. We parameterize this model by a scalar θ, which measures the number of retweets/mentions. Two values for θ are considered: (i) Baseline [$\theta > 0$]: If the ego user retweeted/mentioned the alter, then a Following connection is predicted; (ii) Baseline [$\theta >$ Avg.]: Following connection is predicted if the number of retweets/mentions is greater than the average number of retweets/mentions within the same ego network.

5 Prediction of Following Relationships

In this section, we first evaluate and compare the predictive performance of the selected candidate methods. Then, we scrutinize the performance of the best-performing model, revealing the most important features and investigating the factors affecting the accuracy of predictions.

5.1 Comparative Analysis

Table 4 reports the performance of the models with respect to accuracy, precision, recall, and F1-score. As the performance of the naive Baseline [$\theta > 0$] indicates, interaction alone (i.e., retweeting or mentioning) does not serve as an accurate

Table 4. Prediction results for each candidate method broken down by datasets. The scores represent the weighted average of the two classes, and the best ones are highlighted in bold.

	Twibot-22				CodeSwitchNet			
Method	Accuracy	Precision	Recall	F1-score	Accuracy	Precision	Recall	F1-score
Baseline [θ>0]	0.30	0.30	**1.00**	0.47	0.18	0.18	**1.00**	0.31
Baseline [θ>Avg.]	0.69	0.49	0.37	0.42	0.76	0.35	0.38	0.37
Logistic Regression	0.74	**0.74**	0.74	0.68	0.82	0.78	0.82	0.76
*k*NN	0.64	0.62	0.64	0.63	0.83	0.80	0.83	0.81
Random Forest	0.75	**0.74**	0.76	**0.74**	**0.89**	**0.88**	0.89	**0.88**
XGBoost	**0.76**	**0.74**	0.76	**0.74**	**0.89**	**0.88**	0.89	**0.88**
DGCN	0.70	0.48	0.70	0.57	0.79	0.71	0.79	0.74

predictor of a following connection achieving F1-score of only 47% on TwiBot–22 dataset, and an even lower score on CodeSwitchNet. Due to the threshold value of 0, this method classifies all the edges as Following connections (i.e., the positive class), hence the Recall score of 1. When the threshold value is with respect to the average number of interactions within the ego network, the overall performance of this baseline model slightly improves. This suggests that while the frequency of retweets/mentions is an informative feature, it lacks precision.

Relative to the Baseline, the other models exhibited significantly improved accuracy. Yet, as can be seen from Table 4, their performance varies notably across the two datasets. Specifically, XGBoost achieves 74% on TwiBot–22, compared to 88% on CodeSwitchNet. The difference of around 14% is consistently observed in other models. In some models the difference is larger, for example in DGCN. This disparity can be attributed to the difference in the statistics and characteristics of the datasets (see Table 2). In the following subsection we investigate this observation more closely.

Compared to the statistical ML models, DGCN demonstrated an overall inferior performance. On TwiBot–22, DGCN attained an F1-score of only 57%, closely approaching the score of the baseline model. Note that a similar behavior was observed in [20]. Though convolutional graph neural networks can offer powerful prediction capabilities, these were not originally designed for one-layer ego-centric graphs. Considering this, the low performance on the current datasets at hand was expected.

Among the considered models, ensemble tree-based models, XGBoost and Random Forest, achieved the highest performance metrics, closely following each other. Both yielding an F1-score of 74% on TwiBot–22 and 88% on CodeSwitchNet respectively. In the following section the variation is investigated more closely.

5.2 Performance Scrutiny of XGBoost

To identify the most influential factors that affected the performance of XGBoost, we extracted the feature importance scores, which are depicted in Table 6. The two most prominent features concern the number of retweets and mentions. These observations are consistent with the findings in [2], which suggest that Mentions can quantify the actual strength of Follow connections, thus providing some measure of validation. Additionally, the degree of the ego is also relatively important (Table 5).

Next, in order to shed light on the performance disparity of XGBoost across the two datasets, we partition the ego-users in TwiBot–22 and CodeSwitchNet, and analyze the accuracy of predictions for each category separately. First, we group the ego-users by the degree of their Retweet/Mention networks (i.e. G_{RT}) into 3 classes with divisions at the 25th, 50th, and 75th percentiles. Both in CodeSwitchNet and TwiBot–22, we observe that the performance accuracy of XGBoost increases as the size of ego users' G_{RT} increases. This suggests that extracting Following connections from a larger Retweet/Mention networks is comparably easier.

Table 5. Accuracy broken down by degree category of G_{RT} and G_{F}. Degree categories are calculated based on quantiles for both datasets. The numbers in parentheses indicate the mean and standard deviation.

	G_{RT}			G_F		
	Low degree (102.28 ± 140.05)	Medium degree (209.70 ± 164.21)	High degree (303.45 ± 144.95)	Low degree (27.00 ± 20.40)	Medium degree (176.04 ± 66.45)	High degree (441.29 ± 129.41)
CodeSwitchNet	0.75	0.85	0.90	0.88	0.85	0.79
TwiBot–22	0.68	0.73	0.77	0.76	0.73	0.68

Table 6. Top 5 Feature importance scores extracted from the XGBoost model.

Feature	mentioned_count	retweeted_count	ego_account_age	tweet_retweet_count	ego_follower_following_ratio
Importance	0.439	0.168	0.117	0.044	0.035

Second, we cluster the ego-users by the degree of their Following networks (i.e., G_{F}) into 3 classes with divisions at the 25th, 50th, and 75th percentiles. In contrast to the above pattern, the accuracy of XGBoost seems to be negatively correlated with the size of the network. In both CodeSwitchNet and TwiBot–22, the performance decreases as the size of the Following networks increases. This implies that prediction of Following connections is relatively easier when ego-user follows only a few users whom they mention or retweet.

The above findings indicate that the size of Retweet/Mention networks can play a significant role, and the choice of the predictive model depends largely on the dataset. While for datasets like CodeSwitchNet, with large G_{RT} networks and small G_{F}, ensemble tree models can achieve reasonably high accuracy, for other datasets more advanced predictive models and additional features should be considered. For example, to incorporate features that can capture the interaction nuances between ego-users and alters, the topic and sentiment of the tweets.

6 Concluding Remarks

In this study, we explored the problem of predicting users' Following networks on Twitter from their Retweet/Mention graph. Our approach circumvents the traditional challenges associated with the construction of following networks, particularly on platforms like Twitter that have inherent data access constraints. As demonstrated through evaluations on two different datasets, an ensemble tree-based approach supplied with a moderate number of features can achieve sufficiently accurate results in predicting the Following connections for certain category of ego-networks, thereby reinforcing the potential of this research direction. However, we observe significantly degraded performance for cases when ego-users' Retweet/Mention network is small or when the target Following network is large, which calls for the development of more advanced prediction models.

One promising avenue for future exploration would be integrating a richer, possibly multi-modal, set of features into our models. By intertwining Natural

Language Processing, sentiment analysis, and further contextual information about users, can drastically enhance predictive models' expressiveness, allowing us to capture subtler nuances of user behaviors and preferences.

Disclosure of Interests. The authors have no competing interests to declare that are relevant to the content of this article.

References

1. Ahmad, I., Akhtar, M.U., Noor, S., Shahnaz, A.: Missing link prediction using common neighbor and centrality based parameterized algorithm. Sci. Rep. **10**(1), 364 (2020)
2. Amati, G., et al.: Moving beyond the twitter follow graph. In: 2015 7th International Joint Conference on Knowledge Discovery, Knowledge Engineering and Knowledge Management (IC3K), vol. 1, pp. 612–619. IEEE (2015)
3. Behera, D.K., Das, M., Swetanisha, S., Nayak, J., Vimal, S., Naik, B.: Follower link prediction using the xgboost classification model with multiple graph features. Wireless Pers. Commun. **127**(1), 695–714 (2022)
4. Bild, D.R., Liu, Y., Dick, R.P., Mao, Z.M., Wallach, D.S.: Aggregate characterization of user behavior in twitter and analysis of the retweet graph. ACM Trans. Internet Technol. (TOIT) **15**(1), 1–24 (2015)
5. Feng, S., et al.: Twibot-22: towards graph-based twitter bot detection. Adv. Neural. Inf. Process. Syst. **35**, 35254–35269 (2022)
6. Golbeck, J., Hansen, D.: A method for computing political preference among twitter followers. Soc. Netw. **36**, 177–184 (2014)
7. Grover, A., Leskovec, J.: node2vec: scalable feature learning for networks. In: Proceedings of the 22nd ACM SIGKDD International Conference on Knowledge Discovery and Data Mining, pp. 855–864 (2016)
8. Hours, H., Fleury, E., Karsai, M.: Link prediction in the twitter mention network: impacts of local structure and similarity of interest. In: 2016 IEEE 16th International Conference on Data Mining Workshops, pp. 454–461. IEEE (2016)
9. Jin, H., Xu, G., Cheng, K., Liu, J., Wu, Z.: A link prediction algorithm based on gan. Electronics **11**(13) (2022)
10. Li, T., Wu, Y.J., Levina, E., Zhu, J.: Link prediction for egocentrically sampled networks. J. Comput. Graph. Stat., 1–24 (2023)
11. Liben-Nowell, D., Kleinberg, J.: The link-prediction problem for social networks. journal of the association for information science and technology (2007). Google Scholar Google Scholar Digital Library Digital Library (2007)
12. Martínez, V., Berzal, F., Cubero, J.C.: A survey of link prediction in complex networks. ACM Comput. Surv. **49**(4) (2016)
13. Myers, S.A., Sharma, A., Gupta, P., Lin, J.: Information network or social network? the structure of the twitter follow graph. In: Proceedings of the 23rd International Conference on World Wide Web, pp. 493–498 (2014)
14. Perozzi, B., Al-Rfou, R., Skiena, S.: Deepwalk: online learning of social representations. In: Proceedings of the 20th ACM SIGKDD International Conference on Knowledge Discovery and Data Mining, pp. 701–710 (2014)
15. Pfeffer, J., Matter, D., et al.: Just another day on twitter: a complete 24 hours of twitter data. In: Proceedings of the International AAAI Conference on Web and Social Media, vol. 17, pp. 1073–1081 (2023)

16. Pfeffer, J., Mooseder, A., Lasser, J., Hammer, L., Stritzel, O., Garcia, D.: This sample seems to be good enough! assessing coverage and temporal reliability of twitter's academic api. In: Proceedings of the International AAAI Conference on Web and Social Media vol. 17, pp. 720–729 (2023)
17. Praznik, L., Qudar, M.M.A., Mendhe, C., Srivastava, G., Mago, V.: Analysis of link prediction algorithms in hashtag graphs. In: Çakırtaş, M., Ozdemir, M.K. (eds.) Big Data and Social Media Analytics. LNSN, pp. 221–245. Springer, Cham (2021). https://doi.org/10.1007/978-3-030-67044-3_11
18. Su, T., Macdonald, C., Ounis, I.: Leveraging users' social network embeddings for fake news detection on twitter. arXiv preprint arXiv:2211.10672 (2022)
19. Tong, Z., Liang, Y., Sun, C., Rosenblum, D.S., Lim, A.: Directed graph convolutional network. arXiv preprint arXiv:2004.13970 (2020)
20. Toprak, M., Boldrini, C., Passarella, A., Conti, M.: Harnessing the power of ego network layers for link prediction in online social networks. IEEE Trans. Comput. Soc. Syst. **10**(1), 48–60 (2023)
21. Upadhyaya, A., Chandra, J.: Spotting flares: the vital signs of the viral spread of tweets made during communal incidents. ACM Trans. Web **16**(4), 1–28 (2022)
22. Zhang, M., Chen, Y.: Link prediction based on graph neural networks. Adv. Neural Inform. Process. Syst. **31** (2018)

ClimateMiSt: Climate Change Misinformation and Stance Detection Dataset

YeonJung Choi, Lanyu Shang, and Dong Wang(✉)

School of Information Sciences, University of Illinois Urbana-Champaign, Champaign, IL, USA
{yc55,lshang3,dwang24}@illinois.edu

Abstract. Climate change has been a worldwide concern for more than 50 years, and climate change misinformation has also become a critical issue as it questions the causes and effects of climate change, thereby disrupting climate action. Climate misinformation has been a major obstacle to mitigating climate change and its effects, aggravating the issue and polarizing the public. In this paper, we introduce ClimateMiSt, a new climate change misinformation and stance detection dataset consisting of social media data with manually verified labels. The data is collected from Twitter/X and our dataset contains 146,670 tweets. We implement state-of-the-art baseline models for both misinformation and stance detection on our dataset and discover that GPT-4 outperforms them in both tasks. To the best of our knowledge, ClimateMiSt is the first dataset focused on climate change that includes both veracity and stance annotations collected from a social media platform. Our novel dataset can be used for climate change misinformation and stance detection, and it can further contribute to research in this field.

Keywords: Climage Change · Climate Change Dataset · Misinformation Detection · Stance Detection · Online Social Media

1 Introduction

The emergence of social networking platforms (e.g., Twitter/X, Facebook/Meta) has significantly changed the way people interact and communicate online, leading to a whole new wave of applications and reshaping existing information ecosystems [23, 29]. However, the popularity of social media platforms simultaneously prompted the growth of misinformation, resulting in detrimental societal effects such as undermining public trust and support and putting people's lives at risk [9,19,25,31]. Climate change is no exception as a target of misinformation propagation. Although climate scientists have reached a near-unanimous consensus on anthropogenic climate change [8], widespread public disbelief persists due to misinformation about this scientific conclusion [12,22]. Climate misinformation reduces public acceptance and understanding of climate change, contributes to polarization, and impedes support for mitigation policies [12,20]. In this paper, we propose a novel Climate Change Misinformation and Stance Detection Dataset (ClimateMiSt) that aims to address climate change misinformation issues by facilitating relevant studies.

L. M. Aiello et al. (Eds.): ASONAM 2024, LNCS 15212, pp. 321–330, 2025.
https://doi.org/10.1007/978-3-031-78538-2_28

Since climate change became a matter of "real concern" from a US President's Advisory Committee panel in 1965,[1] numerous efforts have been made worldwide to mitigate its effects, such as United Nations Framework Convention on Climate Change (UNFCCC) and Paris Agreement [5]. However, these efforts have not resolved the climate change issue. On July 27th, 2023, the UN Secretary-General warned that "the era of global boiling has arrived."[2] Recent extreme droughts and heatwaves (including in Europe and the US) and wildfires in California and Canada are all attributed to anthropogenic climate change [15].

Goldberg *et al.* analyze consecutive election cycles from 1990 to 2018 and find that oil and gas companies systematically provide financial support to anti-environmental politicians [16]. In addition, Farrell examines a network of 164 organizations (e.g., think tanks, foundations, etc.) and 4,556 individuals associated with these organizations who participated in the climate change counter-movement. He identifies that organizations with corporate funding are more likely to polarize the climate change issue [14]. He also suggests that science is being privatized due to the increasing influence of corporate wealth on scientific issues. These studies imply that the climate change issue is highly political and that climate change misinformation is created and disseminated by prominent politicians and organizations. That is, climate misinformation is repeated and amplified by people with power, influence, or recognition, from where it reaches a wider public. Considering its unique characteristics, detecting and correcting climate change misinformation is an urgent matter.

Moreover, the general public lacks the expertise and skills to evaluate the veracity of a claim, leading them to rely more on heuristics. Hence, people make judgments based on the character of those who speak about climate change, rather than on the claim itself [7]. This tendency leads people to be more vulnerable to climate misinformation. Considering that falsehood reaches people about six times faster than the truth [28], it is significantly important to accurately identify misleading information on climate change. In addition, studying the stance on climate change is critical as it directly affects policy making, public awareness, and participation. However, there are fewer constructions of climate change misinformation datasets and relevant research compared to other domains. For example, several recent efforts have been made to construct data repositories for climate change [4,11] However, these datasets do not specifically target climate change misinformation: they lack either veracity annotations (e.g., misinformation/non-misinformation) or stance annotations (e.g., favor/against).

To address these limitations, we present ClimateMiSt, a novel dataset specifically targeted for climate misinformation and stance detection tasks. Our goal is to construct a comprehensive dataset for climate misinformation that can be utilized in misinformation or stance detection within the climate change domain. In particular, we extensively collect data from one of the widely used social media platforms and provide veracity and stance annotations for our dataset. Additionally, we experiment with multiple benchmark models to evaluate their performance on our dataset. The evaluation results indicate that GPT-4 achieves the best performance for both misinformation and stance detection tasks. We expect our study to facilitate relevant research in the field, such

[1] https://www.bbc.com/news/science-environment-15874560.

[2] https://news.un.org/en/story/2023/07/1139162.

as advances in misinformation and stance detection models within the climate change domain. This, in turn, can help people distinguish misinformation from factual information and raise awareness of climate change issues. We summarize the key contributions of our paper as follows:

- ClimateMiSt is the first comprehensive dataset for both misinformation and stance detection in the context of climate change, consisting of social media data.
- We implement state-of-the-art baselines, including generative AI models like GPT and Llama, and evaluate their performance on both misinformation and stance detection tasks using our collected dataset.

2 Related Works

2.1 Misinformation Detection

Several studies of climate change focus on either dataset construction [11] or analyses on the misinformation itself [2,6]. Diggelmann *et al.* introduce the CLIMATE-FEVER dataset, designed to verify climate change-related claims based on the methodology of FEVER [11]. However, the dataset does not provide misinformation or stance annotations for English tweets. On the other hand, Al-Rawi *et al.* study public discourses around climate change and global warming by collecting 6.8 million tweets referencing "fake news" and find that discussions about climate change/global warming and fake news are highly polarized [2]. Coan *et al.* investigate the role of misinformation in the climate change debate, suggesting that conservative think tanks and climate contrarian blogs damage the credibility of climate science and scientists through conspiratorial messaging [6].

2.2 Studies on Stance Detection

With the proliferation of online content, many natural language processing tasks, including stance detection, have been extensively investigated. Reveilhac and Schneider present a rule-based stance detection model that is easily replicable across various targets [24]. Gómez-Suta *et al.* propose a two-phase classification system for stance detection that leverages topic modeling features and provides explanations for stance labels using these features [17]. Upadhyaya *et al.* propose a novel framework called Sentiment and Temporal Aided Stance Detection (STASY) to classify tweets' stance on climate change as either "denier" or "believer" [27]. While the other aforementioned studies lack their own datasets for stance detection, Upadhyaya *et al.* create the "CLiCS" dataset for the climate change domain [27]. However, CLiCS consists only of tweets with stance annotations and lacks veracity annotations.

3 Data

In this section, we provide a detailed overview of the data collection and annotation process for ClimateMiSt. In particular, we detail the social media data collection and annotation process in Sect. 3.1. We provide comparison results with other datasets in Sect. 3.2. An overview of the data collection process is illustrated in Fig. 1.

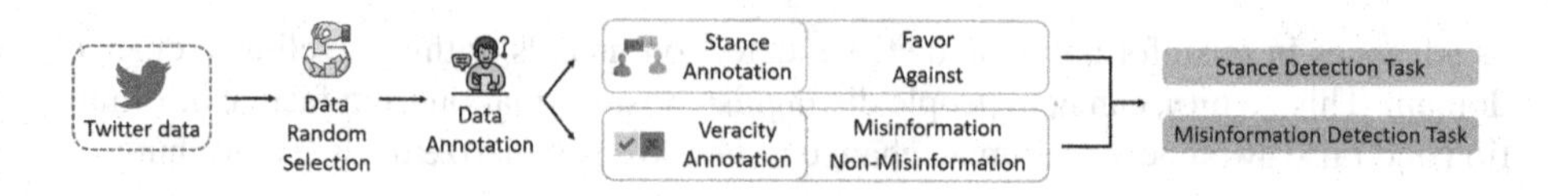

Fig. 1. Overview of Data Collection Process

3.1 Social Media Data

We choose Twitter/X as the social media platform to collect data as Twitter is one of the most popular social media platforms with around 450 million monthly active users.[3] Also, users can follow others and retweet their tweets, allowing information to be easily propagated. Twitter API v2[4] is used to crawl all English tweets that mention "climate change" (excluding retweets), from January 1st, 2022 to September 30th, 2022. For each tweet, we collect a set of attributes, including "author_id", "created_at", "id", "text", "attachments", "url", "verified[5]", etc. A total of 147,957 tweets are collected, with 1,287 being duplicates. In total, we crawl 146,670 unique tweets.

Veracity Annotation. Due to labor constraints, 2,008 tweets are randomly selected from the 146,670 collected tweets and manually annotated as either misinformation (0) or non-misinformation (1). In our study, we define misinformation as false or inaccurate information about climate change issues, and non-misinformation as true or accurate information about them. We exclude any tweets that are 1) too political (extreme right or left inclination), 2) not related to climate change (a mere mention of "climate change"), or 3) lacking sufficient context.

For any tweets, we search for information that could support or rebut the tweet. We look for news articles (e.g., CNN, BBC), reports (e.g., UN, IPCC), or journal papers (e.g., Elsevier, Springer) from reputable sources for verification. If any information is found supporting the statement, the tweet is labeled as non-misinformation. If the information rebuts the statement, the tweet is labeled as misinformation. In total, we annotate 1,140 tweets as misinformation and 868 tweets as non-misinformation.

Stance Annotation. For the same tweets used in the veracity annotation, we also manually annotate them as either "against (0)" or "favor (1)." Regardless of the veracity of each tweet, if the writer denies the existence of (anthropogenic) climate change, we label it as "against." If the writer accepts it, we label it as "favor."

In total, we annotate 980 tweets as "against" and 1,028 tweets as "favor." Among all "favor" tweets, about 82% are "non-misinformation," while approximately 18% are "misinformation." In contrast, among "against" tweets, 97.45% are "misinformation," and only 2.55% are "non-misinformation."

[3] https://www.demandsage.com/twitter-statistics/.
[4] https://developer.twitter.com/en/docs/twitter-api.
[5] The 'verified' account refers to the blue Verification badge on Twitter which indicates that an account of public interest is authentic.

We also calculate Cramer's V to measure the association between two categorical variables: veracity labels and stance labels [1]. The value of Cramer's V ranges from 0 (no association between the variables) to 1 (complete association between the variables). The Cramer's V between veracity labels and stance labels is 0.8017 (with a p-value of 1.274e-28), indicating a strong and significant association between the two variables.

In summary, we collect a total of 146,670 tweets and randomly select 2,008 of them for annotation. We apply two different labels: veracity and stance. Our dataset includes 1,140 tweets labeled as misinformation and 868 as non-misinformation. Additionally, there are 980 tweets labeled "against" and 1,028 labeled "favor" in terms of stance.

3.2 Comparison with Existing Datasets

We compare the properties of ClimateMiSt with those of other existing climate change misinformation datasets in Table 1. One important observation is that only our dataset includes political stance information. While every study in Table 1 provides climate change misinformation data, each has distinct differences from our dataset.

Table 1. Comparison of Properties of Different Climate Change Misinformation Datasets

Dataset	Data Source	Language	Political Stance	Total Amount (Annotated Total)	Time Range
CLIMATE-FEVER [11]	Social Media	English	✗	1,535 (-)	✗
Twitter-COMMs [4]	Social Media	English	✗	212,665 (-)	06/01/2016 - 09/31/2021
Al-Rawi *et al.* [2]	Social Media	English	✗	over 6.8 million (-)	11/27/2019 - 02/14/2020
Coan *et al.* [6]	Social Media	English	✗	249,413 (-)	1998 - 2020
ClimateMiSt (Ours)	Social Media	English	✓	146,670 (5,353)	01/01/2022 - 09/30/2022

CLIMATE-FEVER contains 7,675 annotated claim-evidence pairs, including 1,535 verifiable claims, with labels for each claim-evidence pair [11]. However, the evidence sentences are only retrieved from Wikipedia, which may raise credibility and validity issues. Moreover, the labels refer to the relationship between the claim and the evidence, not the claim's veracity. In addition, this dataset lacks stance annotations and does not include any experiments on misinformation or stance detection tasks. Similarly, Twitter-COMMs is a large-scale multi-modal dataset with a total of 884,331 tweets, including 212,665 tweets related to climate change [4]. However, this dataset does not include any veracity or stance annotations, nor does it involve misinformation or stance detection tasks. Al-Rawi *et al.* and Coan *et al.* also collect large amounts of data from different sources to study contrarian claims or fake news about climate change and analyze the characteristics of these discourses [2, 6]. However, they do not include any annotations for misinformation or stance detection tasks. Compared to these datasets, ClimateMiSt is a comprehensive climate change dataset designed for both misinformation and stance detection tasks. It is manually labeled with veracity and stance annotations, and we have experimented with our dataset on both misinformation and stance detection tasks.

4 Experiment

In this section, we perform misinformation and stance detection on our dataset using several state-of-the-art baseline models. The evaluation results show that GPT-4 achieves the best performance in both tasks.

4.1 Baselines and Experiment Setup

The baselines consist of generative large language models (LLMs) as well as models specifically targeted for misinformation detection and text classification. We select GPT-4[6] and Llama 3[7] as our generative LLMs, employing few-shot learning by providing three to four sample tweets for each label in every task. We use supervised learning for all other models, which include: dEFEND (Explainable FakE News Detection) [26], SHINE (HIerarchical heterogeNEous graph representation learning method for STC) [30], TextING (Text classification method for INductive word representations via Graph neural networks) [32], BERT Text Classification (Bidirectional Encoder Representations from Transformers) [10], and RoBERTa Text Classification (Robustly Optimized BERT-Pretraining Approach) [21].

Table 2. Hyperparameters for Baseline Models

Model	(Graph/Text) Embedding Dimension	Batch Size	Epoch	Learning Rate	Dropout Rate	Max Length
dEFEND	100	20	50	–	–	–
SHINE	1000	–	–	1e-3	0.7	–
TextING	300	4096	200	0.005	0.5	–
BERT	768	16	50	1e-6	0.5	512
RoBERTa	768	16	50	1e-5	0.5	150

For supervised classifiers, the entire annotated dataset is divided into training, development, and test sets in an 8:1:1 ratio, resulting in 1,606 training samples, 201 development samples, and 201 test samples. We run the benchmark models on Ubuntu 20.04 using four NVIDIA A16 and four NVIDIA L40s. Additionally, we use PyTorch version 1.13.1 and TensorFlow version 1.12.0. Specific parameters for each benchmark model are detailed in Table 2.

4.2 Misinformation Detection Performance Analysis

The misinformation detection results on our dataset are shown on the left side of Table 3. We observe that the best-performing model is GPT-4, achieving a classification F1 score of 0.9590, while RoBERTa performs the best among all non-LLMs. GPT-4 outperforms RoBERTa by 2.22%, 2.15%, 1.97%, and 2.36% in terms of F1 score, accuracy, precision, and recall, respectively.

[6] https://openai.com/research/gpt-4.

[7] https://llama.meta.com/llama3/.

Table 3. Baseline Model Performances with Our Dataset on Misinformation Detection (left) and Stance Detection (right)

	Model	F1 score	Accuracy	Precision	Recall
GenerativeLLMs	**GPT-4**	**0.9590**	**0.9602**	**0.9627**	**0.9561**
	Llama 3	0.9267	0.9303	0.9462	0.9176
Supervised Classifiers	dEFEND	0.8010	0.8010	0.8010	0.8010
	SHINE	0.8448	0.8458	0.8454	0.8538
	TextING	0.8831	0.8657	0.8870	0.8793
	BERT	0.8977	0.9005	0.8992	0.8965
	RoBERTa	0.9382	0.9403	0.9441	0.9341

	Model	F1 score	Accuracy	Precision	Recall
GenerativeLLMs	**GPT-4**	**0.8953**	**0.8955**	**0.9002**	**0.8958**
	Llama 3	0.8123	0.8159	0.8443	0.8166
Supervised Classifiers	dEFEND	0.7214	0.7214	0.7214	0.7214
	SHINE	0.8396	0.8408	0.8433	0.8387
	TextING	0.7831	0.7960	0.7957	0.7708
	BERT	0.7904	0.7910	0.7940	0.7908
	RoBERTa	0.8752	0.8756	0.8803	0.8753

4.3 Stance Detection Performance Analysis

The stance detection results are displayed on the right side of Table 3. Similar to the misinformation detection results, the best-performing model for stance detection is GPT-4, with a classification F1 score of 0.8953. Other text classification models, such as RoBERTa and SHINE, also show decent performance in terms of both accuracy and F1 score. One noticeable point is that the overall performance of all baseline models has decreased in the stance detection task. This is possibly due to the characteristics of stance annotation, which is strongly related to the writers' opinions, whereas veracity annotation is contingent on facts.

5 Discussion and Future Work

Experiments on our dataset indicate that generative LLMs achieve the best performance compared to other supervised classifiers, even with few-shot learning. This can be attributed to the billions to trillions of parameters and the massive training data used in training generative LLMs. Hence, generative LLMs could be used for future automatic data annotation, thereby enhancing misinformation detection tasks.

Our study can facilitate relevant climate change misinformation tasks, including 1) constructing datasets for both misinformation and stance detection on climate change, and 2) implementing and evaluating benchmark baselines, including generative AI models such as GPT-4 and Llama 3. This will help people distinguish misinformation from factual information about climate change issues, thereby gradually alleviating the climate change problem.

Moreover, our research can be further investigated in a few ways. Our dataset currently contains only textual data (single modality), but modern data often comprises multiple modalities (e.g., text, image, video, or audio). Incorporating multimodal features into our dataset will further contribute to extending relevant studies and enhancing the performance of both misinformation and stance detection. For example, leveraging other modalities such as image or audio, which exist alongside text, can improve detection performance by providing additional information (e.g., visual representation of images [13], MFCC from audio [18]) for classifying veracity and stance. Moreover, we could further improve model performance by developing human-AI hybrid solutions that incorporate human intelligence (e.g., crowdsourcing) to identify and verify posts likely to be misclassified by LLMs [3]. Additionally, zero-shot learning on generative LLMs could be explored to investigate its impact on misinformation and stance detection performance.

6 Conclusion

Climate change misinformation has undermined mitigation efforts and contributed to public polarization. It is crucial to address this misinformation immediately, but climate change misinformation datasets have not been extensively studied so far. Through this study, we provide a novel climate change dataset specifically designed for both misinformation detection and stance detection tasks, consisting of social media data. We experiment with several state-of-the-art baseline models for misinformation detection and stance detection, and provide benchmark performance results. Both the misinformation detection and stance detection results show that generative LLMs, specifically GPT-4, excel in both tasks. We hope that our study opens the door to further in-depth research on climate change misinformation detection.

Acknowledgement. This research is supported in part by the National Science Foundation under Grant No. IIS-2202481, CHE-2105032, IIS-2130263, CNS-2131622, CNS-2140999. The views and conclusions contained in this document are those of the authors and should not be interpreted as representing the official policies, either expressed or implied, of the U.S. Government. The U.S. Government is authorized to reproduce and distribute reprints for Government purposes notwithstanding any copyright notation here on.

References

1. Akoglu, H.: User's guide to correlation coefficients. Turkish J. Emerg. Med. **18**(3), 91–93 (2018)
2. Al-Rawi, A., O'Keefe, D., Kane, O., Bizimana, A.J.: Twitter's fake news discourses around climate change and global warming. Front. Commun. **6** (2021). https://doi.org/10.3389/fcomm.2021.729818
3. Arous, I., Dolamic, L., Yang, J., Bhardwaj, A., Cuccu, G., Cudré-Mauroux, P.: MARTA: leveraging human rationales for explainable text classification. In: AAAI Conference on Artificial Intelligence, pp. 5868–5876 (2021)
4. Biamby, G., Luo, G., Darrell, T., Rohrbach, A.: Twitter-COMMs: Detecting climate, covid, and military multimodal misinformation. CoRR abs/2112.08594 (2021). https://arxiv.org/abs/2112.08594
5. Blau, J.: The Paris Agreement: climate change, solidarity, and human rights. Springer (2017)
6. Coan, T., Boussalis, C., Cook, J., Nanko, M.O.: Computer-assisted detection and classification of misinformation about climate change (2021). https://doi.org/10.31235/osf.io/crxfm, osf.io/preprints/socarxiv/crxfm
7. Cook, J., Ellerton, P., Kinkead, D.: Deconstructing climate misinformation to identify reasoning errors. Environ. Res. Lett. **13**, 024018 (2018). https://doi.org/10.1088/1748-9326/aaa49f
8. Cook, J., et al.: Consensus on consensus: a synthesis of consensus estimates on human-caused global warming. Environ. Res. Lett. **11**(4), 48002 (2016). https://doi.org/10.1088/1748-9326/11/4/048002

9. Cui, L., Seo, H., Tabar, M., Ma, F., Wang, S., Lee, D.: DETERRENT: knowledge guided graph attention network for detecting healthcare misinformation. In: Proceedings of the 26th ACM SIGKDD International Conference on Knowledge Discovery & Data Mining, pp. 492–502. KDD '20, Association for Computing Machinery, New York, NY, USA (2020). https://doi.org/10.1145/3394486.3403092
10. Devlin, J., Chang, M., Lee, K., Toutanova, K.: BERT: pre-training of deep bidirectional transformers for language understanding. CoRR **abs/1810.04805** (2018). http://arxiv.org/abs/1810.04805
11. Diggelmann, T., Boyd-Graber, J.L., Bulian, J., Ciaramita, M., Leippold, M.: CLIMATE-FEVER: A dataset for verification of real-world climate claims. CoRR abs/2012.00614 (2020). https://arxiv.org/abs/2012.00614
12. Ding, D., Maibach, E., Zhao, X., Roser-Renouf, C., Leiserowitz, A.: Support for climate policy and societal action are linked to perceptions of scientific agreement. Nat. Clim. Change **1**, 462–466 (2011). https://doi.org/10.1038/NCLIMATE1295
13. Fang, Y., et al.: EVA: exploring the limits of masked visual representation learning at scale. In: Proceedings of the IEEE/CVF Conference on Computer Vision and Pattern Recognition, pp. 19358–19369 (2023)
14. Farrell, J.: Corporate funding and ideological polarization about climate change. Proc. Nat. Acad. Sci. **113**(1), 92–97 (2016) https://doi.org/10.1073/pnas.1509433112
15. Fountain, H.: Hotter summer days mean more sierra Nevada wildfires, study finds. The New York Times (2021). https://www.nytimes.com/2021/11/17/climate/climate-change-wildfire-risk.html?smid=url-share
16. Goldberg, M.H., Marlon, J.R., Wang, X., van der Linden, S., Leiserowitz, A.: Oil and gas companies invest in legislators that vote against the environment. Proc. Nat. Acad. Sci. **117**(10), 5111–5112 (2020). https://doi.org/10.1073/pnas.1922175117
17. Gómez-Suta, M., Echeverry-Correa, J., Soto-Mejía, J.A.: Stance detection in tweets: a topic modeling approach supporting explainability. Expert Syst. Appl. **214**(C) (2023). https://doi.org/10.1016/j.eswa.2022.119046
18. Hamza, A., et al.: Deepfake audio detection via MFCC features using machine learning. IEEE Access **10**, 134018–134028 (2022)
19. Islam, M.S., et al.: COVID-19-related infodemic and its impact on public health: a global social media analysis. Am. J. Trop. Med. Hyg. **103**(4), 1621 (2020)
20. van der Linden, S., Leiserowitz, A., Rosenthal, S., Maibach, E.: Inoculating the public against misinformation about climate change. Global Challenges **1**(2), 1600008 (2017). https://doi.org/10.1002/gch2.201600008
21. Liu, Y., et al.: RoBERTa: A robustly optimized BERT pretraining approach. CoRR abs/1907.11692 (2019). http://arxiv.org/abs/1907.11692
22. Oreskes, N., Conway, E.: Defeating the merchants of doubt. Nature **465**, 686–7 (2010) https://doi.org/10.1038/465686a
23. Reis, J.C.S., Correia, A., Murai, F., Veloso, A., Benevenuto, F.: Explainable machine learning for fake news detection. In: Proceedings of the 10th ACM Conference on Web Science, pp. 17–26. WebSci '19, Association for Computing Machinery, New York, NY, USA (2019). https://doi.org/10.1145/3292522.3326027
24. Reveilhac, M., Schneider, G.: Replicable semi-supervised approaches to state-of-the-art stance detection of tweets. Inf. Process. Manag. **60**(2), 103199 (2023). https://doi.org/10.1016/j.ipm.2022.103199
25. Shang, L., Chen, B., Vora, A., Zhang, Y., Cai, X., Wang, D.: SocialDrought: a social and news media driven dataset and analytical platform towards understanding societal impact of drought. In: Proceedings of the International AAAI Conference on Web and Social Media. vol. 18, pp. 2051–2062 (2024)

26. Shu, K., Cui, L., Wang, S., Lee, D., Liu, H.: dEFEND: explainable fake news detection. In: Proceedings of the 25th ACM SIGKDD International Conference on Knowledge Discovery & Data Mining, pp. 395–405 (2019)
27. Upadhyaya, A., Fisichella, M., Nejdl, W.: Towards sentiment and temporal aided stance detection of climate change tweets. Inf. Process. Manag. **60**(4), 103325 (2023)
28. Vosoughi, S., Roy, D., Aral, S.: The spread of true and false news online. Science **359**(6380), 1146–1151 (2018). https://doi.org/10.1126/science.aap9559
29. Wang, D., Kaplan, L., Abdelzaher, T.F.: Maximum likelihood analysis of conflicting observations in social sensing. ACM Trans. Sens. Netw. (ToSN) **10**(2), 1–27 (2014)
30. Wang, Y., Wang, S., Yao, Q., Dou, D.: Hierarchical heterogeneous graph representation learning for short text classification. In: Proceedings of the 2021 Conference on Empirical Methods in Natural Language Processing, pp. 3091–3101. Association for Computational Linguistics, Online and Punta Cana, Dominican Republic (2021). https://doi.org/10.18653/v1/2021.emnlp-main.247
31. Zhang, D., Wang, D., Vance, N., Zhang, Y., Mike, S.: On scalable and robust truth discovery in big data social media sensing applications. IEEE Trans. Big Data **5**(2), 195–208 (2018)
32. Zhang, Y., Yu, X., Cui, Z., Wu, S., Wen, Z., Wang, L.: Every document owns its structure: Inductive text classification via graph neural networks. In: Proceedings of the 58th Annual Meeting of the Association for Computational Linguistics (2020)

Hate Speech Classification in Text-Embedded Images: Integrating Ontology, Contextual Semantics, and Vision-Language Representations

Surendrabikram Thapa[1(✉)], Surabhi Adhikari[2], Imran Razzak[3], Roy Ka-Wei Lee[4], and Usman Naseem[5]

[1] Virginia Tech, Blacksburg, VA, USA
surendrabikram@vt.edu
[2] Columbia University, New York, NY, USA
[3] University of New South Wales, Sydney, NSW, Australia
[4] Singapore University of Technology and Design, Singapore, Singapore
[5] Macquarie University, Sydney, NSW, Australia

Abstract. The growing influence of text-embedded images in online communication demands effective strategies for identifying hate speech. The use of hate speech in different contexts makes it necessary to study it in a particular context. Simultaneously, identifying hate speech targets is a crucial research domain as it can offer insights into propagation, impacts, and potential interventions against hate speech. In this article, we address the problem of hate speech detection and target identification in text-embedded images by presenting a comprehensive approach that combines textual and visual cues to accurately detect hate speech and targets within the context of the Russia-Ukraine Crisis. Leveraging a dataset of 4,723 text-embedded images centered around this crisis, we integrate features from the knowledge graph, ontological insights to indicate the presence of hate speech presence, TF-IDF, Named Entity Recognition (NER), and robust vision-language representations. We also provide the rationale behind using different features in our implementation. Our method surpasses existing baselines and methodologies, suggesting the importance of each feature we employ in decision-making.

Keywords: Hate Speech · Multimodal analysis · Vision-Language

1 Introduction

Social media platforms have revolutionized the way we share and consume information, fostering unprecedented connectivity and enabling us to engage with diverse perspectives from around the world. The power of these platforms to rapidly disseminate content has reshaped the way we communicate, learn, and form opinions. However, alongside their undeniable benefits, there exists a flip

L. M. Aiello et al. (Eds.): ASONAM 2024, LNCS 15212, pp. 331–342, 2025.
https://doi.org/10.1007/978-3-031-78538-2_29

side to this technological advancement – the issue of hate speech. Hate speech, characterized by harmful, discriminatory, or prejudiced language, poses a significant challenge to maintaining a safe and respectful online environment [1]. In recent years, higher internet penetration and the prevalence of multimodal content have introduced new dimensions to online communication. This fusion of text and images, often seen in the form of text-embedded images, while enhancing expression, has also given rise to challenges, notably the proliferation of hate speech within such medium. These text-embedded images, easily shareable across social media platforms, possess the potential to rapidly disseminate toxic ideologies to a global audience, necessitating urgent and innovative strategies to address this critical issue. However, the scale and pace of content sharing makes manual monitoring insufficient. Hence, automated techniques are necessary to immediately detect and counter hate speech in text-embedded images. These solutions alleviate human moderators' load while fostering a safer online environment conducive to positive interactions [1].

Hate speech becomes especially dangerous amid political tensions between nations, and it's even more concerning during times of invasion [1]. Such speech can fuel division and escalate unwanted situations. To tackle this, research is crucial, specifically to identify hate speech in these contexts and determine its targets. In this research, we propose models for hate speech classification and target identification particularly in the context of Russia-Ukraine crisis.

To build a robust classification system, we exploit the full potential of available textual and visual information within text-embedded content. Our approach is grounded in leveraging both textual and visual dimensions, wherein we employ multimodal-information embeddings from CLIP [18] to jointly understand vision and linguistic concepts. To further enhance our understanding, we incorporate insights from knowledge graphs and ontological data, enriching our analysis with contextual depth. The identification of specific targets is also facilitated by NER, enabling us to pinpoint specific entities involved. In parallel, we integrate traditional features like TF-IDF, for word significance. Through this amalgamation, our model aims to offer a robust solution to moderating hate speech.

2 Related Works

Recently, the efforts to detect instances of hate speech within social media have gained considerable attention, predominantly focusing on textual content. However, efforts dedicated to the classification of hate speech within text-embedded images, a significant aspect of current social media communication, remain relatively constrained. In recent times, a rise in academic interest can be seen, particularly concerning the identification of hate speech within memes or images featuring embedded text. Memes, characterized by their amalgamation of images and text, often intended for humor, have emerged as a popular area of exploration. Moreover, the category of text-embedded images extends beyond memes, encompassing an array of textual-visual content forms, including screenshots extracted from television headlines. In these instances, images provide contextual foundations, complemented by accompanying text that conveys information within that established context. While the study of memes has recently gained attention in academic and industry research, the nuanced examination of hate speech within text-embedded images demands equal scholarly consideration.

In addition to the necessity of conducting further research on text-embedded images, there is a pressing need to conduct research in specific contexts and applications. For instance, there exists a pressing demand for studies focused on hate speech within specific contexts like invasion. Also, while the exploration of memes and other forms of multimodal textual-visual content has largely been concentrated within the broader landscape of general social media platforms, the efforts to curate dedicated datasets and undertake research tailored to those distinct contexts remain relatively limited. Recent efforts, however, have shown promising steps in understanding multimodal textual-visual data within specific domains. For example, Pramanick et al. [17] studied harmful memes and their targets during the US election pandemic with curation of the related dataset that involved labeling US election-related memes to indicate harmful content and identifying the specific targets of these harmful memes. Similarly, Naseem et al. [16] introduced a dataset encompassing 10,244 memes that critique vaccines. They also proposed specific models to capture the context within such datasets.

2.1 Feature Extraction for Hate Speech

In the context of multimodal classification tasks, features play an important role [3]. Word references and lexicons serve as straightforward methods for feature extraction in text analysis. Tokenization is a foundational step in both traditional and deep learning models, often using dictionaries/lexicons [14]. Meanwhile, TF-IDF, a widely used technique, assigns importance to terms based on their significance across documents [24]. Features play a critical role in machine learning, hence requiring more robust features to help in better classification. In recent literature, it has been found that TF-IDF is widely used in the analysis of hate speech in the internet. Recent studies demonstrate that integrating knowledge enhances NLP task performance by enriching models with semantic information. In context of hate speech, Maheshappa et al. [13] showed that incorporating insights from knowledge graph helps to improve hate speech detection. In the realm of hate speech detection in text-embedded images, multimodal models like CLIP [18], GroupViT [25] has shown promising directions [1]. Similarly, the incorporation of WordNet information [6] is significant in the classification of hate speech. Furthermore, NER features have played a pivotal role in hate speech and target classification [5,15].

In our implementation, for precise identification of hate speech pertaining to the Russia-Ukraine crisis, we use wordnet features along with information about the presence of hate speech. We also leverage features from TF-IDF and NER along with the vision-language representation offered by the CLIP model. Thus, our work bridges a crucial research void by providing a tailored approach to hate speech detection in the specific landscape of the Russia-Ukraine crisis.

3 Dataset

Since our objective is to classify between the hate speech and non-hate speech dataset in the context of Russia-Ukraine crisis, we use CrisisHateMM, the dataset prepared by Bhandari et al. [1] on our implementation. The dataset comprises 4,723 text-embedded images, focusing on the Russia-Ukraine crisis. The dataset had nearly an equal ratio of hate and non-hate speech data with 2,058 images containing no instances of hate speech, while the remaining 2,665 exhibited elements of hate speech. Within this subset of 2,665 images with hate speech, 2,428 specifically had instances of targeted hate speech. The targeted hate speech means that the hate was targeted at some individual, organization, or community. For our first task of hate speech identification, we took all images in the dataset whereas for the second task of target identification, we only took 2,428 images with directed hate speech. We only took text-embedded images with directed hate speech for the identification of targets of hate speech.

4 Methodology

Given an image, **I**, the objective is to determine whether the image contains hate speech content. Additionally, the goal is also to identify whether the hateful image targets one of the specified categories, namely, individual, organization, or community. Figure 1 illustrates an overview of the proposed method.

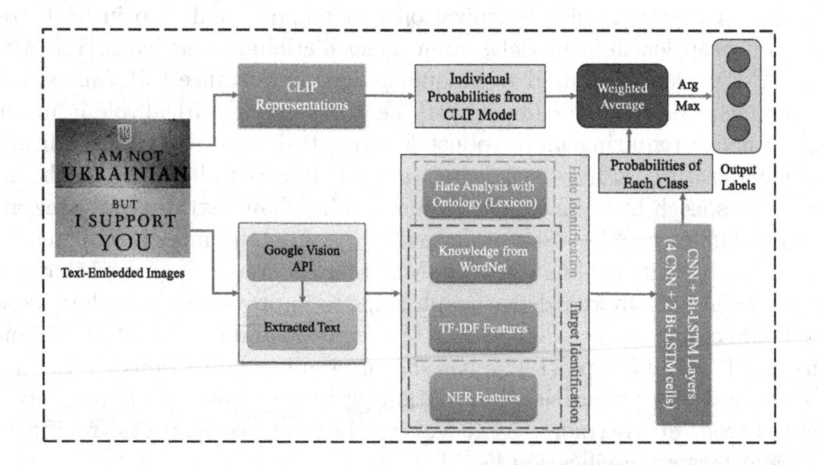

Fig. 1. Our proposed framework leveraging various vision-language representations.

4.1 Textual Features

The textual features from the text-embedded images were extracted using Google Vision API. In this section, we outline the textual features integrated into our methodology. These features collectively contribute to our analysis aimed at accurate hate speech and its target detection. It is important to note that we perform standard preprocessing techniques before calculating the features.

Ontological Insights for Presence of Hate Speech. Leveraging ontology is a significant aspect of our approach. To harness the power of ontology, we utilize the expanded lexicon of abusive words provided by Wiegand et al. [23]. This lexicon comprises an expanded list of 8,478 words, encompassing abusive aspects of a language. Among these, 2,989 words are classified as abusive, while the remaining 5,489 words are categorized as non-abusive. Our utilization of this lexicon involves analyzing text from text-embedded images to investigate their content. For each text-embedded image, we represent the ontological information using a two-dimensional vector. The first value denotes the number of words within the text in image that match the abusive lexicon, and the second value represents the number of words that match with the non-abusive lexicon. This vectorized representation serves as a representation to uncover patterns and relationships in the data, enabling us to effectively recognize and classify the content within text-embedded images. We used this information only for the classification of hate speech and did not use this feature in the target identification model.

TF-IDF Features. TF-IDF features are integral components of our approach. As shown in Eq. 1, it considers both the frequency of a term within a document (term frequency) and its scarcity across the entire corpus (inverse document frequency).

This helps in highlighting significant terms that can aid in hate speech identification. Bhandari et al. [1] showed that, in context of Russia-Ukraine crisis, there are some words in hate speech which are more significant than the others. TF-IDF helps to leverage this information.

$$\text{TF-IDF}(t, d, D) = \text{TF}(t, d) \times \text{IDF}(t, D) \tag{1}$$

where, $\text{TF}(t, d)$ represents the Term Frequency of term t in document d, which measures how often term t appears in document d. $\text{IDF}(t, D)$ represents the Inverse Document Frequency of term t across the entire corpus D, which measures how unique or rare the term is across the corpus.

Leveraging Knowledge Graph Features. We harness the power of the WordNet knowledge graph [6] to enhance our approach. Specifically, we used the hyponymy and hypernymy relationships within WordNet. This enables us to establish connections between specific instances (hyponyms) and their more general concepts (hypernyms), aiding in a deeper understanding of semantic nuances. To identify the top relevant keywords in hate speech, non-hate speech, and specific targets, we employ the SAGE topic model [4]. For hate speech classification, we select 16 keywords, encompassing 8 nonoverlapping words in each category that hold greater relevance in hate speech and non-hate speech posts. Likewise, in target identification, we identify 21 words, including the top 7 non-overlapping words in each target class. In our approach, for each word, we determine the hyponyms and hypernyms, capped at a maximum of 10 per word. By aggregating the resulting hyponyms and hypernyms, we create a combined dictionary. Subsequently, we employ vectorization techniques, essentially assessing whether the text within the text-embedded images contains words from this combined dictionary. This helps to reveal underlying patterns and semantic associations.

Named Entity Recognition (NER). In our target classification process, we also capitalize on NER features. In the context of NER, our approach focuses on leveraging specific entity categories, namely NORP (Nationalities, Religious or Political Groups), PER (Persons), and ORG (Organizations). By recognizing and counting instances of these entities, we extract meaningful information regarding the individuals, groups, and affiliations mentioned within text-embedded images. To achieve this, we employ a counting mechanism for each of the NORP, PER, and ORG entities present within the text. This results in a three-dimensional vector, wherein each dimension corresponds to the count of NORP, PER, and ORG entities, respectively. The resulting vector thus encapsulates the prevalence of these specific entities within the text for reliable target detection.

4.2 Vision-Language Representations

In our methodology, we leverage the powerful capabilities of vision-language representations to enhance our hate speech detection framework. Specifically, we employ the CLIP (Contrastive Language-Image Pretraining) model, which bridges the gap between textual and visual information, facilitating a comprehensive analysis of text-embedded images. The CLIP model possesses the unique ability to understand both images and text in a shared embedding space. It thus provides our framework with a more holistic understanding of the content within text-embedded images. The classifications made by the CLIP model are later incorporated into a weighted ensemble, synergizing with the insights drawn from other textual features. This approach enhances the overall accuracy of hate speech detection and target classification, resulting in a robust and comprehensive solution that aligns with the intricacies of text-embedded image analysis.

4.3 CNN + Bi-LSTM

Convolutional Neural Networks (CNNs) are adept at capturing local text features, while Recurrent Neural Networks (RNNs) excel at capturing long-term dependencies. Combining these architectures can yield improved performance across a spectrum of NLP tasks, including sentiment analysis and text classification [26]. In our experimentation, we employ a model featuring four convolutional layers in series and two Bidirectional Long Short-Term Memory (BiLSTM) cells.

In our model design, we feed word embeddings into the convolutional layer. Following every two convolutional layers, we apply max-pooling with a window size of three to capture salient information. To mitigate overfitting, L2 regularization is incorporated into both networks. The activation function used for BiLSTM cells is Tanh. Subsequent to the first BiLSTM cell, we employ batch normalization [20]. The optimization is handled by the Adam optimizer. The output of the BiLSTM cell is connected to a fully connected layer with ReLU as the activation function. As shown in Fig. 1, we provide the model with a combination of features including ontological features, TF-IDF, and knowledge graph insights for hate speech identification. Similarly, for target identification, we integrate TF-IDF, knowledge graph information, and NER features. This comprehensive approach leverages multiple aspects of linguistic and semantic understanding, contributing to our model's ability to accurately detect hate speech instances and identify specific targets within text-embedded images. The probability of each classes is taken to form a weighted average and make final decision.

4.4 Weighted Average and Final Predictions

In our methodology's final stages, we employ a weighted average technique to amalgamate the predictions from both the CNN+BiLSTM model and the CLIP model. The weighted average is executed by summing up the predicted probabilities from both models for each class and subsequently dividing the sum by two. Effectively, this process computes the average probability for each class, facilitating a comprehensive and balanced assessment. To ensure the adaptability and effectiveness of the weighted average, we introduce adaptive weights based on the performance of the models on the validation set as shown in Eq. 2. This approach allows us to dynamically adjust the contributions of each model's predictions based on their respective capabilities and accuracies. Adaptive weighting aims to give more weight to the more accurate model. The weights are normalized so that they sum up to 1. Our train-test-validation split follows a ratio of 70/15/15. The weighted average of probabilities is obtained and the label is assigned to the one with the highest probability.

$$\text{Weighted Average} = \frac{\sum_{i=1}^{n} w_i \cdot \text{Prob}_{\text{Model}_i}}{\sum_{i=1}^{n} w_i} \tag{2}$$

where, n is the number of models (2 for hate speech detection, 3 for target identification). w_i are the adaptive weights for each model based on validation performance. $\text{Prob}_{\text{Model}_i}$ is the predicted probabilities from the i-th model.

5 Experiments

In our experiments, we have performed various experiments with different unimodal and multimodal models.

Table 1. Performance Comparison of our model different unimodal and multimodal algorithms on CrisisHateMM dataset

Modality	Model	Hate Classification			Target Classification		
		Accuracy ↑	F1-score ↑	MMAE ↓	Accuracy ↑	F1-score ↑	MMAE ↓
Textual	BERT	0.779	0.767	0.240	0.629	0.427	0.998
	DistilBERT	0.754	0.750	0.247	0.637	0.423	1.008
	DistilRoBERTa	0.777	0.769	0.233	0.654	0.440	0.919
	CNN + BiLSTM	0.781	0.773	0.227	0.687	0.619	0.520
Visual	DenseNet-161	0.741	0.739	0.259	0.538	0.425	0.774
	Visformer_small	0.741	0.739	0.257	0.451	0.407	0.772
	MViTv2_base	0.731	0.726	0.276	0.576	0.422	0.657
	VGG19	0.686	0.679	0.305	0.525	0.395	0.785
Multimodal	CLIP	0.798	0.786	0.204	0.684	0.615	0.579
	GroupViT	0.792	0.785	0.214	0.598	0.451	0.763
	MOMENTA	0.787	0.772	0.229	0.688	0.629	0.531
	DisMultiHate	0.771	0.756	0.244	0.624	0.588	0.591
	CogVLM	0.701	0.677	0.349	0.493	0.411	0.740
	MultimodalGPT	0.663	0.610	0.451	0.478	0.383	0.729
	Our Approach	**0.848**	**0.833**	**0.126**	**0.778**	**0.753**	**0.261**

Unimodal Models: We used the following textual and visual models:

- **Textual Models:** Among textual models, we used BERT [9], DistillBERT [19], DistilRoBERTa [19] and our CNN + BiLSTM model.
- **Visual Models:** In order to assess the performance of unimodal visual models, we used DenseNet-161 [8], Visformer [2], Multiscale Vision Transformers (MViTv2) [11], and VGG-19 [21].

Multimodal Models: These models are important to gauge how jointly learning multiple modalities can help in detection of hate speech and targets. In the case of multimodal models, we use CLIP [18], and GroupViT (Grouping Vision Transformer) [25]. We also experimented with hate speech-specific models such as MOMENTA [17] and DisMultiHate [10]. We also use recent large vision language models (LVLMs) like CogVLM [22], and MultimodalGPT [7].

5.1 Experimental Settings

For baselines, we assessed the model performance by using accuracy, F1-score (macro), and MMAE (Macro Mean Absolute Error) as performance metrics.

Text Preprocessing: The text retrieved from OCR was preprocessed along with the image filtering criteria. We removed non-alphanumeric elements, including special characters, hyperlinks, symbols, and non-English characters that may contribute to noise in the data, which could ultimately distort analysis results. Further, non-English words were removed using the English corpus from the NLTK library [12].

6 Results and Analysis

6.1 Comparison With Baselines

Table 1 shows the overall performance of our proposed model against the state-of-the-art baselines. The performance of our model is higher than all the baselines used in our study. As seen in Table 1, among the unimodal models, the textual models have relatively higher accuracy than the visual models. However, this performance is expected as the memes or text-embedded images have more information within the text than the visual modality alone. Furthermore, it is seen that the non-LVLM multimodal models used in our baselines mostly outperform the unimodal models. This is also as expected as such multimodal models can better understand the context by leveraging both textual and visual information. Among the multimodal and unimodal models used, our proposed model performs the highest with an F1-score of 0.833 in the case of hate speech classification and an F1-score of 0.753 in the case of the targets of hate speech identification. The results show that our method was able to perform the best with a nearly 5-point increase in the F1-score in hate speech classification as compared to the second-best performing model. Our model was able to achieve an F1-score of over 83%. Additionally, there was a huge jump in performance in target classification with an increase of nearly 14 points in terms of F1-score. Our model was able to achieve an F1-score of 75.3%. This shows that our features along with the weighted average technique yielded better results.

6.2 Ablation Analysis

We conducted an ablation analysis to evaluate the impact of individual components on the performance of our proposed framework for hate speech and target classification tasks. The results are summarized in Table 2.

For hate speech classification, our proposed framework achieved an accuracy of 0.848 and an F1-score of 0.833, with performance degradation observed when individual components were removed, particularly a decrease of accuracy to 0.811 when excluding lexicon-based features. For target classification, the framework achieved an accuracy of 0.778 and an F1-score of 0.753, with the most significant accuracy decrease to 0.703 observed when excluding NER features, highlighting the importance of these components. These results highlight the effectiveness of incorporating multiple components, including lexicon-based features, semantic knowledge, textual representations, and entity recognition, in our proposed framework for hate speech classification and target classification tasks. Thus, our model benefits from the collective advantages of different components used in our framework.

Table 2. Results of the proposed framework with and without individual components used.

Task	Model	Accuracy	F1-score
Hate Classification	Proposed	**0.848**	**0.833**
	Proposed - Lexicon	0.811	0.809
	Proposed - WordNet	0.820	0.815
	Proposed - TFIDF	0.802	0.788
Target Classification	Proposed	**0.778**	**0.753**
	Proposed - WordNet	0.755	0.732
	Proposed - TFIDF	0.751	0.717
	Proposed - NER	0.703	0.686

7 Conclusion

In this paper, we have delved into the critical domain of hate speech identification within text-embedded images, an increasingly influential mode of communication in today's online landscape. Through rigorous experimentation, we show the importance of a diverse set of features, including ontological insights, TF-IDF analysis, knowledge graph utilization, NER entity recognition, and vision-language representations. Our findings demonstrate the potency of combining these features, enhancing the accuracy of hate speech classification and target identification. The adaptive weighted ensemble technique further bolsters our approach, enabling the fusion of diverse models and their outputs, leveraging the strengths of each. While our approach presents promising results, we acknowledge areas for further refinement. In particular, the challenge of identifying satirical yet hateful content remains, suggesting avenues for deeper explorations into nuanced expressions of hate speech. Moreover, our focus on the Russia-Ukraine crisis context prompts consideration of applying our methodology to other contexts and domains, each with its distinct linguistic and visual characteristics. In conclusion, our research fills a critical void in the study of hate speech within text-embedded images, offering a multifaceted methodology that harnesses the synergy of various features and models. As the digital landscape continues to evolve, our findings contribute to fostering a more respectful and inclusive online environment, where hate speech's harmful impact is mitigated, and meaningful interactions prevail.

References

1. Bhandari, A., Shah, S.B., Thapa, S., Naseem, U., Nasim, M.: CrisisHateMM: multimodal analysis of directed and undirected hate speech in text-embedded images from Russia-Ukraine conflict. In: Proceedings of the IEEE/CVF Conference on Computer Vision and Pattern Recognition, pp. 1993–2002 (2023)

2. Chen, Z., Xie, L., Niu, J., Liu, X., Wei, L., Tian, Q.: Visformer: the vision-friendly transformer. In: Proceedings of the IEEE/CVF International Conference on Computer Vision, pp. 589–598 (2021)
3. Chhabra, A., Vishwakarma, D.K.: A literature survey on multimodal and multilingual automatic hate speech identification. Multimedia Systems, pp. 1–28 (2023)
4. Eisenstein, J., Ahmed, A., Xing, E.P.: Sparse additive generative models of text. In: Proceedings of the 28th International Conference on Machine Learning (ICML-11), pp. 1041–1048 (2011)
5. ElSherief, M., Kulkarni, V., Nguyen, D., Wang, W.Y., Belding, E.: Hate lingo: a target-based linguistic analysis of hate speech in social media. In: Proceedings of the International AAAI Conference on Web and Social Media. vol. 12 (2018)
6. Fellbaum, C.: WordNet. In: Theory and applications of ontology: computer applications, pp. 231–243. Springer (2010)
7. Gong, T., et al.: Multimodal-GPT: A vision and language model for dialogue with humans. arXiv preprint arXiv:2305.04790 (2023)
8. Huang, G., Liu, Z., Van Der Maaten, L., Weinberger, K.Q.: Densely connected convolutional networks. In: Proceedings of the IEEE Conference on Computer Vision and Pattern Recognition, pp. 4700–4708 (2017)
9. Kenton, J.D., Chang, M.W., Toutanova, L.K.: BERT: pre-training of deep bidirectional transformers for language understanding. In: Proceedings of NAACL-HLT, pp. 4171–4186 (2019)
10. Lee, R.K.W., Cao, R., Fan, Z., Jiang, J., Chong, W.H.: Disentangling hate in online memes. In: Proceedings of the 29th ACM International Conference on Multimedia, pp. 5138–5147 (2021)
11. Li, Y., Wu, C.Y., Fan, H., Mangalam, K., Xiong, B., Malik, J., Feichtenhofer, C.: MViTv2: improved multiscale vision transformers for classification and detection. In: Proceedings of the IEEE/CVF Conference on Computer Vision and Pattern Recognition, pp. 4804–4814 (2022)
12. Loper, E., Bird, S.: NLTK: The Natural Language Toolkit. arXiv preprint cs/0205028 (2002)
13. Maheshappa, P., Mathew, B., Saha, P.: Using knowledge graphs to improve hate speech detection. In: Proceedings of the 3rd ACM India Joint International Conference on Data Science & Management of Data (8th ACM IKDD CODS & 26th COMAD), pp. 430–430 (2021)
14. Mekki, A., Zribi, I., Ellouze, M., Belguith, L.H.: Tokenization of Tunisian Arabic: a comparison between three machine learning models. ACM Transactions on Asian and Low-Resource Language Information Processing (2023)
15. Montariol, S., Riabi, A., Seddah, D.: Multilingual auxiliary tasks training: bridging the gap between languages for zero-shot transfer of hate speech detection models. In: Findings of the Association for Computational Linguistics: AACL-IJCNLP 2022, pp. 347–363 (2022)
16. Naseem, U., Kim, J., Khushi, M., Dunn, A.G.: A multimodal framework for the identification of vaccine critical memes on twitter. In: Proceedings of the 16th ACM International Conference on Web Search and Data Mining, pp. 706–714 (2023)
17. Pramanick, S., Sharma, S., Dimitrov, D., Akhtar, M.S., Nakov, P., Chakraborty, T.: MOMENTA: a multimodal framework for detecting harmful memes and their targets. In: Findings of the Association for Computational Linguistics: EMNLP 2021, pp. 4439–4455 (2021)
18. Radford, A., et al.: Learning transferable visual models from natural language supervision. In: International Conference on Machine Learning, pp. 8748–8763. PMLR (2021)

19. Sanh, V., Debut, L., Chaumond, J., Wolf, T.: DistilBERT, a distilled version of BERT: smaller, faster, cheaper and lighter. arXiv preprint arXiv:1910.01108 (2019)
20. Santurkar, S., Tsipras, D., Ilyas, A., Madry, A.: How does batch normalization help optimization? Adv. Neural Inf. Process. Syst. **31** (2018)
21. Simonyan, K., Zisserman, A.: Very deep convolutional networks for large-scale image recognition. arXiv preprint arXiv:1409.1556 (2014)
22. Wang, W., et al.: CogVLM: Visual expert for pretrained language models. arXiv preprint arXiv:2311.03079 (2023)
23. Wiegand, M., Ruppenhofer, J., Schmidt, A., Greenberg, C.: Inducing a lexicon of abusive words–a feature-based approach. In: Proceedings of the 2018 Conference of the North American Chapter of the Association for Computational Linguistics: Human Language Technologies, Volume 1 (Long Papers), pp. 1046–1056 (2018)
24. Wu, H.C., Luk, R.W.P., Wong, K.F., Kwok, K.L.: Interpreting TF-IDF term weights as making relevance decisions. ACM Trans. Inf. Syst. (TOIS) **26**(3), 1–37 (2008)
25. Xu, J., et al.: GroupViT: semantic segmentation emerges from text supervision. In: Proceedings of the IEEE/CVF Conference on Computer Vision and Pattern Recognition, pp. 18134–18144 (2022)
26. Zhao, N., Gao, H., Wen, X., Li, H.: Combination of convolutional neural network and gated recurrent unit for aspect-based sentiment analysis. IEEE Access **9**, 15561–15569 (2021)

VLP: A Label Propagation Algorithm for Community Detection in Complex Networks

Sharon Boddu(✉), Maleq Khan, and Mais Nijim

Department of Electrical Engineering and Computer Science, Texas A&M University-Kingsville, Kingsville, USA
sharonbdd25@gmail.com, {maleq.khan,mais.nijim}@tamuk.edu

Abstract. Community detection is a commonly encountered problem in social network analysis and many other areas. A community in a graph or network is a subgraph containing vertices that are closely connected to other vertices within the same subgraph but have fewer connections to the other vertices. Community detection is useful in analyzing complex systems and recognizing underlying patterns and structures that govern them. There are several algorithms that currently exist for community detection, ranging from simple and fast approaches, such as the label propagation algorithm (LPA), to more complex and time-consuming methods, such as the state-of-the-art Louvain method. We propose a new method called vector label propagation (VLP), which is a generalization of the LPA approach. The VLP algorithm significantly enhances the quality of the detected communities compared to LPA while being much faster than the Louvain method. For example, on the Twitter network, VLP has a normalized mutual information (NMI) score of 0.82, while LPA has an NMI score of 0.47. With rigorous experimentations, we demonstrate that the VLP algorithm is significantly faster than state-of-the-art algorithms such as Louvain and Infomap. On the Twitter network, VLP is 2.8 times faster than Louvain.

Keywords: Community detection · graph algorithms · network analysis · graph mining

1 Introduction

Complex networks and systems can be analyzed by representing them as graphs in which entities are denoted by vertices and the relationships between entities by edges [4]. The study and analysis of complex networks has revealed various structural properties that provide insight into the organization and behavior of such systems [4,14,15,17]. Examples of such structural properties include the small world effect [4], skewed power-law degree distributions [17], scale-free phenomenon [15], and community structure [8,14]. Community structure refers to the existence of differentiated heterogeneous modules known as communities

L. M. Aiello et al. (Eds.): ASONAM 2024, LNCS 15212, pp. 343–353, 2025.
https://doi.org/10.1007/978-3-031-78538-2_30

in the underlying structure of the system [8]. A community refers to a group of vertices within a graph or network that are densely connected to each other but only sparsely connected with vertices outside of the community [8]. The terms networks and graphs are used interchangeably in this paper.

Community detection is a problem of interest in many disciplines, including computer science, physics, sociology, and biology [7]. In many real-world networks, such as social networks and citation networks, communities form spontaneously and reflect the underlying functional or organizational structure of the system being analyzed [14]. For example, in social networks, communities can represent real-world social circles, such as friends, family, or colleagues [17]. Identifying and analyzing communities in real-world graphs can provide valuable insights into the underlying phenomena and help to deliver better solutions to complex problems. For example, in VLSI circuits, community detection can be used to group the functional units of a circuit, optimize the layout, and minimize the required chip area [7].

In recent years, many community detection algorithms have been proposed [2–5,9,11,12,14,15]. Raghavan et al. [19] proposed a novel algorithm called the Label Propagation Algorithm (LPA), which has since become a popular method for community detection in complex networks. LPA works by iteratively spreading labels among vertices until a consensus is reached [19].

We propose a generalization of the label propagation method called Vector Label Propagation (VLP). VLP identifies communities by maintaining a vector of candidate community labels of length k for each vertex in the graph. Each vertex's label vector is updated based on the label vectors of its neighbors. To assess VLP's performance, we conducted extensive experiments on various real-world networks. To evaluate the quality of the detected communities, we use two goodness measures: modularity and normalized mutual information (NMI). VLP improves the quality of detected communities compared to LPA and addresses known issues with LPA, such as solution instability and label oscillations [19]. Furthermore, our experiments show that VLP is faster than state-of-the-art methods, such as Louvain [2] and Infomap [5], while detecting communities of comparable quality. For example, VLP is 3.2 times faster than Louvain on the Orkut graph. Furthermore, VLP shows a slightly improved NMI score of 0.60 compared to the Louvain method's score of 0.58.

The remainder of the paper is organized as follows. In Sect. 2, we discuss preliminary concepts and related work. We present our VLP algorithm in Sect. 3. The experimental results are given in Sect. 4, and we conclude in Sect. 5.

2 Preliminaries and Related Work

In this section, we describe some of the notation and preliminary concepts used in this paper. A graph $G = (V, E)$ consists of a set of vertices or nodes V and a set of edges $E \subseteq V \times V$. The number of vertices and edges is denoted by $n = |V|$ and $m = |E|$, respectively. We assume that all edges are undirected. If $(u, v) \in E$, u and v are called neighbors. $N(v)$ denotes the set of neighbors of vertex v; that is, $N(v) = \{u : (u, v) \in E\}$. The degree of v, denoted by $d(v)$, is given by $d(v) = |N(v)|$. A community $C \subseteq V$ refers to a group of vertices in a network that has a relatively high number of internal edges within the vertices in C and relatively fewer external edges to the vertices in the rest of the graph. Community detection involves a process of identifying a community partition or set of communities $P = \{C_1, C_2, \ldots, C_k\}$ in a graph $G(V, E)$, where each $C_i \subseteq V$. If the communities $C_1, C_2, \ldots, C_k$ are disjoint, they are called non-overlapping communities. Otherwise, they are called overlapping communities.

When the true community structure (also called ground truth) is known in advance, solutions of community detection algorithms can be evaluated by comparing the detected communities against the ground truth communities. Mutual information (MI) is an information-theoretic measure that quantifies the correlation between two sets of communities [7]. However, MI has limitations when differentiating communities derived from hierarchical splitting. Normalized mutual information (NMI) overcomes this issue by normalizing the MI score using the partitions' entropy [7]. NMI is widely used to evaluate the performance of community detection algorithms when the ground truth is known [7].

In the absence of known ground truth, detected communities can be evaluated using quality scores, such as modularity [14], which measure the strength of the community structure [7]. Modularity is defined as the deviation of the internal edges from the expected number of edges in a random graph [7,14]. Higher modularity scores indicate a better community structure, as vertices are more densely connected within the community and sparsely connected to the rest of the network [14].

Community detection algorithms can be broadly categorized into partitioning-based, clustering-based, optimization-based, and label-propagation-based algorithms. Partitioning-based community detection algorithms group the vertices of a graph into some k groups of nearly equal size and a small number of edges between the groups. Popular approaches include bisection-based algorithms [15] and edge removal-based algorithms [16]. Clustering-based community detection algorithms are often stochastic in nature [11], and examples include Infomap, WalkTrap, and spectral clustering [5,11,15]. Optimization-based community detection algorithms seek to identify communities by maximizing or minimizing some global quality score, such as modularity. Modularity maximization produces communities of high quality by repeatedly maximizing the gain in modularity [2,3,14,15]. Louvain [2] is a state-of-the-art modularity-based community detection algorithm [11] and is widely adopted due to its speed and good community quality. However, modularity-based algorithms are not effective in detecting smaller communities [6].

Label Propagation Algorithm: Raghavan et al. [19] proposed a community detection algorithm called the label propagation algorithm (LPA) (Algorithm 1) as described below. Each vertex v is assigned a label $L(v)$, and these labels are updated iteratively so that, at the end, all vertices in the same community have the same label. We have the initial labels $L(v) = v$ for all $v \in V$. In each

Algorithm 1 Label Propagation Algorithm

1: **for all** $v \in V$ **do**
2: $\quad L(v) \leftarrow v$
3: $Converged \leftarrow false$
4: **while** not $Converged$ **do**
5: $\quad Converged \leftarrow true$
6: $\quad$ **for all** $v \in V$ **do**
7: $\quad\quad \mathrm{NL}(v) \leftarrow \{L(v)\}$
8: $\quad\quad$ **for all** $u \in N(v)$ **do**
9: $\quad\quad\quad \mathrm{NL}(v) \leftarrow \mathrm{NL}(v) \cup \{L(u)\}$
10: $\quad\quad \mu(v) \leftarrow \mathrm{mode}(\mathrm{NL}(v))$
11: $\quad$ **for all** $u \in N(v)$ **do**
12: $\quad\quad$ **if** $\mu(v) \neq L(v)$ **then**
13: $\quad\quad\quad Converged \leftarrow false$
14: $\quad\quad\quad L(v) \leftarrow \mu(v)$
15: **end while**

iteration, the label $L(v)$ of each vertex v is updated to a new label that is the most frequent label among neighbors $N(v)$. Let $\mathrm{NL}(v)$ be a multiset containing the labels of v and its neighbors; that is $\mathrm{NL}(v) = \{L(v)\} \cup \{L(u)|u \in N(v)\}$. Then the new label of v is simply the most frequent element in $\mathrm{NL}(v)$, that is, $mode(\mathrm{NL}(v))$. When there is a tie for the most frequent label, one of them is chosen arbitrarily. After some iterations, the algorithm converges, and no vertices change their labels from the previous iteration. At this point, the algorithm is terminated, and we say that the community labels have stabilized or that the consensus has been reached.

3 Vector Label Propagation

In this section, we present our community detection algorithm called vector label propagation (VLP), which is a generalization of the label propagation algorithm (LPA) described in Sect. 2. Our experiments show that VLP significantly improves the quality of communities compared to LPA. VLP is also much faster than state-of-the-art algorithms such as Louvain [2] and Infomap [5], and at the same time, it detects communities with quality comparable to that of the Louvain and Infomap algorithms.

3.1 Overview of VLP

Vector label propagation (VLP) uses the label propagation technique to propagate the most dominant labels selected from the neighboring vertices. To enhance the community detection ability of the algorithm, we generalize this technique by maintaining k candidate community labels for each vertex, where k is a parameter of the algorithm. The candidate labels are the k most dominant or frequent labels in the local neighborhood. On the other hand, LPA tracks only the most frequent label for each vertex. Our experimental results (given in Sect. 4) show that it is enough to use a small value of k, such as $k = 3$, to significantly improve the quality of the detected communities compared to LPA. VLP stores the candidate labels in label vectors, which are vectors consisting of pairs as follows: $\mathbb{L} = [(\ell_0, c_0), (\ell_1, c_1), \ldots, (\ell_s, c_s)]$, where s is the size or length of the vector. The i-th pair (ℓ_i, c_i) consists of a candidate label $\ell_i = \mathbb{L}.\ell_i$ and a label confidence

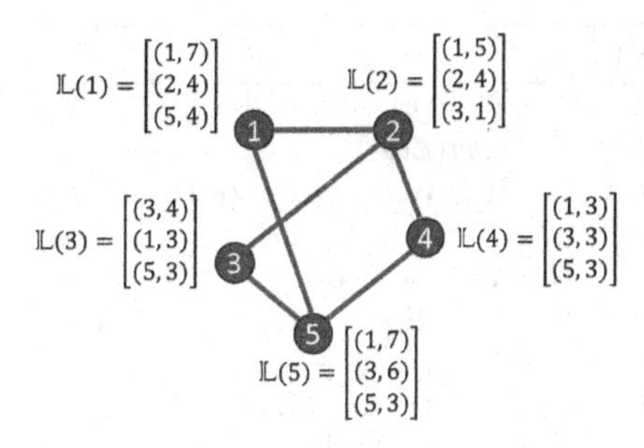

Fig. 1. A graph with five vertices showing label vectors of length $k = 3$

$$\begin{bmatrix}(1,7)\\(2,4)\\(5,4)\end{bmatrix} \oplus \begin{bmatrix}(3,4)\\(1,3)\\(5,3)\end{bmatrix} = \begin{bmatrix}(1,10)\\(5,8)\\(2,4)\\(3,4)\end{bmatrix}$$

Fig. 2. Label-wise sum

$$\mathbb{Z}(5) = \begin{bmatrix}(1,7)\\(3,6)\\(5,3)\end{bmatrix} \oplus \begin{bmatrix}(1,7)\\(2,4)\\(5,4)\end{bmatrix} \oplus \begin{bmatrix}(3,4)\\(1,3)\\(5,3)\end{bmatrix} \oplus \begin{bmatrix}(1,3)\\(3,3)\\(5,3)\end{bmatrix}$$

$$\mathbb{Z}(5) = \begin{bmatrix}(1,20)\\(2,4)\\(3,13)\\(4,0)\\(5,13)\end{bmatrix} \quad \overset{sorted}{\mathbb{Z}(5)} = \begin{bmatrix}(1,20)\\(3,13)\\(5,13)\\(2,4)\\(4,0)\end{bmatrix} \quad \mathbb{L}(5) = \begin{bmatrix}(1,20)\\(3,13)\\(5,13)\end{bmatrix}$$

Fig. 3. Updating label vector of vertex 5

$c_i = \mathbb{L}.c_i$. The confidence c_i represents the count or frequency of ℓ_i. VLP maps each vertex v to a label vector $\mathbb{L}(v)$ of length k with each of the k elements in the vector corresponding to one candidate label. The k pairs in $\mathbb{L}(v)$ are sorted by confidence in descending order. At the end of the algorithm, a vertex v is assigned to the label $L(v) = \mathbb{L}(v).\ell_0$. The label vectors of the vertices in an example graph are shown in Fig. 1.

The VLP algorithm initializes the label vectors $\mathbb{L}(v)$ for each vertex v as follows: $\ell_0 \leftarrow v$, $c_0 \leftarrow 1$, and for $1 \leq i < k$, $\ell_i \leftarrow$ NULL, and $c_i \leftarrow 0$. Similarly to LPA, the algorithm proceeds in multiple iterations. In each iteration, the label vector $\mathbb{L}(v)$ of each vertex v is updated using the following three steps.

1. Aggregation: The label vectors $\mathbb{L}(u)$ of neighbors $u \in N(v)$, along with $\mathbb{L}(v)$, are aggregated into a single label vector $\mathbb{Z}$ using a label-wise sum, denoted by $\oplus$, in which the confidences corresponding to the same labels are added together. The details of the aggregation step are discussed in Sect. 3.2.
2. Sorting: The aggregated label vector $\mathbb{Z}$ is sorted by confidence values c_i in descending order.
3. Top k selection: The label vector $\mathbb{L}(v)$ is updated to the first k pairs in $\mathbb{Z}$, which are the k most frequent candidate labels.

After some iterations of the VLP algorithm, no vertices change their most frequent candidate label $\mathbb{L}(v).\ell_0$ from the previous iteration. At this point, the most frequent label $\mathbb{L}(v).\ell_0$ is chosen to be the final community label $L(v)$, and the algorithm is terminated.

3.2 Aggregating Label Vectors

The aggregation of the label vectors is performed using a *label-wise sum*. A *label-wise sum*, denoted by $\oplus$, of two label vectors $\mathbb{X}$ and $\mathbb{Y}$ is another label vector $\mathbb{Z} = \mathbb{X} \oplus \mathbb{Y}$. The label vector $\mathbb{Z}$ accumulates the total confidences of all distinct labels that appear in $\mathbb{X}$ and/or $\mathbb{Y}$ as follows:

Algorithm 2 Vector Label Propagation VLP(k, G)

```
1: for all v ∈ V do
2:     𝕃(v).ℓ0 ← v
3:     𝕃(v).c0 ← 1
4:     for all i = 1, 2, ...k − 1 do
5:         𝕃(v).ℓi ← NULL
6:         𝕃(v).ci ← 0
7: Converged ← false
8: while not Converged do
9:     Converged ← true
10:    for all v ∈ V do
11:        ℤ(v) ← 𝕃(v)
12:        for all u ∈ N(v) do
13:            ℤ(v) ← ℤ(v) ⊕ 𝕃(u)
14:        sort(ℤ(v))
15:        if 𝕃(v).ℓ0 ≠ ℤ(v).ℓ0 then
16:            Converged ← false
17:    for all v ∈ V do
18:        for all i = 1, 2, . . . , k − 1 do
19:            𝕃(v).ℓi ← ℤ(v).ℓi
20:            𝕃(v).ci ← ℤ(v).ci
21:    for all v ∈ V do
22:        L(v) = 𝕃(v).ℓ0
23: end while
```

- If ℓ appears as a pair (ℓ, c') in only one of $\mathbb{X}$ and $\mathbb{Y}$, then we have the pair $(\ell, c) = (\ell, c')$ in $\mathbb{Z}$.
- If ℓ appears as pairs (ℓ, c') in $\mathbb{X}$ and (ℓ, c'') in $\mathbb{Y}$, then we have the pair $(\ell, c) = (\ell, c' + c'')$ in $\mathbb{Z}$.

An example of label-wise sum is shown in Fig. 2. For each vertex v, label vectors of its neighbors are aggregated using the label-wise sum as follows: $\mathbb{Z} = \mathbb{L}(v) \oplus \mathbb{L}(u_1) \oplus \mathbb{L}(u_2) \oplus \ldots \oplus \mathbb{L}(u_{d(v)})$, where $u_1, u_2, \ldots u_{d(v)}$ are neighbors of v. For example, consider the vertex 5 in Fig. 1, which has neighbors 1, 3, and 4. Figure 3 illustrates how vertex 5 updates its label vector. The label vectors $\mathbb{L}(1)$, $\mathbb{L}(3)$, and $\mathbb{L}(4)$ of the neighbors of the vertex 5 and its own label vector $\mathbb{L}(5)$ are aggregated into $\mathbb{Z}(5) = \mathbb{L}(5) \oplus \mathbb{L}(1) \oplus \mathbb{L}(3) \oplus \mathbb{L}(4)$. $\mathbb{Z}(5)$ has 5 pairs, which are sorted by confidence values. The updated label vector $\mathbb{L}(5)$ is set to the most frequent $k = 3$ labels in $\mathbb{Z}(5)$.

3.3 Pseudocode of the VLP Algorithm

The pseudocode of the VLP algorithm is given in Algorithm 2. The label vectors are initialized in Lines 1–6 and iteratively updated in the while loop in Lines 8–23. The label vectors of the neighbors are aggregated in Lines 11–13. The aggregated label vector $\mathbb{Z}(v)$ is sorted in Line 14. The first k pairs, corresponding to the most frequent k labels, in $\mathbb{Z}(v)$ are selected as the new label vector $\mathbb{L}(v)$ in Lines 18–20. In the end (Lines 21–22), the most frequent label $\mathbb{L}(v).\ell_0$ is chosen as the final label $L(v)$. We implemented our algorithm in C++ and used *std::unordered_map* of the Standard Template Library (STL) to implement the label vectors. This allows us to perform label-wise sum efficiently using *std::accumulate*.

3.4 Weighted Aggregation of Label Vectors

We enhance label vector aggregation by giving more importance to label vectors from vertices with more common neighbors, in a mechanism called *weighted aggregation*. When a vertex v aggregates the label vectors of neighbors, we assign a weight $\lambda = J(u, v)$ to neighbors $u \in N(v)$ corresponding to the Jaccard similarity index given by $J(u, v) = \frac{|N(u) \cap N(v)|}{|N(u) \cup N(v)|}$. In weighted label vector aggregation, the label confidences in vectors $\mathbb{L}(u)$ of the neighbors $u \in N(v)$ are scaled up or down by a factor of λ as follows. If $\mathbb{L}(u) = [(\ell_0, c_0), (\ell_1, c_1), \ldots, (\ell_{k-1}, c_{k-1})]$, then $\lambda\mathbb{L}(u) = [(\ell_0, \lambda c_0), (\ell_1, \lambda c_1), \ldots, (\ell_{k-1}, \lambda c_{k-1})]$. Now, $\mathbb{Z} \oplus \lambda\mathbb{L}(u)$ is computed by performing a label-wise sum on two label vectors $\mathbb{Z}$ and $\lambda\mathbb{L}(u)$. To use the weighted aggregation mechanism, we replace Line 13 in Algorithm 2 with $\mathbb{Z} \leftarrow \mathbb{Z} \oplus \lambda\mathbb{L}(u)$, where $\lambda = J(u, v)$ is the Jaccard index of the edge. Note that the Jaccard index weight of the edges needs to be calculated only once at the beginning of the algorithm, as they depend only on the graph's topological structure. The weight also serves as a damping factor for confidence values, which can be useful when the algorithm goes through too many iterations or the confidence values become too high. The experimental results show that the weighted label vector aggregation detects better quality communities than the unweighted aggregation.

4 Experimental Results

In this section, we rigorously evaluate the performance of our VLP algorithm experimentally on a wide variety of real-world graphs. Our experiments test the runtime and the quality (measured in NMI) of the detected communities. The performance of our algorithm is compared to recent state-of-the-art algorithms [2,5,13,15,19]. The experimental setup and the nature of the graph datasets used to evaluate the VLP algorithm are detailed below, along with the choice of parameters used in the algorithm.

4.1 Experimental Setup

We conducted our experiments on the Bridges-2 supercomputer at the Pittsburg Supercomputing Center (PSC), which was available to us via the Advanced Cyberinfrastructure Coordination Ecosystem: Services and Support (ACCESS) [18]. On the Bridges-2 system, we used the Regular Memory (RM) nodes having 2xAMD EPYC 7742 CPUs (2.25–3.40GHz and 256MB L3 cache) with 64 cores each and 256GB RAM on each node. Our implementations were made in the C++ language and compiled using gcc version 8.2.0.

4.2 Dataset

Our graph data set consists of undirected, unweighted graphs with known ground truth communities. The data set is chosen to represent a wide variety of graph types and graph sizes. These networks are taken from the Stanford Large Network Dataset Collection [10] and the GAP Benchmark Suite [1]. Table 1 lists the graphs used in our experiments and the number of vertices, the number of edges, and the average degree of the graphs.

Table 1. List of graphs used in experiments.

Graph name	Nodes	Edges	Avg-Deg
Karate	34	78	4.6
Dolphins	62	159	5.1
Football	115	613	10.7
Polbooks	105	441	8.4
web-Google	0.9M	5.1M	11.3
web-BerkStan	0.7M	7.6M	21.7

Graph name	Nodes	Edges	Avg-Deg
soc-LiveJournal1	4.8M	68.8M	28.6
com-Orkut	3.1M	117.2M	75.6
twitter	61.6M	1.4B	61.7
com-FriendSter	65.6M	1.8B	55.3
uk2002	39.5M	1.9B	45.5

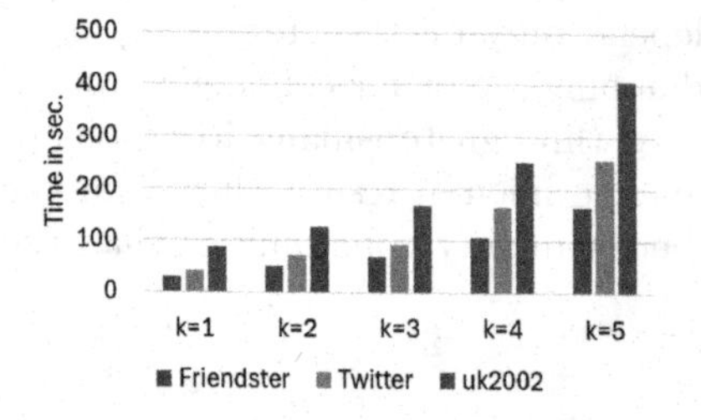

Fig. 4. Effect of parameter k on the runtime of the algorithm in the Friendster, Twitter, and uk2002 networks.

Fig. 5. Effect of parameter k on NMI scores of detected communities in the Friendster, Twitter, and uk2002 networks.

4.3 Performance

In order to determine the best value of parameter k for Algorithm 2, a comparative analysis of the algorithm runtime and the quality score of the identified communities is performed for various values of k. The runtimes (in seconds) for our VLP algorithm on three graphs - Friendster, Twitter, and uk2002 - are presented in Fig. 4. The normalized mutual information score (NMI) is used to compare the detected communities with the ground truth, and the results are presented in Fig. 5. The objective is to have an algorithm that has a fast runtime and a high NMI score. After careful consideration, $k = 3$ is chosen as it demonstrated a satisfactory quality score while still being reasonably fast. $k = 3$ is used for all subsequent experiments.

To find the effect of weighted aggregation, we compare VLP with and without weighted label vector aggregation. Figure 6 shows the quality of the detected communities, measured using NMI, on various graphs. Using weighted label vector aggregation shows a consistent improvement in the quality of the detected communities. For example, on the Orkut graph, using weighted aggregation improves the NMI score from 0.456 to 0.603. In the rest of our experiments, we use VLP with weighted aggregation.

We compare the performance of VLP using weighted label vector aggregation with the following state-of-the-art algorithms: *i*) Label Propagation (LPA) [19], *ii*) LPAm+ [13], *iii*) Infomap [5], *iv*) ICSD+ [15], and *v*) Louvain [2]. In

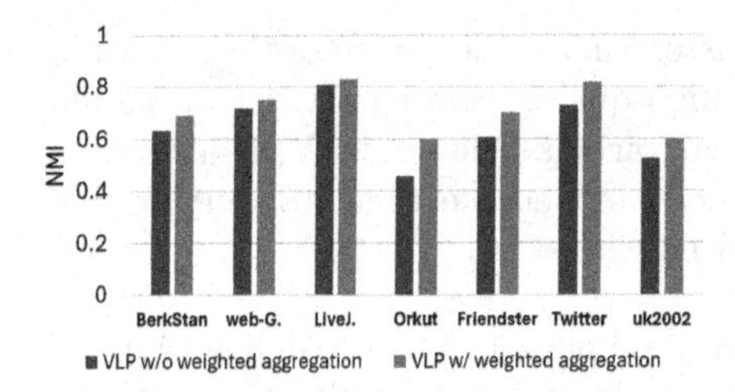

Fig. 6. Effect of using weighted aggregation on NMI scores of detected communities in various networks.

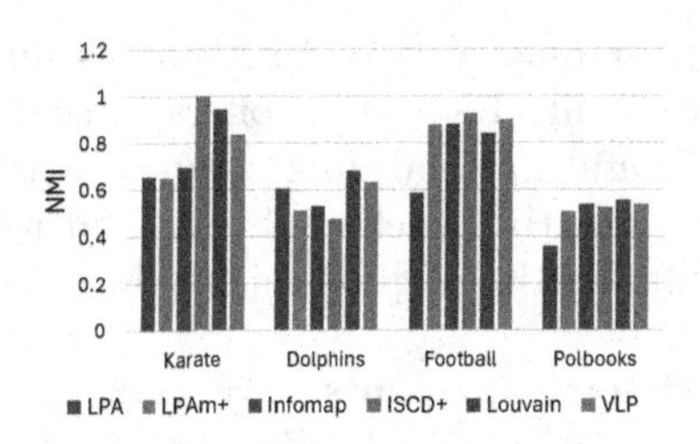

Fig. 7. NMI scores of detected communities in four small networks.

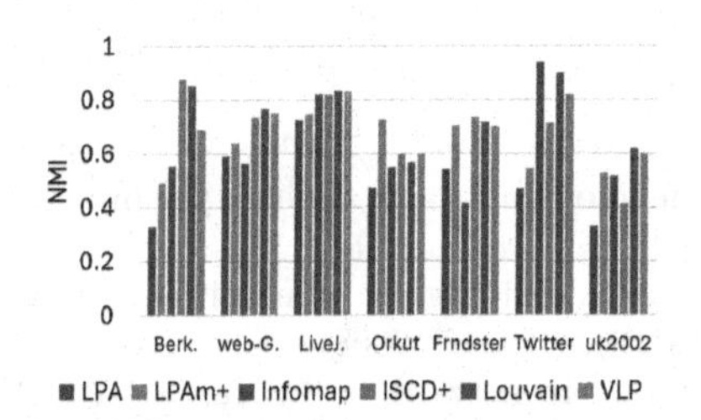

Fig. 8. NMI scores for communities detected in various medium and large networks by different algorithms.

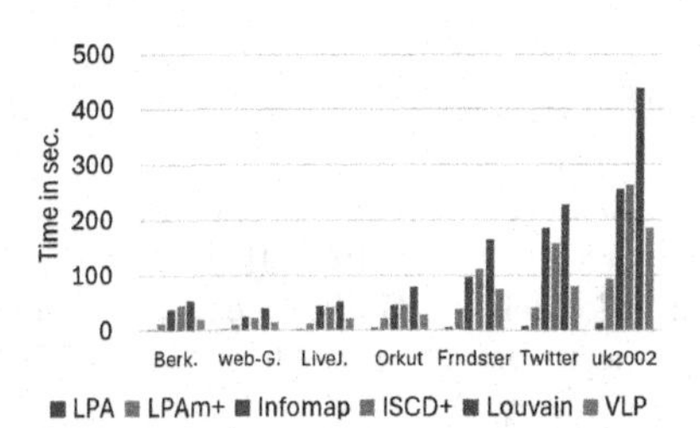

Fig. 9. The runtime of various algorithms, including our VLP algorithm, on various medium and large networks.

Fig. 7, we compare the NMI scores on four small networks - Karate, Dolphins, Football, and Polbooks - which have widely been referenced in the community detection literature. Figure 8 compares the NMI scores for seven medium to large networks. VLP showed an average improvement of 1.52 times in NMI scores over LPA. Moreover, compared to the algorithm with the best NMI score for each network, VLP was only worse by 0.072 NMI points on average. Our VLP algorithm outperforms LPA by a wide margin and performs comparably to state-of-the-art algorithms.

Figure 9 illustrates the runtime comparison of VLP with several state-of-the-art algorithms on seven medium to large networks. VLP outperformed Louvain and Infomap on all graphs and was considerably faster. For example, on large graphs such as Friendster, Twitter, and UK2002, VLP was 2.2 times, 2.8 times, and 2.4 times faster than Louvain and 1.3 times, 2.3 times, and 1.4 times faster than Infomap, respectively. In summary, the experimental results demonstrate that the proposed VLP algorithm is significantly faster than the state-of-the-art algorithms while maintaining comparable quality of the communities.

5 Conclusion

We have introduced a community detection algorithm called vector label propagation (VLP), which is a generalization of LPA. VLP detects communities with higher quality than LPA and is much faster than other state-of-the-art algorithms like Louvain. It offers efficient and high-quality community detection and is useful for analyzing complex networks in various fields. Our VLP algorithm will greatly benefit analysts and researchers and enable them to quickly find communities of good quality in real-world networks.

Acknowledgements. The work was partially supported by NSF BIGDATA grant IIS-1633028 and used PSC Bridges-2 RM at Pittsburg Supercomuting Center (PSC) through allocation CIS230052 from the Advanced Cyberinfrastructure Coordination Ecosystem: Services & Support (ACCESS) program [18], which is supported by NSF grants #2138259, #2138286, #2138307, #2137603, and #2138296.

References

1. Beamer, S., Asanović, K., Patterson, D.: The gap benchmark suite. arXiv preprint arXiv:1508.03619 (2015)
2. Blondel, V., Guillaume, J., Lambiotte, R.: Fast unfolding of communities in large networks. J. Stat. Mech. Theory Exp. **2008**(10), P10008 (2008)
3. Clauset, A.: Finding local community structure in networks. Phy. Rev. E **72**(2), 026132 (2005)
4. Deweese, K., Gilbert, J., Lugowski, A., Reinhardt, S., Kepner, J.: Graph clustering in sparql. In: SIAM Workshop on Network Science, vol. 34, pp. 930–941. Citeseer (2013)
5. Edler, D., Guedes, T., Zizka, A., Rosvall, M., Antonelli, A.: Infomap bioregions: interactive mapping of biogeographical regions from species distributions. Syst. Biol. **66**(2), 197–204 (2017)
6. Fortunato, S., Barthelemy, M.: Resolution limit in community detection. Proc. Nat. Acad. Sci. **104**(1), 36–41 (2007)
7. Fortunato, S., Hric, D.: Community detection in networks: a user guide. Phys. Rep. **659**, 1–44 (2016)
8. Girvan, M., Newman, M.: Community structure in social and biological networks. Proc. Nat. Acad. Sci. **99**(12), 7821–7826 (2002)
9. Lancichinetti, A., Fortunato, S.: Community detection algorithms: a comparative analysis. Phys. Rev. E **80**(5), 056117 (2009)
10. Leskovec, J., Krevl, A.: SNAP Datasets: Stanford large network dataset collection. http://snap.stanford.edu/data (2014)
11. Leskovec, J., Lang, K.: Empirical comparison of algorithms for network community detection. In: Proceedings of the 19th International Conference on WWW, pp. 631–640 (2010)
12. Leung, I., Hui, P., Lio, P., Crowcroft, J.: Towards real-time community detection in large networks. Phys. Rev. E **79**(6), 066107 (2009)
13. Liu, X., Murata, T.: Advanced modularity-specialized label propagation for detecting communities. Phys. A: Stat. Mech. and its Apps. **389**(7), 1493–1500 (2010)
14. Newman, M.: Modularity and community structure in networks. Proc. Nat. Acad. Sci. **103**(23), 8577–8582 (2006)
15. Newman, M.: Spectral methods for community detection and graph partitioning. Phys. Rev. E **88**(4), 042822 (2013)
16. Newman, M., Girvan, M.: Finding and evaluating community structure in networks. Phys. Rev. E **69**(2), 026113 (2004)

17. Papadopoulos, S., Kompatsiaris, Y., Vakali, A.: Community detection in social media: performance and applications. Data Min. Know. Dis. **24**, 515–554 (2012)
18. Parashar, M., Friedlander, A., Gianchandani, E., Martonosi, M.: Transforming science through cyberinfra. Commun. ACM **65**(8), 30–32 (2022)
19. Raghavan, U., Albert, R., Kumara, S.: Near linear time algorithm to detect community structures in large-scale networks. Phys. Rev. E **76**(3), 036106 (2007)

Author Index

L. M. Aiello et al. (Eds.): ASONAM 2024, LNCS 15212, pp. 355–356, 2025.
https://doi.org/10.1007/978-3-031-78538-2

Printed in the United States
by Baker & Taylor Publisher Services